W9-AFO-433

Lecture Notes in Mathematics

Edited by A. Dold and B. Eckmann

448

Spectral Theory and Differential Equations

Proceedings of the Symposium held at Dundee, Scotland, 1–19 July, 1974

Edited by W. N. Everitt

Springer-Verlag
Berlin · Heidelberg · New York 1975

Prof. William N. Everitt
Department of Mathematics
University of Dundee
Dundee DD1 4HN/Scotland

Library of Congress Cataloging in Publication Data

Symposium on Spectral Theory and Differential Equations, University of Dundee, 1974.
Spectral theory and differential equations.

(Lecture notes in mathematics ; 448)
Bibliography: p.
Includes index.
1. Differential equations--Congresses. 2. Differential operators--Congresses. 3. Spectral theory (Mathematics)--Congresses. I. Everitt, William Norrie. II. Title. III. Series: Lecture notes in mathematics (Berlin) ; 448.
QA3.L28 no. 448 [QA371] 510'.8s [515'.35] 75-6675

AMS Subject Classifications (1970): 26A84, 26A86, 34B15, 34B20, 34B25, 34C15, 34G05, 35B10, 35B45, 35C25, 35D05, 35D10, 35G05, 35J05, 35J10, 35J15, 35J25, 35J45, 35J55, 35P05, 35P10, 35P15, 35P25, 35Q10, 47A05, 47A10, 47A25, 47A55, 47B25, 47B40, 47B44, 47E05, 47F05, 49A05, 49A10, 49G20, 65L10, 65L15, 65L99, 65K05, 76A05, 76A10, 76B10, 76D05, 76D10, 76D15, 78A20, 78A25

ISBN 3-540-07150-4 Springer-Verlag Berlin · Heidelberg · New York
ISBN 0-387-07150-4 Springer-Verlag New York · Heidelberg · Berlin

Offsetdruck: Julius Beltz, Hemsbach/Bergstr.

This volume is dedicated to the life,

work and memory of

KONRAD JÖRGENS

1926-1974

P R E F A C E

These Proceedings form a record of the lectures given to the Symposium on Spectral Theory and Differential Equations held at the University of Dundee, Scotland during the period 1 to 19 July 1974.

The Symposium was organised on behalf of the Science Research Council of the United Kingdom. The Council provided financial support which made it possible to bring together many of the foremost workers in various aspects of the application of spectral analysis to the theory of both ordinary and partial differential equations. Without this support from the Council it would have been impossible to bring the Symposium into being.

The Symposium was attended by 60 mathematicians from the following countries: BRD (Germany), Canada, Japan, Sweden, The Netherlands, the United Kingdom and the United States of America.

Three mathematicians, Professors E. A. Coddington, Tosio Kato and Joachim Weidmann, were each invited to give a series of four lectures and to prepare manuscripts which appear in this volume. Twelve other mathematicians were invited to give single lectures and papers resulting from these lectures are also collected in this volume. Additionally Professor F S Rofe-Beketov, of Kharkov, the Ukraine, USSR, sent a manuscript for inclusion in this collection.

The Symposium was organised, on behalf of the Science Research Council, by the following Committee: W. N. Everitt (Chairman): I. M. Michael and B. D. Sleeman (Organising Secretaries).

On behalf of the Committee I express our keen appreciation to all mathematicians who took part in the work of the Symposium. In particular I thank all those who delivered lectures and supplied manuscripts for this volume. Special thanks are due to Professor Weidmann who took over responsibility for the lectures on the spectral theory of partial differential operators when it became clear that the late Professor Jörgens would not be able to attend the Sympsoium.

The Committee thanks: the University of Dundee for generously providing facilities which made it possible to hold the Symposium in Dundee. The Wardens and Staff of Chalmers Hall and Peterson House for accommodating many of those attending the Symposium; and many colleagues and research students in the Department of Mathematics, for help received. Mr E R Dawson of the Department of Mathematics gave particular advice in the preparation of manuscripts for this volume, and helped in several other ways.

We are grateful to the Staff of the Mathematics Committee of the Science Research Council for much help and advice; in particular Miss Jean Melville and Dr K D Crosbie. Also to Professor D E Edmunds (University of Sussex) in his capacity as Adviser to the Council for the Symposium.

The Committee extends very special gratitude to Mrs Norah Thompson, Secretary in the Department of Mathematics, for her significant contribution to the work of the Symposium. It would have been difficult, at times impossible, to cope with the volume of secretarial work without her help; many of the papers appearing in this volume bear witness to her ability as a typist of mathematical manuscripts.

Finally it is a pleasure for me to thank once again my colleagues Brian Sleeman and Ian Michael for their sustained efforts over several months; and for their patience and cheerfulness in the face of responsibility for so much of the organisation of the Symposium. I am grateful to both of them.

One mathematician was sorely missed from the Symposium. By July of 1974 the tall, commanding figure and personality of Konrad Jörgens was no more; he died in April of that year as a result of a terminal illness which had had a quite unexpected onset in the summer of 1973. Professor Jörgens had earlier accepted our invitation to deliver a series of lectures to the Symposium but as 1974 advanced it became clear that he would be unable to attend. The loss sustained by mathematics, and in particular the study of differential equations, as a result of his untimely death will be felt not only in his own country but in many parts of the world. It is for others in different circumstances to speak and write in greater detail of his outstanding contributions to our discipline. Here my colleagues and I dedicate, so appropriately, this volume to the memory of a mathematician who contributed with such significance to spectral theory and differential equations. Konrad Jörgens was indeed a colleague and a friend to many of us and we are the poorer for losing him at the height of his powers and influence.

October 1974 W N Everitt

C O N T E N T S

Survey papers

Papers

Lectures delivered to the Symposium

Series of four lectures

E. A. Coddington:

Spectral theory of ordinary differential operators

T. Kato:

Quasi-linear equations of evolution, with applications to partial differential equations

J. Weidmann:

Spectral theory of partial differential operators

Single Lectures

J. S. Bradley:

Integral inequalities and spectral theory

L. Collatz:

Optimisation and differential equations

M. S. P. Eastham

Results and problems in spectral theory of periodic differential equations

W D Evans

Sobelev embeddings

M. Giertz:

On the deficiency indices of powers of symmetric differential expressions

H. Kalf:

On the spectral theory of Dirac and Schrödinger operators

S. T. Kuroda:

Scattering theory for differential operators

J. B. McLeod:

Swirling flow

Å. Pleijel:

A survey of spectral theory for pairs of differential operators

K. Schmitt:

Eigenvalue problems for non-linear ordinary differential equations

B. D. Sleeman:

Left-definite multi-parameter eigenvalue problems

W. Velte:

On variational problems with unilateral constraints.

Address list of authors and speakers

J. S. Bradley: Department of Mathematics, The University of Tennessee, KNOXVILLE, Tennessee 37916, USA.

E. A. Coddington: Department of Mathematics, University of California, LOS ANGELES, California 90024, USA.

L. Collatz: Institut für angewandte Mathematik, Universität Hamburg, 2 HAMBURG 13, Rothenbaumchaussee 41, BRD Germany.

M. S. P. Eastham: Department of Mathematics, Chelsea College, Manresa Road, LONDON SW3 6LX, England, UK.

W. D. Evans: Department of Pure Mathematics, University College, CARDIFF CF1 1XL, Wales, UK.

W. N. Everitt: Department of Mathematics, The University, DUNDEE DD1 4HN, Scotland, UK.

M. Giertz: Institutionen für Matematik, Kungliga Tekniska Högskolan, 100 44 STOCKHOLM 70, Sweden.

H. Kalf: Institut für Mathematik, Rhein.-Westf. Techn. Hochschule Aachen, 51 AACHEN, Templergraben 55, BRD Germany.

T. Kato: Department of Mathematics, University of California, BERKELEY, California 94720, USA.

S. T. Kuroda: Department of Pure and Applied Sciences, University of Tokyo, Komaba Meguro-ku, TOKYO, Japan.

J. B. McLeod: The Mathematical Institute, 24-29 St Giles, OXFORD, England, UK.

Å. Pleijel: Department of Mathematics, Uppsala University, Sysslomansgatan 8, 752 23 UPPSALA, Sweden.

F. S. Rofe-Beketov: Department of Mathematics, Low Temperature Physics Institute of the Ukrainian SSR, Pr Lenina 47, KHARKOV 86, USSR.

U.-W. Schmincke: Institut für Mathematik, Rhein.-Westf. Techn. Hochschule Aachen, 51 AACHEN, Templergraben 55, BRD Germany.

K. Schmitt: Department of Mathematics, The University of Utah, SALT LAKE CITY, Utah 84112, USA.

B. D. Sleeman: Department of Mathematics, The University, DUNDEE DD1 4HN, Scotland, UK.

W. Velte: Department of Applied Mathematics, The University of Würzburg, 87 WÜRZBURG, Am Hubland, BRD Germany.

J. Walter: Institut für Mathematik, Rhein.-Westf. Techn. Hochschule Aachen, 51 AACHEN, Templergraben 55, BRD Germany.

J. Weidmann: Fachbereich Mathematik, Universität Frankfurt-am-Main, FRANKFURT-AM-MAIN, Robert-Meyer-Strasse 10, BRD Germany.

R. Wüst: Institut für Mathematik, Rhein.-Westf. Techn. Hochschule Aachen, 51 AACHEN, Templergraben 55, BRD Germany.

SPECTRAL THEORY OF ORDINARY DIFFERENTIAL OPERATORS

Earl A. Coddington

1. Introduction. This is a report on some work which was completed during the last several years, together with some results which were obtained jointly with A. Dijksma during the 1973-1974 academic year. The work of the author was supported in part by the National Science Foundation under NSF Grant GP-33696X.

The classical eigenvalue problem can be exemplified by the problem on $0 \leq x \leq 1$ given by

$$Lf = \lambda f, \ af(0) + bf(1) = 0, \tag{1.1}$$

where L denotes the formal operator $L = id/dx$. In case the boundary condition is $f(0) - f(1) = 0$ we know that there are orthonormal eigenfunctions $\chi_n(x) = \exp(-2\pi inx)$, $n = 0, \pm 1, \ldots,$ with eigenvalues $\lambda_n = 2\pi n$, and there is the eigenvalue expansion

$$f = \sum_{n=-\infty}^{\infty} (f, \chi_n)\chi_n, \ (f, \chi_n) = \int_0^1 f\bar{\chi}_n .$$

For each $f \in \mathfrak{L}^2(0, 1)$ this series converges to f in the metric of $\mathfrak{L}^2(0, 1)$. This is an example of a selfadjoint problem, and, in fact, all problems of the form (1.1) with $|a| = |b| \neq 0$ are selfadjoint ones.

Since $\mathfrak{H} = \mathfrak{L}^2(0, 1)$ is a Hilbert space, and since we shall be concerned with spectral theory in Hilbert spaces, let us interpret the problem (1.1) in the context of this Hilbert space. The f trivially satisfying the boundary condition in (1.1) are those such that $f(0) = f(1) = 0$. Let S_0 be the minimal operator in $\mathfrak{H}$ for L. Thus the domain $\mathfrak{D}(S_0)$ of S_0 is the set of all those $f \in \mathfrak{H}$ such that f is absolutely continuous on $0 \leq x \leq 1$, $f' \in \mathfrak{H}$, and $f(0) = f(1) = 0$, and for $f \in \mathfrak{D}(S_0)$ we have $S_0 f = Lf = if'$. This S_0 is a symmetric operator in $\mathfrak{H}$, $(S_0 f, g) = (f, S_0 g)$ for all $f, g \in \mathfrak{D}(S_0)$. The maximal operator in $\mathfrak{H}$ for L is S_0^*, where the graph $\mathfrak{G}(S_0^*)$ of S_0^* is defined by

$$\mathfrak{G}(S_0^*) = \{\{h, k\} \in \mathfrak{H}^2 = \mathfrak{H} \oplus \mathfrak{H} \,|\, (S_0 f, h) = (f, k), \text{ all } f \in \mathfrak{D}(S_0)\} .$$

This is the graph of an operator (single-valued function), and $S_0^* h = Lh$ for $h \in \mathfrak{D}(S_0^*)$, where $\mathfrak{D}(S_0^*)$ is the set of all $f \in \mathfrak{H}$ which are absolutely continuous on $0 \leq x \leq 1$ and such that $f' \in \mathfrak{H}$. For two fixed complex numbers a, b satisfying $|a| = |b| \neq 0$ define

$$\mathfrak{D}(H) = \{f \in \mathfrak{D}(S_0^*) \,|\, af(0) + bf(1) = 0\},$$

and for $f \in \mathfrak{D}(H)$ let $Hf = Lf$. Then $S_0 \subset H \subset S_0^*$, in the sense that $\mathfrak{G}(S_0) \subset \mathfrak{G}(H) \subset \mathfrak{G}(S_0^*)$, and H is a selfadjoint operator in $\mathfrak{H}$, i.e., $H = H^*$. Moreover, all selfadjoint extensions of S_0 (or selfadjoint restrictions of S_0^*) are of this form.

We now seek to broaden the type of problems for the differential operator L. For example, we could consider a side condition for $f \in \mathfrak{D}(S_0^*)$ of the form

$$af(0) + bf(1) + c \int_0^1 f\overline{\varphi} = 0,$$

where $a,b,c \in \mathbb{C}$ (the complex numbers), and $\varphi \in \mathfrak{H}$, $\|\varphi\| \neq 0$. As before the f trivially satisfying this are the $f \in \mathfrak{D}(S_0)$ such that

$$\int_0^1 f\overline{\varphi} = (f, \varphi) = 0 .$$

This leads immediately to the consideration of a restriction S of S_0 with domain $\mathfrak{D}(S)$ given by

$$\mathfrak{D}(S) = \mathfrak{D}(S_0) \cap \{\varphi\}^{\perp},$$

where $\{\varphi\}$ is the subspace spanned by φ in $\mathfrak{H}$, and $\{\varphi\}^{\perp} = \mathfrak{H} \ominus \{\varphi\}$ is the orthogonal complement of $\{\varphi\}$ in $\mathfrak{H}$. This S is symmetric in $\mathfrak{H}$, and we can seek to determine those selfadjoint H such that $S \subset H$. By analogy with our first example we would expect $S \subset H \subset S^*$. But what is S^*? Identifying it with its graph we would want it to be

$$(1.2) \qquad \{\{h, k\} \in \mathfrak{H}^2 | (Sf, h) = (f, k), \quad \text{all} \quad f \in \mathfrak{D}(S)\}.$$

We note that $\{h, S_0^* h\}$, $h \in \mathfrak{D}(S_0^*)$, belongs to this set. However, $\{0, \varphi\}$ does also, and thus this set is <u>not</u> the graph of an operator in $\mathfrak{H}$, although it is a perfectly nice closed linear manifold in $\mathfrak{H}^2$. We <u>define</u> S^* to be the set (1.2), and it is not difficult to see that

$$S^* = \{\{h, S_0^* h + d\varphi\} | h \in \mathfrak{D}(S_0^*), \ d \in \mathbb{C}\} .$$

Since S^* is not the graph of an operator in $\mathfrak{H}$, we must now expect selfadjoint H satisfying $S \subset H \subset S^*$ which are not operators.

Let us look more closely at the conditions

$$(1.3) \qquad f(0) = 0, \ f(1) = 0, \ (f, \varphi) = 0, \ f \in \mathfrak{D}(S_0^*) .$$

Clearly the maps $f \to f(0)$, $f \to f(1)$, and $f \to (f, \varphi)$ are linear functionals on $\mathfrak{D}(S_0^*)$, and (1.3) just says that f lies in the null spaces of these linear functionals. The last map, $f \to (f, \varphi)$ is clearly continuous, but the evaluation maps $f \to f(0)$, $f \to f(1)$ are not. To be explicit, if $f_n(x) = (2n + 1)^{1/2} x^n$, then $\|f_n\| = 1$ but $f_n(1) = (2n + 1)^{1/2} \to \infty$, $n \to \infty$. However, the maps $\{f, S_0^* f\} \to f(0)$, and $\{f, S_0^* f\} \to f(1)$ <u>are</u> continuous on $\mathfrak{G}(S_0^*)$. For example,

$$f(1) = \int_0^1 (xf)' dx = \int_0^1 [f(x) + xf'(x)] dx$$

$$= (f, \sigma) + (Lf, \tau) = (\{f, S_0^* f\}, \{\sigma, \tau\}),$$

where $\sigma(x) = 1$, $\tau(x) = ix$. Thus we see that the conditions (1.3) can be considered as restrictions on $\mathfrak{G}(S_0^*)$, or $\mathfrak{G}(S_0)$, whereby elements in $\mathfrak{G}(S_0^*)$ or $\mathfrak{G}(S_0)$ are required to lie in the null spaces of a finite set of continuous linear functionals on $\mathfrak{G}(S_0^*)$, or $\mathfrak{G}(S_0)$. We propose to identify operators with their graphs, and to look at all selfadjoint problems which arise from a formally symmetric ordinary differential operator L in this way, characterize these problems, and give an eigenfunction expansion result for each of them.

Such problems include more general ones than those considered in the above examples. Suppose μ is a function of bounded variation on $0 \leq x \leq 1$, and consider the condition

$$(1.4) \qquad \int_0^1 f d\bar{\mu} = 0.$$

If μ' exists, and is in $\mathfrak{H}$, then this condition becomes $(f, \mu') = 0$; if μ is a jump function with jumps at 0 and 1, the condition reduces to $af(0) + bf(1) = 0$, for some $a, b \in \mathbb{C}$. In general, for $f \in \mathfrak{D}(S_0^*)$,

$$\int_0^1 f d\bar{\mu} = f(1)\bar{\mu}(1) - f(0)\bar{\mu}(0) - \int_0^1 f'\bar{\mu} ,$$

and this clearly shows that the map $\{f, S_0^* f\} \to \int_0^1 f d\bar{\mu}$ is continuous on $\mathfrak{G}(S_0^*)$. Conditions such as (1.4) do not exhaust those which we consider. If $\tau \in C[0, 1]$, $\tau \notin BV[0, 1]$, then there is no $\mu \in BV[0, 1]$ such that $(if', \tau) = \int_0^1 f d\bar{\mu}$ for all $f \in \mathfrak{D}(S_0)$, and the condition $(if', \tau) = 0$, for $f \in \mathfrak{D}(S_0)$, is not of the type considered in the examples above.

2. Subspaces. Let $\mathfrak{H}$ be a Hilbert space over the complex numbers $\mathbb{C}$, and let $\mathfrak{H}^2 = \mathfrak{H} \oplus \mathfrak{H}$, considered as a Hilbert space. A subspace T is a closed linear manifold in $\mathfrak{H}^2$. We shall denote elements in T by $\{f, g\}$, where $f, g \in \mathfrak{H}$. The domain $\mathfrak{D}(T)$ and range $\mathfrak{R}(T)$ of T are given by

$$\mathfrak{D}(T) = \{f \in \mathfrak{H} | \{f, g\} \in T, \text{ some } g \in \mathfrak{H}\},$$

$$\mathfrak{R}(T) = \{g \in \mathfrak{H} | \{f, g\} \in T, \text{ some } f \in \mathfrak{H}\},$$

and we let

$$T(f) = \{g \in \mathfrak{H} | \{f, g\} \in T\} , \; f \in \mathfrak{D}(T).$$

The subspace T is an operator in $\mathfrak{H}$ if $T(0) = \{0\}$, and then we write $T(f) = Tf$. We consider subspaces as linear relations with

$$\alpha T = \{\{f, \alpha g\} | \{f, g\} \in T, \ \alpha \in \mathbb{C}\},$$

$$T + S = \{\{f, g + k\} | \{f, g\} \in T, \{f, k\} \in S\},$$

$$T^{-1} = \{\{g, f\} | \{f, g\} \in T\},$$

$$\nu(T) = \{f \in \mathfrak{H} | \{f, 0\} \in T\} = T^{-1}(0) .$$

The <u>algebraic sum</u> of two subspaces T and S is

$$T \dot{+} S = \{\{f + h, g + k\} | \{f, g\} \in T, \{h, k\} \in S\} ;$$

this sum is <u>direct</u> if $T \cap S = \{\{0, 0\}\}$. The <u>orthogonal sum</u> of T and S is denoted by $T \oplus S$, and this is $T \dot{+} S$ when $T \perp S$ in $\mathfrak{H}^2$. The <u>orthogonal complement</u> of T in $\mathfrak{H}^2$ is denoted by $T^{\perp} = \mathfrak{H}^2 \ominus T$. The <u>adjoint</u> of a subspace T is the subspace T^* defined by

$$T^* = \{\{h, k\} \in \mathfrak{H}^2 | (g, h) = (f, k), \ \text{all} \ \{f, g\} \in T\}.$$

We have $T^* = (- T^{-1})^{\perp}$, and it is easy to see that T^* has all the expected properties, $T^{**} = T$, $(T^{-1})^* = (T^*)^{-1}$, $S \subset T$ implies $T^* \subset S^*$, etc. For any subspace T we may write $T = T_s \oplus T_{\infty}$, where

$$T_{\infty} = \{\{f, g\} \in T | f = 0\}, \ T_s = T \ominus T_{\infty}.$$

The subspace T_s is called the <u>operator part</u> of T; it is a closed operator in $\mathfrak{H}$ with $\mathfrak{D}(T_s) = \mathfrak{D}(T)$ which is dense in $(T^*(0))^{\perp}$, and $\mathfrak{R}(T_s) \subset (T(0))^{\perp}$.

A subspace S is said to be a <u>symmetric subspace</u> if $S \subset S^*$, and a <u>selfadjoint subspace</u> H is one satisfying $H = H^*$. Fundamental to the study of selfadjoint H is the following result, originally due to Arens.

THEOREM 2.1. <u>If</u> $H = H_s \oplus H_{\infty}$ <u>is a selfadjoint subspace in</u> $\mathfrak{H}^2$ <u>then</u> H_s <u>is a densely defined selfadjoint operator in</u> $(H(0))^{\perp}$.

This allows a spectral analysis of H, once the components H_s, H_∞ of H are identified, and a spectral analysis of H_s is known.

For a given symmetric subspace S we are interested in those selfadjoint extensions H of S in $\mathfrak{H}^2$. Such H must satisfy $S \subset H \subset S^*$. We let $M = S^* \ominus S$, and define $M_S(\ell)$ by

$$M_S(\ell) = \{\{h, k\} \in S^* | k = \ell h, \ \ell \in C\}.$$

It can be shown that $\dim M_S(\ell)$ is constant for $\ell \in C^+$, and for $\ell \in C^-$, where $C^{\pm} = \{ \ell \in C | \mathrm{Im}\ell \gtrless 0\}$. The basic facts which we require about selfadjoint extensions of S are summarized in the following theorem.

THEOREM 2.2. *Let* S *be a symmetric subspace in* $\mathfrak{H}^2$. *Then*

(a) $S^* = S \dot{+} M_S(\ell) \dot{+} M_S(\overline{\ell})$, $\ell \in C^+$, *a direct sum*,

(b) S *has a selfadjoint extension* H *in* $\mathfrak{H}^2$ *if and only if*

$$\dim M_S(\ell) = \dim M_S(\overline{\ell}), \ \ell \in C^+; \ \textit{equivalently, if and only if}$$

$$M = M_1 \oplus (- M_1^{-1}), \ H = S \oplus M_1 = S^* \ominus (- M_1^{-1}),$$

for some subspace $M_1 \subset M$,

(c) S *always has selfadjoint extensions in some* $\mathfrak{K}^2$, *where* $\mathfrak{K} \supset \mathfrak{H}$ *for an appropriate Hilbert space* $\mathfrak{K}$, *and these can be characterized*.

3. *The basic operator.* Let S_0 be a symmetric densely defined operator in a Hilbert space $\mathfrak{H}$, and identify this with its graph in $\mathfrak{H}^2$. Let B be a finite-dimensional subspace in $\mathfrak{H}^2$. We define the symmetric operator S by

$$S = S_0 \cap B^{\perp}, \ \dim B < \infty; \tag{3.1}$$

this is our basic operator, which results by restricting the elements of S_0 to lie in the null spaces of a finite set of continuous linear functionals in $\mathfrak{H}^2$.

THEOREM 3.1. *If* S *is defined by* (3.1), *then* $S^* = S_0^* \dotplus (-B^{-1})$. *The sum is direct if and only if* $B \cap S_0^{\perp} = \{\{0, 0\}\}$.

Without loss of generality we can, and do, assume $B \cap S_0^{\perp} = \{\{0, 0\}\}$; for if $\hat{B} = B \ominus (B \cap S_0^{\perp})$, then $S = S_0 \cap \hat{B}^{\perp}$, and $\hat{B} \cap S_0^{\perp} = \{\{0, 0\}\}$. This corresponds to the fact that the space of continuous linear functionals on S_0 is isomorphic to $\mathfrak{H}^2/S_0^{\perp}$.

A *special case* of (3.1) occurs when $B = \mathfrak{H}_0 \oplus \{0\}$, where $\mathfrak{H}_0$ is a finite-dimensional subspace of $\mathfrak{H}$. Then $B^{\perp} = \mathfrak{H}_0^{\perp} \oplus \mathfrak{H}$, $S = S_0 \cap (\mathfrak{H}_0 \oplus \{0\})^{\perp} = S_0 \cap (\mathfrak{H}_0^{\perp} \oplus \mathfrak{H})$, $S^* = S_0^* \dotplus (\{0\} \oplus \mathfrak{H}_0)$, $(S^*)_\infty = \{0\} \oplus \mathfrak{H}_0$. We note that $\mathfrak{D}(S) = \mathfrak{D}(S_0) \cap \mathfrak{H}_0^{\perp}$, and $Sf = S_0 f$ for $f \in \mathfrak{D}(S)$. Thus S is obtained from S_0 only by restricting $\mathfrak{D}(S_0)$.

The general case (3.1) can be reduced to this special case. We are led to this if we inquire about the nature of $(S^*)_\infty$. From Theorem 3.1 we have

$$S^* = \{\{h, S_0^* h\} + \{\tau, -\sigma\} \mid h \in \mathfrak{D}(S_0^*), \{\sigma, \tau\} \in B\},$$

and thus $\{0, \varphi\} \in S^*$ if and only if $0 = h + \tau$, $\varphi = S_0^* h - \sigma$, for some $h \in \mathfrak{D}(S_0^*)$, $\{\sigma, \tau\} \in B$. Thus $\tau = -h \in \mathfrak{D}(S_0^*)$ and $\varphi = -S_0^* \tau - \sigma$. Let

$$(3.2) \quad \begin{cases} B_2 = \{\{\sigma, \tau\} \in B \mid \tau \in \mathfrak{D}(S_0^*)\}, \ B_1 = B \ominus B_2, \\ \mathfrak{H}_0 = \{\varphi \in \mathfrak{H} \mid \varphi = -S_0^* \tau - \sigma, \ \{\sigma, \tau\} \in B_2\} . \end{cases}$$

Then we have $S = S_1 \cap B_2^{\perp}$, with $S_1 = S_0 \cap B_1^{\perp}$, and the following result is valid.

THEOREM 3.2. *Let* S *be given by* (3.1), *where* $B \cap S_0^{\perp} = \{\{0, 0\}\}$, *and let* $B_1, B_2, \mathfrak{H}_0$ *be defined by* (3.2). *Then*

(a) $(S^*)_\infty = \{0\} \oplus \mathfrak{H}_0$,

(b) $S_1 = S_0 \cap B_1^{\perp}$ *is a densely defined symmetric operator in* $\mathfrak{H}$,

(c) $S = S_1 \cap B_2^{\perp} = S_1 \cap (\mathfrak{H}_0 \oplus \{0\})^{\perp}$,

(d) $S^* = S_1^* \dotplus (\{0\} \oplus \mathfrak{H}_0)$, *a direct sum*.

Thus the general case (3.1) for S_0 is reduced to the special case for S_1.

4. Selfadjoint extensions in $\mathfrak{H}^2$. Let S be as in (3.1) with $B \cap S_0^\perp = \{\{0, 0\}\}$. We ask for the possible selfadjoint extensions of S in $\mathfrak{H}^2$. If

$$M(\ell) = M_S(\ell),\ M_0(\ell) = M_{S_0}(\ell),$$

then we have the following result.

THEOREM 4.1. *For* $\ell \in C_0 = C^+ \cup C^-$,

$$\dim M(\ell) = \dim M_0(\ell) + \dim B,$$

and therefore S *has selfadjoint extensions in* $\mathfrak{H}^2$ *if and only if* S_0 *does*.

Suppose $\dim M_0(\ell) = \dim M_0(\overline{\ell})$, $\ell \in C^+$. What do all selfadjoint extensions of S in $\mathfrak{H}^2$ look like? The answer for the general case is a little involved. We have indicated in Theorem 3.2 how the general case can be reduced to the special case, and so we present the answer in the special case. In Section 5 we shall show how this finds application in the case of an S_0 which is a minimal ordinary symmetric differential operator. An example will be given to illustrate what takes place in the general case for an ordinary differential operator.

In order to present the main result we need some notation which has proved useful. For matrices $F = (F_{kj})$, $G = (G_{kj})$, having the same number of rows, with elements in $\mathfrak{H}$, we define the "matrix inner product" (F, G) to be the matrix whose i, j-th element is

$$(F, G)_{ij} = \sum_k (F_{kj}, G_{ki}) .$$

For example, if the elements of F, G are in $\mathfrak{H} = C$, then $(F, G) = G^*F$, whereas if the elements of F, G are in $\mathfrak{H} = \mathfrak{L}^2(0, 1)$ then $(F, G) = \int_0^1 G^*F$. This matrix inner product has the following properties:

$$(G, F) = (F, G)^*,$$

$$(F, F) \geq 0, \text{ and } (F, F) = 0 \text{ if and only if } F = 0,$$

$$(F_1 + F_2, G) = (F_1, G) + (F_2, G),$$

$$(FC, G) = (F, G)C, \ (F, GD) = D^*(F, G) ,$$

where C, D are matrices of constants (elements in $\mathbb{C}$). A true inner product is given by $F \cdot G = \text{trace}(F, G)$. Finally, we denote by $(F : G)$ the matrix whose columns are obtained by placing the columns of G next to those of F in the order indicated.

The basic operator S in $\mathfrak{H}$ in the special case is described as follows:

$$(4.1) \qquad S = S_0 \cap (\mathfrak{H}_0 \oplus \{0\})^{\perp}, \ \mathfrak{H}_0 \subset \mathfrak{H}, \ \dim \mathfrak{H}_0 = p < \infty,$$

where S_0 is a densely defined symmetric operator in $\mathfrak{H}$, and where we now assume that

$$(4.2) \qquad \dim M_0(\ell) = \dim M_0(\bar{\ell}) = \omega, \quad \ell \in \mathbb{C}^+.$$

Then $S^* = S_0^* \dot{+} (\{0\} \oplus \mathfrak{H}_0)$, a direct sum, and

$$\dim M(\ell) = p + \omega.$$

For matrices F, G, with elements in $\mathfrak{D}(S_0^*)$ and having the same number of rows, we define

$$\langle F, G \rangle = (S_0^* F, G) - (F, S_0^* G).$$

THEOREM 4.2. <u>Let</u> S <u>be given by</u> (4.1) <u>and</u> (4.2), <u>and let</u> H <u>be a selfadjoint extension of</u> S <u>in</u> $\mathfrak{H}^2$ <u>with</u> $\dim H(0) = s$. <u>Let</u> $\Phi_0 = (\varphi_1, \ldots, \varphi_s)$ <u>be an orthonormal basis for</u> $H(0)$, <u>and let</u> $\Phi_1 = (\varphi_{s+1}, \ldots, \varphi_p)$ <u>be such that</u> $\Phi = (\Phi_0 : \Phi_1)$ <u>is an orthonormal basis for</u> $\mathfrak{H}_0$. <u>Then there exist</u>

(a) $\gamma = (\gamma_{s+1}, \ldots, \gamma_p) \in \mathfrak{D}(S_0^*)$,

(b) $\delta = (\delta_{p+1}, \ldots, \delta_{p+\omega}) \in \mathfrak{D}(S_0^*)$, <u>linearly independent</u> mod $\mathfrak{D}(S_0)$, <u>and</u> $\langle \delta, \delta \rangle = 0$,

<u>and a matrix</u> E <u>such that if</u>

(c) $\psi = \Phi_1[E - (1/2) \langle \gamma, \gamma \rangle]$, $E = E^*$, $E_{jk} \in \mathbb{C}$,

(d) $\zeta = -\Phi_1 \langle \delta, \gamma \rangle$,

<u>then</u>

$$H = \{\{h, S_0^* h + \varphi\} \mid h \in \mathfrak{D}(S_0^*),\ \varphi \in \mathfrak{H}_0\},$$

<u>where</u>

(i) $(h, \Phi_0) = 0$,

(ii) $\langle h, \delta \rangle - (h, \zeta) = 0$,

(iii) $\varphi = \Phi_0 c + \Phi_1[(h, \psi) - \langle h, \gamma \rangle]$,

c <u>an</u> $s \times 1$ <u>matrix with elements in</u> $\mathbb{C}$,

(iv) $H_s h = S_0^* h - \Phi_0(S_0^* h, \Phi_0) + \Phi_1[(h, \psi) - \langle h, \gamma \rangle]$.

<u>Conversely, if</u> Φ <u>is an orthonormal basis for</u> $\mathfrak{H}_0$, γ, δ <u>exist satisfying</u> (a), (b), <u>and</u> ψ, ζ <u>are defined by</u> (c), (d), <u>then the</u> H <u>given by</u> (i) - (iv) <u>is a selfadjoint extension of</u> S <u>with</u> Φ_0 <u>a basis for</u> $H(0)$.

5. *Ordinary differential operators.* Let ι be an arbitrary open interval $a < x < b$ on the real line $\mathbb{R}$; $a = -\infty$, $b = +\infty$, or both, are not ruled out. We consider a formally symmetric ordinary differential expression

$$L = \sum_{k=0}^{n} p_k D^k = \sum_{k=0}^{n} (-1)^k D^k \bar{p}_k, \quad D = d/dx,$$

where the $p_k \in C^k(\iota)$ and $p_n(x) \neq 0$, $x \in \iota$. In the Hilbert space $\mathfrak{H} = \mathfrak{L}^2(\iota)$ we let S_0 be the <u>minimal operator</u> for L This is defined as the closure in $\mathfrak{H}^2$ of the set

$$\{\{f, Lf\} \in \mathfrak{H}^2 \mid f \in C_0^\infty(\iota)\},$$

where $C_0^\infty(\iota)$ denotes the set of all infinitely differentiable functions on ι with compact support. This S_0 is symmetric, and $\mathfrak{D}(S_0^*)$ is the set of all $f \in \mathfrak{H}$ such that $f \in C^{n-1}(\iota)$, $f^{(n-1)}$ is absolutely continuous on each closed subinterval of ι, and $Lf \in \mathfrak{H}$. For $f \in \mathfrak{D}(S_0^*)$ we have $S_0^* f = Lf$; the operator S_0^* is the <u>maximal operator</u> for L in $\mathfrak{H}$.

For $f, g \in \mathfrak{D}(S_0^*)$ we have Green's formula:

$$\int_x^y \bar{g}Lf - f\overline{Lg} = [fg](y) - [fg](x),$$

where $[fg](x) = \tilde{g}^*(x)B(x)\tilde{f}(x)$, $B \in C(\iota)$ is an invertible, skew-hermitian $n \times n$ matrix-valued function, (not to be confused with the subspace B in (3.1)), and $\tilde{f}(x)$ denotes the matrix whose rows are $f(x), f'(x), \ldots, f^{(n-1)}(x)$. It follows from Green's formula that the limits

$$\lim_{y \to b} [fg](y) = [fg](b), \quad \lim_{x \to a} [fg](x) = [fg](a),$$

exist, and

$$\langle f, g \rangle = (Lf, g) - (f, Lg) = [fg](b) - [fg](a) .$$

The right side of this expression depends only on f, g and their first $n - 1$ derivatives in the neighborhood of a and b. We see that Theorem 4.2 applies to an S generated by the minimal operator S_0 for L and a subspace $\mathfrak{H}_0 \subset \mathfrak{H}$. In that theorem (i), (ii) show that $\mathfrak{D}(H)$ is given by boundary-integral conditions, and (iii), (iv) show that such terms enter into the operator part of H as well.

We note several particular cases of Theorem 4.2. Some of the selfadjoint extensions H are operators $(s = 0)$. Then Φ_0 is not present, $\Phi = \Phi_1$, and

$$Hh = Lh + \Phi[(h, \psi) - \langle h, \gamma \rangle].$$

If $\gamma = 0$, $E = 0$, then $\psi = 0$, $\zeta = 0$, and $Hh = Lh = S_0^* h$, which means that $S_0 \subset H \subset S_0^*$. In this case

$$\mathfrak{D}(H) = \{h \in \mathfrak{D}(S_0^*) \mid \langle h, \delta \rangle = 0\},$$

where δ satisfies (b). This is the familiar characterization of the selfadjoint extensions H of S_0 by means of a set of ω selfadjoint boundary conditions $\langle h, \delta \rangle = 0$.

In the regular case when $\imath$ is a finite interval, and $p_k \in C^k(\bar{\imath})$, $p_n(x) \neq 0$, $x \in \bar{\imath}$, where $\bar{\imath}$ is the closure of $\imath$, we have

$$\mathfrak{D}(S_0^*) = \{f \in \mathfrak{H} | f \in C^{n-1}(\bar{\imath}),\ f^{(n-1)} \in AC(\bar{\imath}),\ Lf \in \mathfrak{H}\},$$

$$\mathfrak{D}(S_0) = \{f \in \mathfrak{D}(S_0^*) | \tilde{f}(a) = \tilde{f}(b) = 0\}.$$

Here $AC(\bar{\imath})$ denotes the set of absolutely continuous functi^ns on $\bar{\imath}$. Consequently

$$\langle f, g\rangle = \tilde{g}^*(b)B(b)\tilde{f}(b) - \tilde{g}(a)B(a)\tilde{f}(a).$$

Since $\dim M_0(\ell) = n$, $\ell \in C$, we see that we always have selfadjoint extensions H of S in $\mathfrak{H}^2$ in this case, and the description of such H can be made more explicit. For example, the condition (ii) becomes

$$M\tilde{h}(a) + N\tilde{h}(b) + (h, \zeta) = 0,$$

for some $n \times n$ matrices of constants M, N, and (b) is equivalent to the conditions : $\operatorname{rank}(M : N) = n$, $MB^{-1}(a)M^* - NB^{-1}(b)N^* = 0$.

We now illustrate the general case, of an S given by (3.1), by a simple example of a first order regular ordinary differential operator. Let $L = iD$ and let S_0 be the minimal operator for L in $\mathfrak{H} = \mathfrak{L}^2[0, 1]$. Suppose τ is a function of bounded variation on $0 \leq x \leq 1$, which is not a constant function. Let

$$\mathfrak{D}(S) = \left\{f \in \mathfrak{D}(S_0) \ \Big|\ \int_0^1 f d\bar{\tau} = 0\right\}$$

and $S \subset S_0$. Since, for $f \in \mathfrak{D}(S_0)$,

$$\int_0^1 f d\bar{\tau} = f(1)\bar{\tau}(1) - f(0)\bar{\tau}(0) - \int_0^1 f'\bar{\tau} = -(f', \tau),$$

we see that

$$\mathfrak{D}(S) = \{f \in \mathfrak{D}(S_0) | (S_0 f, \tau) = (if', \tau) = 0\}.$$

If $B = \{0\} \oplus \{\tau\}$, where $\{\tau\}$ is the subspace spanned by τ in $\mathfrak{H}$, then we have $S = S_0 \cap B^{\perp}$ as in (3.1). The nontriviality condition $B \cap S_0^{\perp} = \{\{0, 0\}\}$ is valid since τ is not a constant. We have

$$S^* = S_0^* \dot{+} (-B^{-1}) = \{\{h, ih'\} + \{c\tau, 0\} | h \in \mathfrak{D}(S_0^*), c \in \mathbb{C}\},$$

and for any $\ell \in \mathbb{C}$, $\dim M_0(\ell) = 1$, $\dim B = 1$, and so $\dim M(\ell) = 2$. There are two cases: either (1) $\tau \in \mathfrak{D}(S_0^*)$, or (2) $\tau \notin \mathfrak{D}(S_0^*)$. In case (1) we have $B = B_2$, $\mathfrak{H}_0 = S^*(0) = \{S_0^*\tau\} = \{i\tau'\}$, and this is really the special case. In case (2) we see that $B = B_1$, $\mathfrak{H}_0 = S^*(0) = \{0\}$, and S^* is an operator. We describe all the selfadjoint extensions H of S in $\mathfrak{H}^2$ for each of these cases.

In case (1) there are two sub-cases corresponding to (i) $H(0) = \{0\}$, or (ii) $H(0) = \{i\tau'\}$. In case (1) (i) we have

$$H = \{\{h, ih' + \alpha i\tau'\}\}, \quad h \in \mathfrak{D}(S_0^*), \tag{5.1}$$

and where

$$mh(0) + nh(1) + i(\bar{d}n - \bar{c}m)(h, i\tau') = 0,$$

$$\alpha = ch(0) + dh(1) + [e + (i/2)(|d|^2 - |c|^2)](h, i\tau').$$

Here $c, d \in \mathbb{C}$ are arbitrary, $e \in \mathbb{R}$, and $m, n \in \mathbb{C}$ satisfy $|m| = |n| \neq 0$. In case (1) (ii) we have H is given by (5.1), where now $\alpha \in \mathbb{C}$ is arbitrary, and

$$(h, i\tau') = 0, \quad mh(0) + nh(1) = 0, \quad |m| = |n| \neq 0.$$

Of course H is an operator in case (1) (i), but it is not one in case (1) (ii).

In case (2) the selfadjoint extensions H of S are described by

$$H = \{\{h + \alpha\tau, ih'\}\}, \quad h \in \mathfrak{D}(S_0^*), \tag{5.2}$$

where $\alpha \in \mathbb{C}$ satisfies

$$m_1h(0) + n_1h(1) + a_1(ih', \tau) + f_1\alpha = 0,$$

$$m_2h(0) + n_2h(1) + a_2(ih', \tau) + f_2\alpha = 0,$$

and $m_j, n_j, a_j, f_j \in C$ are such that

$$\text{rank}\begin{bmatrix} m_1 & n_1 & a_1 & f_1 \\ m_2 & n_2 & a_2 & f_2 \end{bmatrix} = 2,$$

$$m_j\bar{m}_k - n_j\bar{n}_k = i(a_jf_k - f_j\bar{a}_k),\ j,\ k = 1,\ 2.$$

At first glance it does not appear from (5.2) that all such H are operators. But note that $h + \alpha\tau = 0$ implies $\alpha\tau \in \mathfrak{D}(S_0^*)$, and since $\tau \notin \mathfrak{D}(S_0^*)$ in case (2) we have $\alpha = 0$, and then $h = 0$ also. Thus $ih' = 0$, and H is indeed an operator.

A more general example can be obtained as follows. Let L be a formally symmetric ordinary differential operator L of order n on some finite interval $a \leq x \leq b$. Let S_0 be the minimal operator for L in $\mathfrak{H} = \mathfrak{L}^2[a, b]$. Suppose we define a new symmetric operator $S \subset S_0$, with

$$\mathfrak{D}(S) = \left\{ f \in \mathfrak{D}(S_0) \;\middle|\; \int_a^b f^{(j-1)}\, d\bar{\mu}_{ij} = 0 \right\},$$

where the μ_{ij}, $i = 1, \ldots, r_j$, $j = 1, \ldots, n$, are functions of bounded variation on $a \leq x \leq b$. It can be shown that there exists a $\tau = (\tau_1, \ldots, \tau_p)$, with $\tau_j \in \mathfrak{H}$, such that

$$\mathfrak{D}(S) = \{f \in \mathfrak{D}(S_0) | (S_0f, \tau) = 0\}.$$

Thus, if $B = \{0\} \oplus \{\tau_1, \ldots, \tau_p\}$, then $S = S_0 \cap B^\perp$, which is just (3.1) for this B.

6. Generalized resolvents and spectral families. Let us return to the abstract case of the basic operator S in $\mathfrak{H}$ defined by $S = S_0 \cap B^{\perp}$, where S_0 is a densely defined symmetric operator in $\mathfrak{H}$, and B is a finite-dimensional subspace of $\mathfrak{H}^2$, $\dim B = p$. We assume the non-triviality condition $B \cap S_0^{\perp} = \{\{0, 0\}\}$, and so $S^* = S_0^* \dotplus (-B^{-1})$, a direct sum. Moreover, $\dim M(\ell) = \dim M_0(\ell) + \dim B$ for $\ell \in C_0$. Let $\dim M_0(\ell) = \omega^{\pm}$, $\ell \in C^{\pm}$, and put $q^{\pm} = \omega^{\pm} + p$. Here we do not assume that $\omega^+ = \omega^-$, so that S need not have any selfadjoint extensions in $\mathfrak{H}^2$. However, from Theorem 2.2 (c) we know that S always has selfadjoint extensions in $\mathfrak{K}^2$ for some Hilbert space $\mathfrak{K} \supset \mathfrak{H}$. Let $H = H_s \oplus H_\infty$ be such a subspace extension of S in $\mathfrak{K}^2$. Then H_s is a selfadjoint operator in $H(0)^{\perp} = \mathfrak{K} \ominus H(0)$, and has a spectral resolution

$$H_s = \int_{-\infty}^{\infty} \lambda dE_s(\lambda).$$

Here $E_s = \{E_s(\lambda) | \lambda \in \mathbb{R}\}$ is the spectral family of projections in $H(0)^{\perp}$ for H_s. The resolvent R_H of H is the operator-valued function $R_H(\ell) = (H - \ell I)^{-1}$, $\ell \in C_0$, where I is the identity operator. For each $\ell \in C_0$, $R_H(\ell)$ is an operator which is defined on all of $\mathfrak{K}$, and it has the following properties:

$$(6.1)\quad \begin{cases} \text{(i)} & \|R_H(\ell)\| \leq 1/|\mathrm{Im}\,\ell|, \\ \text{(ii)} & R_H^*(\ell) = R_H(\bar{\ell}), \\ \text{(iii)} & R_H(\ell) - R_H(m) = (\ell - m) R_H(\ell) R_H(m), \\ \text{(iv)} & R_H \text{ is analytic in the uniform topology.} \end{cases}$$

We can write

$$(6.2)\qquad R_H(\ell) = \int_{-\infty}^{\infty} \frac{dE(\lambda)}{\lambda - \ell}, \quad \ell \in C_0,$$

where $E(\lambda)f = E_s(\lambda)f$, $f \in H(0)^{\perp}$, and $E(\lambda)f = 0$ for $f \in H(0)$. Thus $E(\lambda) = E_s(\lambda) \oplus O_0$, where O_0 is the zero operator on $H(0)$. The set $E = \{E(\lambda) | \lambda \in \mathbb{R}\}$ is called the <u>spectral family</u> of projections in $\mathfrak{K}$ for the selfadjoint subspace H.

Let P denote the orthogonal projection of $\mathfrak{K}$ onto $\mathfrak{H}$, and let R be the operator-valued function defined on $\mathfrak{H}$ by

$$R(\ell)f = PR_H(\ell)f, \quad f \in \mathfrak{H}, \quad \ell \in \mathbb{C}_0.$$

This R is called the <u>generalized resolvent</u> of S corresponding to the extension H. It has the properties:

$$(6.3) \quad \begin{cases} \text{(i)} & \|R(\ell)\| \leq 1/|\mathrm{Im}\ell|, \\ \text{(ii)} & R^*(\ell) = R(\overline{\ell}), \\ \text{(iii)} & \dfrac{\mathrm{Im}(R(\ell)f, f)}{\mathrm{Im}\ell} \geq \|R(\ell)f\|^2, \; f \in \mathfrak{H}, \\ \text{(iv)} & R \text{ is analytic in the uniform topology,} \\ \text{(v)} & \text{if } T(\ell) = \{\{R(\ell)f, \ell R(\ell)f + f\} | f \in \mathfrak{H}\}, \text{ then} \\ & S \subset T(\ell) \subset S^*. \end{cases}$$

The relation (6.2) implies that

$$(6.4) \qquad (R(\ell)f, f) = (R_H(\ell)f, f) = \int_{-\infty}^{\infty} \frac{d(F(\lambda)f, f)}{\lambda - \ell}, \quad f \in \mathfrak{H}, \; \ell \in \mathbb{C}_0,$$

where $F(\lambda)f = PE(\lambda)f$, $f \in \mathfrak{H}$. The set $F = \{F(\lambda) | \lambda \in \mathbb{R}\}$ is the <u>generalized spectral family</u> of S corresponding to the extension H.

The central problem for a given symmetric S in $\mathfrak{H}^2$ is to determine explicitly all the F. We shall indicate how this can be done in the case of an S generated by a minimal ordinary differential operator S_0. This leads to eigenfunction expansion results for all such F. The proof which we shall sketch depends on an inversion of (6.4). Let $\Delta = \{\nu \in \mathbb{R} | \mu < \nu \leq \lambda\}$, and $F(\Delta) = F(\lambda) - F(\mu)$, where λ, μ are continuity points of F. Then we have

$$(6.5) \qquad (F(\Delta)f, f) = \lim_{\varepsilon \to +0} \frac{1}{\pi} \int_\Delta \operatorname{Im}(R(\nu + i\varepsilon)f, f)d\nu, \quad f \in \mathfrak{H},$$

which is the inversion of the relation (6.4).

7. Generalized resolvents in the case of an ordinary differential operator. Let S_0 be the minimal operator for a formally symmetric ordinary differential operator L in the Hilbert space $\mathfrak{H} = \mathfrak{L}^2(\iota)$, as defined at the beginning of Section 5. Let B be a subspace of $\mathfrak{H}^2$, $\dim B = p < \infty$, $B \cap S_0^{\perp} = \{\{0, 0\}\}$, and define $S = S_0 \cap B^{\perp}$. We shall use the notations of Section 6. Let H be a selfadjoint subspace extension of S in $\mathfrak{K}^2$, with the corresponding generalized resolvent R.

THEOREM 7.1. *For each* $\ell \in C_0$ *the operator* $R(\ell)$ *is an integral operator of Carleman type*

$$R(\ell)f(x) = \int_a^b K(x, y, \ell)f(y)dy, \quad f \in \mathfrak{H},$$

with kernel K *satisfying* $K(x, y, \ell) = \overline{K}(y, x, \overline{\ell})$.

We briefly sketch the proof. There exists a right inverse $G(\ell)$ of $S_0^* - \ell I$, which is an integral operator of Carleman type

$$G(\ell)f(x) = \int_a^b G(x, y, \ell)f(y)dy, \quad f \in \mathfrak{H}, \ \ell \in C_0,$$

satisfying

$$(7.1) \qquad \begin{cases} \text{(i)} & \|G(\ell)\| \le 1/|\operatorname{Im}\ell|, \\ \text{(ii)} & G^*(\ell) = G(\overline{\ell}), \\ \text{(iii)} & (S_0^* - \ell I)G(\ell)f = f, \ f \in \mathfrak{H}, \ \ell \in C_0, \\ \text{(iv)} & G \text{ is analytic in the uniform topology.} \end{cases}$$

For $f \in \mathfrak{H}$ we have $\{R(\ell)f, \ell R(\ell)f + f\} \in S^*$ and $\{G(\ell)f, \ell G(\ell)f + f\} \in S_0^* \subset S^*$, by virtue of (6.3) (v) and (7.1) (iii). Thus $A(\ell) = R(\ell) - G(\ell)$ satisfies

$\{A(\ell)f, \ell A(\ell)f\} \in M(\ell)$, or $A(\ell)f \in \nu(S^* - \ell I) = \mathfrak{D}(M(\ell))$. Let $\alpha(\ell) = (\alpha_1(\ell), \ldots, \alpha_{q^\pm}(\ell))$, $\ell \in C^\pm$, be an orthonormal basis for $\nu(S^* - \ell I)$. Then $A(\ell)f = \alpha(\ell)(A(\ell)f, \alpha(\ell)) = \alpha(\ell)(f, A(\overline{\ell})\alpha(\ell)) = \alpha(\ell)a^\pm(\ell)(f, \alpha(\overline{\ell}))$, where $a^\pm(\ell) = (A(\ell)\alpha(\overline{\ell}), \alpha(\ell))$, $\ell \in C^\pm$. Therefore we have

$$A(\ell)f(x) = \int_a^b A(x, y, \ell)f(y)dy, \quad f \in \mathfrak{H}, \ \ell \in C_0,$$

where the kernel is given by

$$A(x, y, \ell) = \alpha(x, \ell)a^\pm(\ell)\alpha^*(y, \overline{\ell}), \quad \ell \in C^\pm,$$

$$(a^+(\ell))^* = a^-(\overline{\ell}), \quad \ell \in C^+.$$

This yields Theorem 7.1 with $R(\ell) = G(\ell) + A(\ell)$ and $K(x, y, \ell) = G(x, y, \ell) + A(x, y, \ell)$.

In order to motivate the form that F takes, we consider this R in a little more detail and express the kernel K in terms of other functions which are smoother than $\alpha(\ell)$. The relations $S \subset S_0 \subset S_0^* \subset S^*$ imply $S_0^* - \ell I \subset S^* - \ell I$. Thus it is clear that $\nu(S_0^* - \ell I) \subset \nu(S^* - \ell I)$. Indeed,

$$\nu(S^* - \ell I) = \nu(S_0^* - \ell I) \dot{+} N_B(\ell),$$

a direct sum, where

$$N_B(\ell) = \{G(\ell)(\sigma + \ell\tau) + \tau \,|\, \{\sigma, \tau\} \in B\}.$$

Note that if $v = G(\ell)(\sigma + \ell\tau)$, then (7.1) (iii) implies that $S_0^* v = \ell v + \sigma + \ell\tau$, and hence $\{v + \tau, \ell(v + \tau)\} = \{v, S_0^* v\} + \{\tau, -\sigma\} \in S^* = S_0^* \dot{+} (-B^{-1})$. Let $\theta^1(\ell)$ be a $1 \times \omega^\pm$ matrix $(\ell \in C^\pm)$ whose elements form a basis for $\nu(S_0^* - \ell I)$, and let $\theta^2(\ell) = G(\ell)(\sigma + \ell\tau) + \tau$, where now

$$\{\sigma, \tau\} = \{\{\sigma_1, \tau_1\}, \ldots, \{\sigma_p, \tau_p\}\}$$

is a basis for B. Then $\theta(\ell) = (\theta^1(\ell) : \theta^2(\ell))$ is a basis for $\nu(S^* - \ell I)$. We remark that

$$(L - \ell)\theta^1(\ell) = 0,$$

$$(L - \ell)v(\ell) = \sigma + \ell\tau, \quad v(\ell) = G(\ell)(\sigma + \ell\tau).$$

Now let $c \in \imath$ be fixed and let

$$s^1(x, \ell) = (s_1(x, \ell), \ldots, s_n(x, \ell)),$$

$$u(x, \ell) = (u_{n+1}(x, \ell), \ldots, u_{n+p}(x, \ell)),$$

be the unique functions satisfying

$$(7.2) \qquad \begin{aligned} &(L - \ell)s^1(\ell) = 0, \quad \tilde{s}^1(c, \ell) = I_n, \quad \ell \in C, \\ &(L - \ell)u(\ell) = \sigma + \ell\tau, \quad \tilde{u}(c, \ell) = 0_n^p, \quad \ell \in C, \end{aligned}$$

and put

$$(7.3) \qquad s(\ell) = (s^1(\ell) : s^2(\ell)), \quad s^2(\ell) = u(\ell) + \tau.$$

Here I_n is the $n \times n$ identity matrix and 0_n^p is the $n \times p$ zero matrix. If $w(\ell) = (s^1(\ell) : u(\ell))$ then $\tilde{w}$ is continuous on $\imath \times C$, and for each fixed $x \in \imath$ it is entire in ℓ. The basis $\theta(\ell)$ can be expressed in terms of $s(\ell)$ as $\theta(\ell) = s(\ell)d(\ell)$, for some matrix $d(\ell)$, and then the kernel A can be expressed in terms of $s(\ell)$ to yield

$$A(x, y, \ell) = s(x, \ell)a(\ell)s^*(y, \overline{\ell}), \quad \ell \in C_0,$$

for some $(n + p) \times (n + p)$ matrix $a(\ell)$. We now write $G(x, y, \ell) = K_0(x, y, \ell) + G_1(x, y, \ell)$, where K_0 is a fundamental solution for $L - \ell$ given by

$$K_0(x, y, \ell) = \overline{K}_0(y, x, \overline{\ell}) = (1/2)s^1(x, \ell)\mathfrak{s}^{-1}(s^1(y, \overline{\ell}))^*, \quad x \geq y,$$

where $\mathfrak{s} = [s^1(\ell)s^1(\overline{\ell})](x)$ is independent of x and ℓ by virtue of Green's formula. This K_0 has the appropriate jump in the $(n - 1)$st derivative at $x = y$, which is the same as that for G. Thus G_1 is smooth in x, y, and we may write

$$G_1(x, y, \ell) = s^1(x, \ell)g(\ell)(s^1(y, \overline{\ell}))^*,$$

where $g(\ell)$ is some $n \times n$ matrix. We now have

$$(7.4) \qquad K(x, y, \ell) = K_0(x, y, \ell) + K_1(x, y, \ell),$$

where

$$(7.5) \qquad K_1(x, y, \ell) = G_1(x, y, \ell) + A(x, y, \ell) = s(x, \ell)\Psi(\ell)s^*(y, \overline{\ell}),$$

with the $(n + p) \times (n + p)$ matrix $\Psi(\ell)$ given by

$$\Psi(\ell) = a(\ell) + \begin{bmatrix} g(\ell) & 0_n^p \\ 0_p^n & 0_p^p \end{bmatrix}.$$

We note that K_0 is entire in ℓ but has a jump in its $(n - 1)$st derivative at $x = y$, whereas K_1 is nice as a function of x, y but as a function of ℓ has possible discontinuities across the real axis, $\operatorname{Im}\ell = 0$, due to the matrix $\Psi(\ell)$.

8. Eigenfunction expansions. We can now state the main result which describes the generalized spectral family F of S corresponding to the selfadjoint extension H.

THEOREM 8.1. *Let* H *be any selfadjoint subspace extension of* S *in* $\mathfrak{K}^2$, $\mathfrak{H} \subset \mathfrak{K}$, *with corresponding generalized spectral family* F. *If* $s(x, \ell)$ *is defined by* (7.2), (7.3), *then there exists an* $(n + p) \times (n + p)$ *matrix-valued function* ρ *on* $\mathbb{R}$ *which is hermitian, nondecreasing, of bounded variation on each finite interval, and such that*

$$(8.1) \qquad F(\Delta)f = \int_{\Delta} s(\nu)d\rho(\nu)\hat{f}(\nu), \quad f \in C_0(\iota),$$

where

$$\hat{f}(\nu) = (f, s(\nu)) = \int_a^b s^*(x, \nu)f(x)dx.$$

We remark that ρ is essentially unique, in the sense that if ρ_1, ρ_2 are two matrix-valued functions on $\mathbb{R}$ such that

$$(F(\Delta)f,\ g) = \int_\Delta \hat{g}^*(\nu)d\rho_j(\nu)\hat{f}(\nu), \quad j = 1,\ 2,$$

for all $f,\ g \in C_0(\iota)$ and all right-closed intervals Δ, then

$$\int_\Delta d\rho_1(\nu) = \int_\Delta d\rho_2(\nu)$$

for all such Δ. Thus, properly normalized, ρ is called the <u>spectral matrix</u> for H and F.

The proof of Theorem 8.1 follows from the decomposition (7.4) and the properties of the matrix-valued function Ψ. It can be shown that Ψ is analytic for $\ell \in C_0$, $\Psi^*(\ell) = \Psi(\bar{\ell})$, and that

$$\frac{\mathrm{Im}\Psi(\ell)}{\mathrm{Im}\ell} \geq 0, \quad \mathrm{Im}\Psi = (\Psi - \Psi^*)/2i.$$

This last inequality results from (6.3)(iii) applied to appropriate $f \in \mathfrak{H}$. These properties of Ψ imply the representation

$$\Psi(\ell) = \alpha + \ell\beta + \int_{-\infty}^{\infty} \frac{\lambda\ell + 1}{\lambda - \ell}\, d\sigma(\lambda), \quad \ell \in C_0,$$

where α, β are hermitian constant matrices, $\beta \geq 0$, and σ is a hermitian nondecreasing matrix-valued function of bounded variation. In turn, this implies that

$$\rho(\lambda) = \lim_{\varepsilon \to +0} \frac{1}{\pi} \int_0^\lambda \mathrm{Im}\Psi(\nu + i\varepsilon)d\nu$$

exists, is nondecreasing and of bounded variation on each finite subinterval; in fact,

$$d\rho(\nu) = (1 + \nu^2)d\sigma(\nu).$$

Now, using the inversion formula (6.5) we obtain Theorem 8.1 with this ρ by writing $R(\ell) = R_0(\ell) + R_1(\ell)$, for $f \in C_0(\iota)$, with kernels K_0, K_1, respectively. In the limit as $\varepsilon \to +0$ the part with $R_0(\ell)$ tends to zero and the part with $R_1(\ell)$ yields the term (8.1) with ρ; see (7.4) and (7.5).

Let ζ, η be functions from $\mathbb{R}$ to $\mathbb{C}^{n+p}$, considered as $(n+p) \times 1$ matrices, and let

$$(\zeta, \eta) = \int_{-\infty}^{\infty} \eta^*(\nu) d\rho(\nu) \zeta(\nu).$$

Since ρ is nondecreasing, $(\zeta, \zeta) \geq 0$, and we can define $\|\zeta\| = (\zeta, \zeta)^{1/2}$, and

$$\hat{\mathfrak{H}} = \mathfrak{L}^2(\rho) = \{\zeta \mid \|\zeta\| < \infty\}.$$

THEOREM 8.2. <u>Let</u> $H, s(\ell), \rho$ <u>be as in Theorem</u> 8.1. <u>For</u> $f \in \mathfrak{H}$,

$$\hat{f}(\nu) = \int_a^b s^*(x, \nu) f(x) dx$$

<u>converges in norm in</u> $\hat{\mathfrak{H}}$, <u>and</u>

$$F(\infty) f = \int_{-\infty}^{\infty} s(\nu) d\rho(\nu) \hat{f}(\nu),$$

<u>where this integral converges in norm in</u> $\mathfrak{H}$. <u>We have</u>

$$(F(\infty) f, g) = (\hat{f}, \hat{g}),$$

<u>and the map</u> V <u>of</u> $\mathfrak{H}$ <u>into</u> $\hat{\mathfrak{H}}$ <u>defined by</u> $Vf = \hat{f}$ <u>is a contraction</u>, $\|Vf\| \leq \|f\|$. <u>It is an isometry</u>, $\|Vf\| = \|f\|$, <u>for</u> $f \in \mathfrak{H} \cap H(0)^{\perp} = \mathfrak{H} \ominus PH(0)$, <u>and</u>

$$f = \int_{-\infty}^{\infty} s(\nu) d\rho(\nu) \hat{f}(\nu), \quad f \in \mathfrak{H} \ominus PH(0).$$

This result may be viewed as an eigenfunction expansion corresponding to the selfadjoint extension H and the generalized spectral family F. We explore it in a little more detail. Note that $F(\infty) = PP_s$, where P_s is the projection of $\mathfrak{K}$ onto $H(0)^{\perp}$. The map V implies a splitting of $\mathfrak{H}$ and $V\mathfrak{H} \subset \hat{\mathfrak{H}}$. Let

$$\mathfrak{H}_0 = \{f \in \mathfrak{H} \mid F(\infty) f = f\}, \quad \mathfrak{H}_1 = \{f \in \mathfrak{H} \mid F(\infty) f = 0\},$$

where this $\mathfrak{H}_0$ is not to be confused with the $\mathfrak{H}_0$ appearing in Theorem 4.2.

THEOREM 8.3. _We have_

$$\mathfrak{H}_0 = \mathfrak{H} \cap H(0)^{\perp} = \{f \in \mathfrak{H} \mid \|Vf\| = \|f\|\},$$

$$\mathfrak{H}_1 = \mathfrak{H} \cap H(0) = \{f \in \mathfrak{H} \mid Vf = 0\},$$

and

$$\mathfrak{H} = \mathfrak{H}_0 \oplus \mathfrak{H}_1 \oplus \mathfrak{H}_2, \quad V\mathfrak{H} = V\mathfrak{H}_0 \oplus V\mathfrak{H}_2.$$

It is easy to see that if $\mathfrak{D}(S)$ is dense in $\mathfrak{H}$, or if H is an operator, then $\mathfrak{H} = \mathfrak{H}_0$. Thus nontrivial $\mathfrak{H}_1$, $\mathfrak{H}_2$ can exist only for nondensely defined S and a subspace (nonoperator) extension H in a $\mathfrak{K}$ properly containing $\mathfrak{H}$. Simple examples exist illustrating the occurrence of such nontrivial $\mathfrak{H}_1$, $\mathfrak{H}_2$.

THEOREM 8.4. _We have_ $V\mathfrak{H}_0 = \hat{\mathfrak{H}}$ _if and only if_ F _is the spectral family for a selfadjoint subspace extension_ H _of_ S _in_ $\mathfrak{H}^2$ _itself_.

If H is in $\mathfrak{H}^2$ we must have $\omega^+ = \omega^-$. If $\omega^+ = \omega^- = n$, the order of the differential expression L, then the elements of $s(\ell)$, given by (7.2), (7.3), are in $\mathfrak{H} = \mathfrak{L}^2(\iota)$ for all $\ell \in C_0$. This implies that $R(\ell)$ is of Hilbert-Schmidt type, and hence the spectrum of H is a pure point spectrum. Thus, in this case, ρ consists of step functions only. In particular, this is true for selfadjoint H in the regular case.

Some References to Recent Work

1. E. A. Coddington, Extension theory of formally normal and symmetric subspaces, Mem. Amer. Math. Soc. No. 134 (1973).

2. E. A. Coddington, Selfadjoint subspace extensions of nondensely defined symmetric operators, Bull. Amer. Math. Soc. 79 (1973), 712-715; complete version with proofs to appear in Advances in Math.

3. E. A. Coddington, Eigenfunction expansions for nondensely defined operators generated by symmetric ordinary differential expressions, Bull. Amer. Math. Soc. 79 (1973), 964-968; complete version will appear in Advances in Math. under the title: Selfadjoint problems for nondensely defined ordinary differential operators and their eigenfunction expansions.

4. E. A. Coddington and A. Dijksma, Selfadjoint subspaces and eigenfunction expansions for ordinary differential subspaces, to appear.

5. A. Dijksma and H. S. V. de Snoo, Eigenfunction expansions for nondensely defined differential operators, to appear in J. Diff. Equations.

6. A. M. Krall, Differential-boundary operators, Trans. Amer. Math. Soc. 154 (1971), 429-458.

7. O. Vejvoda and M. Tvrdý, Existence of solutions to a linear integro-boundary-differential equation with additional conditions, Ann. di Mat. Pura ed Appl. (Ser. 4) 89 (1971), 169-216.

8. H. J. Zimmerberg, Linear integro-differential-boundary-parameter problems, to appear in Ann. di Mat. Pura ed Appl.

Quasi-linear equations of evolution, with applications to partial differential equations

Tosio Kato

Introduction

The purpose of this paper is to present a unified treatment of the Cauchy problem (local in time) for various quasi-linear partial differential equations that appear in mathematical physics, based on the theory of abstract equations of evolution. We shall prove two rather general theorems on abstract equations, one on the existence and uniqueness, and the other on the continuous dependence on the data, of the solution. Then we shall show that these theorems are applicable to different kinds of quasi-linear differential equations such as

symmetric hyperbolic systems of the first order,
wave equations,
Korteweg-de Vries equation,
Navier-Stokes and Euler equations,
equations for compressible fluids,
magnetohydrodynamic equations,
coupled Maxwell and Dirac equations, etc.

Of course these equations have been studied by many authors, but so far independently of one another. In our method, however, they can be handled as a straightforward application of the abstract theorems; we have only to verify several conditions, and this can be done rather mechanically, although there might be some computational problems involved.

In this paper we have endeavored to make the abstract theorems as simple as possible, while retaining sufficient generality to cover many applications. But we shall also indicate how they can be generalized to more powerful (though more complicated) ones.

Our results are <u>local in time</u>; we are concerned with solutions that exist possibly for only a short time, depending on the data. (It is not likely that the global existence of solutions for different types of equations as shown above can be handled in a unified way.) In view of the fact that the construction of a local solution is usually a major part of the work, even when a global solution exists, it is hoped that our results are nevertheless of sufficient general interest.

The abstract equation of evolution we consider has the form

$$\text{(Q)} \qquad du/dt + A(t,u)u = f(t,u), \quad 0 \leq t \leq T, \quad u(0) = \phi.$$

Here the unknown u takes values $u(t)$ in a Banach space X, $A(t,y)$ is a linear operator in X depending on t and (certain) $y \in X$, and $f(t,y)$ depends on t and (certain) $y \in X$. It will be noted that we regard (Q) as a general "hyperbolic" equation, i.e., we assume that $-A(t,y)$ is the generator of a C_0-semigroup but not necessarily of an analytic semigroup. This makes for a wider range of application, although it may lead to a less sharp result when the equation is in fact "parabolic".

The results we prove for (Q) are based on the results for linear "hyperbolic" equation

$$\text{(L)} \qquad du/dt + A(t)u = f(t), \quad 0 \leq t \leq T, \quad u(0) = \phi.$$

These results, which have been recently obtained by the author, are summarized in Part 1 without proof, in a form slightly more general than required in this paper (so that they are useful in generalizing the main theorems).

In Part 2 we prove the basic theorems for (Q) and give a simple illustrative example. We shall also indicate how they can be generalized so as to cover a wider range of application.

Part 3 is devoted to the application of the abstract theorems to various differential equations mentioned above. In most cases the theorems proved in Part 2 are directly applicable, but there are some equations in which we have to use the generalized ones indicated above.

Part 1. Linear equations of evolution

1. Introduction.

In this part we consider the Cauchy problem for the linear equation

$$\text{(L)} \qquad du/dt + A(t)u = f(t), \quad 0 \leq t \leq T, \quad u(0) = \phi,$$

in a Banach space X. Equations of this form have been studied by many authors by different methods (for general expositions see the books by Lions [1], Yosida [2], Krein [3], Friedman [4], Carroll [5], Pazy [6]). In this paper we shall use the method based on the semigroup theory.

Such a method was first used by the author [7], under the rather strong assumptions that $-A(t)$ generates a contraction semigroup and has a domain independent of t. Many attempts have been made to weaken these assumptions. It turned out that this can be done rather easily if one assumes that $-A(t)$ generates an analytic semigroup (the "parabolic" case), but this assumption is too restrictive to be useful in a unified treatment of nonlinear equations we have in mind. Most of the expositions listed above lean toward the parabolic case. (There are problems in which parabolicity appears to be essential, e.g. viscous compressible fluids, but we shall not consider them in this paper.)

Recently, in [8, 9] we have been able to prove rather general theorems on (L) which are applicable to the general "hyperbolic" case, eliminating the restriction on the domain of $A(t)$ mentioned above. These theorems will be used in the sequel as a basis for solving quasi-linear equations of evolution. In this part we shall introduct these theorems without proof.

First we introduce some definitions and notation. In what follows we consider real Banach spaces $X, Y, \ldots$, with the norms denoted by $\| \|_X$, $\| \|_Y$, ... or simply by $\| \|$ if there is no ambiguity. (In some spplications we consider spaces of complex-valued functions, but we shall regard them as real Banach spaces.)

We denote by $B(X,Y)$ the set of all bounded linear operators on X to Y, with the associated norm denoted by $\| \|_{X,Y}$ of simply by $\| \|$. We write $B(X)$ for $B(X,X)$ and $\| \|_X$ for $\| \|_{X,X}$. The domain of an operator T is denoted by $D(T)$.

$G(X)$ denotes the set of all negative generators of C_0-semigroups on X. (For semigroup theory see Hille-Phillips [10], Yosida [2], Kato [11]). More precisely, we denote by $G(X,M,\beta)$ the set of all linear operators A in X such that $-A$ generates a C_0-semigroup $\{e^{-tA}\}$ with $\|e^{-tA}\| \leq Me^{\beta t}$, $0 \leq t < \infty$. In particular, A is m-accretive if $A \in G(X,1,0)$, in which case $\{e^{-tA}\}$ is a contraction semigroup. A is quasi-m-accretive if $A \in G(X,1,\beta)$.

Let $\{A(t)\}_{0 \leq t \leq T}$ be a family of operators belonging to $G(X)$. $\{A(t)\}$ is said to be stable if there are M, β such that

$$\left\| \prod_{j=1}^{k} (A(t_j) + \lambda)^{-1} \right\| \leq M(\lambda - \beta)^k, \quad \lambda > \beta, \tag{1.1}$$

for every finite family $0 \leq t_1 \leq \ldots\ldots \leq t_k \leq T$, $k = 1,2,\ldots$. The pair M, β will be called the stability index for $\{A(t)\}$. In (1.1) the operator product on the left is time-ordered: $A(t_j)$ is to the left of $A(t_i)$ if $t_j > t_i$. It can be shown (see [8]) that (1.1) is equivalent to

$$\left\| \prod_{j=1}^{k} e^{-s_j A(t_j)} \right\| \leq Me^{\beta(s_1+\ldots+s_k)} \tag{1.2}$$

for all the t_j of the kind described above and for all $s_j \geq 0$, with the product on the left again time-ordered. We note that $\{A(t)\}$ is trivially stable (with stability index 1, β) if $A(t) \in G(X,1,\beta)$.

A more general notion of <u>quasi-stability</u> was introduced in [9], but we shall not need it in this paper.

2. <u>An existence theorem for the evolution operator</u>.

A major part of the study of (L) consists in constructing the <u>evolution operator</u> $U(t,s) \in B(X)$, defined on the triangle

$$\Delta : T \geq t \geq s \geq 0.$$

$\{U(t,s)\}$ is defined as the family of operators such that $u(t) = U(t,s)\emptyset$ is the solution of the homogeneous differential equation $du/dt + A(t)u = 0$ with the initial value $u(s) = \emptyset$. Obviously $U(t,s) = e^{-(t-s)A}$ if $A(t) = A = \text{const}$ and $A \in G(X)$. Once $\{U(t,s)\}$ has been constructed, the solution of (L) is given by

$$\text{(S)} \qquad u(t) = U(t,0)\emptyset + \int_0^t U(t,s)f(s)ds$$

at least formally.

To construct the evolution operator, we introduce the follow conditions, which will be assumed throughout Part 1.

(i) $\{A(t)\}_{0 \leq t \leq T}$ is a stable family of operators in $G(X)$, with the stability index M, β.

(ii) There is a Banach space Y, continuously and densely embedded in X, and an isomorphism S of Y onto X, such that

$$SA(t)S^{-1} = A(t) + B(t), \quad B(t) \in B(X), \quad 0 \leq t \leq T, \tag{2.1}$$

where $t \to B(t)$ is strongly measurable as an operator-valued function (i.e. $t \to B(t)x$ is strongly measurable as an X-valued function for each $x \in X$) and where $t \to \|B(t)\|_X$ is upper integrable on $[0,T]$.

(iii) $Y \in D(A(t))$, $0 \leq t \leq T$, so that $A(t) \in B(Y,X)$. (More precisely, the restriction of $A(t)$ to Y belongs to $B(Y,X)$.) $t \to A(t) \in B(Y,X)$ is continuous in norm.

REMARK. Condition (2.1) should be satisfied in the strict sense including the domain relation. $\|B(t)\|_X$ need not be measurable in t. This is why we assumed upper integrability.

THEOREM 1. Under the assumptions (i) to (iii), there exists a unique evolution operator $\{U(t,s)\}$ defined on the triangle Δ, with the following properties.

(a) U is strongly continuous on Δ to $B(X)$, with $U(s,s) = 1$.

(b) $U(t,s)U(s,r) = U(t,r)$.

(c) $U(t,s)Y \subset Y$, and U is strongly continuous on Δ to $B(Y)$.

(d) $dU(t,s)/dt = - A(t)U(t,s)$, $dU(t,s)/ds = U(t,s)A(s)$, which exist in the strong sense in $B(Y,X)$ and are strongly continuous on Δ to $B(Y,X)$.

This theorem is a special case of Theorem I of [9], in which (i), (ii) are replaced by more general conditions (in particular, S is allowed to depend on t). It was proved previously in [8] under a stronger assumption that $B(t)$ be strongly continuous. It appears that weakening of the strong continuity of $B(t)$ to strong measurability is essential in applications to nonlinear problems.

3. Estimates for $U(t,s)$.

We shall need estimates for

$$|||U|||_{\infty,X} = \sup_{t,s\in\Delta} \|U(t,s)\|_X \tag{3.1}$$

and $|||U|||_{\infty,Y}$ defined similarly. We have

$$\|U\|_{\infty,X} \leq Me^{\beta T}, \tag{3.2}$$

$$\|U\|_{\infty,Y} \leq \|S\|_{Y,X}\|S^{-1}\|_{X,Y}\, Me^{\beta T+M\|B\|_{1,X}}, \tag{3.3}$$

where

$$\|B\|_{1,X} = \int_0^T (*)\, \|B(t)\|_X dt, \tag{3.4}$$

where $(*)$ indicates the upper integral.

For the proof of (3.2-3), see [9]. Actually the estimate for $\|U\|_{\infty,Y}$ given in [9] was more crude than (3.3), but a simple refinement in the computation easily leads to (3.3).

4. Solution of (L).

With the evolution operator $\{U(t,s)\}$ constructed by Theorem 1, the solution of (L) is formally given by (S). For the u given by (S) to be differentiable, however, we need further assumptions on $\emptyset$ and f. In any case we may call (S) a mild solution of (L). The following theorem is proved in [9].

THEOREM 2. Let u be given by (S).

(a) If $\emptyset \in X$ and $f \in L^1[0,T;X]$, then $u \in C[0,T;X]$.

(b) If $\emptyset \in Y$ and $f \in L^1[0,T;Y]$, then $u \in C[0,T;Y]$.

(c) If $\emptyset \in Y$ and $f \in C[0,T,X] \cap L^1[0,T;Y]$, then $u \in C[0,T;Y] \cap C^1[0,T;X]$ and u satisfies (L).

Furthermore, we have the estimates

$$\|u\|_{\infty,X} \leq \|U\|_{\infty,X}(\|\emptyset\|_X + \|f\|_{1,X}) \tag{4.1}$$

$$\|u\|_{\infty,Y} \leq \|U\|_{\infty,Y}(\|\emptyset\|_Y + \|f\|_{1,Y}), \tag{4.2}$$

$$\|du/dt\|_{\infty,X} \leq \|f\|_{\infty,X} + \|A\|_{\infty,Y,X}(\|\emptyset\|_Y + \|f\|_{1,Y}). \tag{4.3}$$

Here we have used the obvious notation such as

$$||f||_{\infty, X} = \sup_{0 \leq t \leq T} \|f(t)\|_X, \quad ||f||_{1,X} = \int_0^T \|f(t)\|_X dt,$$

$$||A||_{\infty, Y, X} = \sup_{0 \leq t \leq T} \|A(t)\|_{Y,X}.$$

5. Perturbation theory.

In applications to non-linear equations, it is important to know how the solution of (L) changes when $A(t)$, $f(t)$, and ϕ are subjected to small changes. Consider another equation

$$(L') \qquad du'/dt + A'(t)u' = f'(t), \quad 0 \leq t \leq T, \quad u'(0) = \phi'.$$

We assume that conditions (i) to (iii) are satisfied for the family $\{A'(t)\}$ with the same Y and S, so that the evolution operator $\{U'(t,s)\}$ exists and (L') is solvable. We denote by M', β' the stability index for $\{A'(t)\}$, and by $B'(t)$ the operator corresponding to $B(t)$ for (L).

THEOREM 3. Let $\phi \in Y$, $f \in L^1[0,T;Y]$, $\phi' \in X$ and $f' \in L^1[0,T;X]$. If u, u' are the mild solutions of (L), (L'), respectively, we have

$$||u' - u||_{\infty, X} \leq ||U'||_{\infty, X}[\|\phi' - \phi\|_X + ||f' - f||_{1,X} + ||(A' - A)u||_{1,X}]; \tag{5.1}$$

note that $u \in C[0,T;Y]$ by Theorem 2 so that $[A'(t) - A(t)]u(t)$ is in X.

THEOREM 4. Let $\phi, \phi' \in Y$ and $f, f' \in L^1[0,T;Y]$. Then

$$||u' - u||_{\infty, Y} \leq K'[\|\phi' - \phi\|_Y + ||f' - f||_{1,Y} + ||(B' - B)Su||_{1,X} + ||h||_{\infty, X}], \tag{5.2}$$

where

$$h(t) = [U'(t,0) - U(t,0)]S\phi + \int_0^t [U'(t,s) - U(t,s)]g(s)ds,$$

$$g(s) = Sf(s) - B(s)Su(s); \qquad (5.3)$$

note that $S\phi \in X$ and $g(s) \in X$ because $u(s) \in Y$ by Theorem 2. Here K' denotes a constant depending only on $\|U'\|_{\infty,X}$ and $\|U'\|_{\infty,Y}$.

These theorems are special cases of Theorems III, IV of [9].

6. A Convergence theorem.

Suppose we have a sequence of equations in X:

$$(L^n) \qquad du^n/dt + A^n(t)u^n = f^n(t), \quad 0 \leq t \leq T, \quad u^n(0) = \phi^n,$$

$n = 1,2,\ldots$. Suppose the A^n satisfy conditions (i), (ii), (iii) uniformly in n, by which we mean that the stability index M, β for $\{A^n(t)\}$ can be chosen independent of n and that $\|B^n\|_{1,X}$ is bounded in n (B^n is defined from A^n as in (2.1)). It is assumed that X, Y, S are common to all (L^n).

Then the evolution operator $\{U^n(t,s)\}$ for $\{A^n(t)\}$ exists.

THEOREM 5. In addition to the above assumptions, suppose that

$$A^n(t) \to A(t) \text{ strongly in } B(Y,X) \text{ for a.e. } t, \qquad (6.1)$$

$$\int_E \|A^n(t)\|_{Y,X}dt \to 0 \text{ as } |E| \to 0 \text{ uniformly in } n, \qquad (6.2)$$

where $|\ |$ denotes the Lebesgue measure. Then we have

$$U^n(t,s) \to U(t,s) \text{ strongly in } B(X), \text{ uniformly in } t,s. \qquad (6.3)$$

If in addition $\phi^n \to \phi$ in X and $f^n \to f$ in $L^1[0,T;X]$, then $u^n \to u$ in $C[0,T;X]$, where u^n, u are the solutions to (L^n), (L), respectively.

For the proof see [9].

Part 2. Quasi-linear equations of evolution

7. Statement of the main theorems.

In this part we consider the Cauchy problem for the quasi-linear equation of evolution

$$(Q) \qquad du/dt + A(t,u)u = f(t,u), \quad 0 \leq t \leq T, \quad u(0) = \phi,$$

in a Banach space X. We may call $A(t,u)u$ the quasi-linear part and $f(t,u)$ the semi-linear part of the given equation, but the division of a given non-linear term into these two parts is not necessarily unique. In general, however, the quasi-linear part is more singular than the semi-linear part.

Roughly, we want to solve (Q) in the following way. For certain functions $t \to v(t) \in X$, we consider the linear equation

$$(L^v) \qquad du/dt + A(t,v(t))u = f(t,v(t)), \qquad u(0) = \phi.$$

If (L^v) has a solution $u = u(t)$, we have defined a mapping $v \to u = \Phi v$. Then we seek a fixed point of Φ, which will be a solution of (Q). To show that Φ has a fixed point, we shall use the contraction mapping theorem.

To this end we introduce the following assumptions.

(X) X is a reflexive Banach space. There is another reflexive Banach space $Y \subset X$, continuously and densely embedded in X. There is an isomorphism S of Y onto X. The norm in Y is chosen so that S becomes an isometry.

(A1) A is a function on $[0,T] \times W$ into $G(X,1,\beta)$, where W is an open ball in Y and β is a real number. In other words,

$$\|e^{-sA(t,y)}\|_X \leq e^{\beta s}, \quad s \in [0,\infty), \quad t \in [0,T], \quad y \in W. \tag{7.1}$$

(A2) For each $t,y \in [0,T] \times W$, we have

$$SA(t,y)S^{-1} = A(t,y) + B(t,y), \tag{7.2}$$

where

$$B(t,y) \in B(X), \qquad \|B(t,y)\|_X \leq \lambda_1, \tag{7.3}$$

with $\lambda_1 > 0$ a constant. (The relation (7.2) should be satisfied in the strict sense, including the domain relation. Thus $x \in X$ is in $D(A(t,y))$ if and only if $S^{-1}x \in D(A(t,y))$ with $A(t,y)S^{-1}x \in Y$.)

(A3) For each $t, y \in [0,T] \times W$, we have $A(t,y) \in B(Y,X)$ (in the sense that $D(A(t,y)) \supset Y$ and the restriction of $A(t,y)$ to Y is in $B(Y,X)$). For each $y \in W$, $t \to A(t,y)$ is continuous in the $B(Y,X)$-norm. For each $t \in [0,T]$, $y \to A(t,y)$ is Lipschitz-continuous in the sense that

$$\|A(t,y) - A(t,z)\|_{Y,X} \leq \mu_1 \|y-z\|_X, \tag{7.4}$$

where μ_1 is a constant.

(A4) Let y_0 be the center of W. Then $A(t,y)y_0 \in Y$ for all $t, y \in [0,T] \times W$, with

$$\|A(t,y)y_0\|_Y \leq \lambda_2, \quad t \in [0,T], \quad y \in W. \tag{7.5}$$

(f1) f is a bounded function on $[0,T] \times W$ to Y:

$$\|f(t,y)\|_Y \leq \lambda_3, \quad t \in [0,T], \quad y \in W. \tag{7.6}$$

For each $y \in W$, $t \to f(t,y)$ is continuous on $[0.T]$ to X.
For each $t \in [0,T]$, $y \to f(t,y)$ is X-Lipschitz continuous:

$$\|f(t,y) - f(t,z)\|_X \leq \mu_2 \|y-z\|_X. \tag{7.7}$$

REMARK 7.1 1. Condition (A4) is trivially satisfied if $y_0 = 0$.

2. In many cases $A(t,y)$ is defined for all $y \in Y$, so that W may be chosen as an arbitrary ball with center 0, although the constants $\beta, \lambda_1, \mu_1, \ldots\ldots$ will depend on the radius of the ball.

3. A sufficient condition for (7.2) and (7.3) to be satisfied is that

$$\|[SA(t,y) - A(t,y)S]S^{-1}w\|_X \leq \lambda_1 \|w\|_X \tag{7.8}$$

is true for all w in a core for $A(t,y)$ (which may depend on t,y). For a proof see [8, §8].

We are now able to state the existence and uniqueness theorem.

THEOREM 6. Let conditions (X), (A1) to (A4), and (f1) be satisfied. If $\emptyset \in W$, then (Q) has a unique solution

$$u \in C[0,T';W] \cap C^1[0,T';X] \quad \text{with} \quad u(0) = \emptyset, \tag{7.9}$$

for some $T' > 0$, $T' \leq T$.

To formulate the continuous dependence of the solution u on the data, we consider, in addition to (Q), a sequence of equations

(Q^n) $\quad du^n/dt + A^n(t,u^n)u^n = f^n(t,u^n), \quad 0 \leq t \leq T, \quad u^n(0) = \emptyset^n, \quad n = 1, 2, \ldots.$

For the functions A^n and f^n, we shall assume not only conditions (A1) to (A4) and (f1) with the same X, Y, S, W as above but we shall need the following additional conditions.

(A5) $$\|B(t,y) - B(t,z)\|_X \leq \mu_3 \|y - z\|_Y,$$

(f2) $$\|f(t,y) - f(t,z)\|_Y \leq \mu_4 \|y - z\|_Y.$$

THEOREM 7. In addition to the assumptions in Theorem 6, assume that (A1) to (A5) and (f1), f(2) are satisfied for (Q^n) uniformly in n (in the sense that all the constants β, λ_1. ..., μ_4 are independent of n). Furthermore, assume that for each $t, y \in [0,T] \times W$

$$A^n(t,y) \to A(t,y) \quad \text{strongly in } B(Y,X), \tag{7.10}$$

$$B^n(t,y) \to B(t,y) \quad \text{strongly in } B(X), \tag{7.11}$$

$$f^n(t,y) \to f(t,y) \quad \text{in } Y, \quad \text{as } n \to \infty. \tag{7.12}$$

If $\emptyset, \emptyset_n \in W$ and $\emptyset_n \to \emptyset$ in the Y-norm as $n \to \infty$, then there is $T'' > 0$, $T'' \leq T$, such that there are unique solutions

$$u^n \in C[0,T'';W] \cap C^1[0,T'';X] \quad \text{with} \quad u^n(0) = \emptyset^n \tag{7.13}$$

to (Q^n), $n = 1,2,\ldots$, and a unique solution u to (Q) in the same class. Moreover, we have

$$u^n(t) \to u(t) \quad \text{in } Y, \text{ uniformly in } t \in [0,T'']. \tag{7.14}$$

REMARK 7.2. The assertion of Theorem 7 is rather strong. A weaker result asserting the convergence in X can be proved under weaker assumptions. In particular, Theorem 7 shows that $u(t)$ depends continuously on $\emptyset = u(0)$ in the Y-norm. But it is impossible to prove a Hölder continuity with any prescribed exponent; for a counterexample see Kato [12].

The proof of these theorems will be given in the following sections. Here we list two simple lemmas that follow immediately from (X).

LEMMA 7.3 If a subset of Y is convex, closed and bounded, then it is also closed in X.

LEMMA 7.4 If a function g on $[0,T]$ to Y is bounded in Y-norm and continuous in X-norm, then g is weakly continuous (hence strongly measurable) as a Y-valued function.

8. An example.

We take a simple example to illustrate the meaning of the assumptions (X), (A1), (A2), etc.

Consider the quasi-linear "wave equation"

$$u_t + a(u)u_x = b(u)u, \quad t \geq 0, \quad -\infty < \infty, \tag{8.1}$$

where $u = u(t,x)$ is a real-valued function and $u_t = \partial u/\partial t$, $u_x = \partial u/\partial x$.

[We put a factor u on the right of (8.1) so that condition (7.6) is satisfied with $f(y) = yb(y)$.] The real-valued functions a and b are assumed, for simplicity, to be as smooth as one likes. It is convenient to introduce the notation

$$\alpha_k(R) = \sup_{|r|\leq R} |a^{(k)}(r)|, \qquad \beta_k(R) = \sup_{|r|\leq R} |b^{(k)}(r)|,$$

$$k = 0,1,2,\ldots. \tag{8.2}$$

To apply Theorems 6 and 7, we choose

$$X = H^0(-\infty,\infty), \qquad Y = H^2(-\infty,\infty), \tag{8.3}$$

where H^s denotes the Sobolev space of order s of the L^2-type, so that $H^0 = L^2$. As an isometric isomorphism of Y onto X, we choose

$$S = 1 - D^2, \qquad D = d/dx. \tag{8.4}$$

In this case we have also $\|S^{-1}\|_X \leq 1$.

We choose W as the ball in Y with center 0 and radius R. For $A(t,y)$ and $f(t,y)$, we naturally take

$$A(t,y) = A(y) = a(y)D, \qquad f(t,y) = f(y) = yb(y). \tag{8.5}$$

[We could as well take $A(y) = a(y)D - b(y)$, $f(y) = 0$.]

To verify condition (A1), we note that $(d/dx)a(y(x)) = a'(y(x))y_x(x)$ is continuous and bounded if $y \in W$. Thus $A(y) \in G(X,1,\beta)$ with $\beta \leq (1/2)\sup|a'(y(x)y_x(x)| \leq cR\alpha_1(R)$, where c is a numerical constant. (Note that the L^∞-norms of y and y_x are smaller than $\|y\|_Y$.)

To verify (A2), it is convenient to compute the <u>commutator</u>

$$[S,A(y)] = -[D^2,a(y)D] = -[D^2,a(y)]D$$

$$= -(D^2a(y))D - 2(Da(y))D^2$$

$$= -[a'(y)y_{xx}+a''(y)y_x^2]D - 2a'(y)y_xD^2. \tag{8.6}$$

Hence

$$[S,A(y)]S^{-1} = -[a'(y)y_{xx}+a''(y)y_x^2]DS^{-1} - 2a'(y)y_x D^2S^{-1}. \qquad (8.7)$$

The computation used here is rather formal, but (8.7) is true when applied to a sufficiently smooth function w ($w \in Y$, for example). By Remark 7.1,3, we see that (7.2) is satisfied with $B(y)$ equal to the operator on the right of (8.7).

This operator $B(y)$ is in $B(X)$. Indeed, the last term on the right of (8.7) is in $B(X)$ since D^2S^{-1} is in $B(X)$ and $a'(y)y_x$ is in L^∞ as shown above, with the L^∞-norm $\leq cR\,\alpha_1(R)$. The same is true with $a''(y)y_x^2DS^{-1}$, where $\|a''(y)y_x^2\|_{L^\infty} \leq cR^2\alpha_2(R)$. For the term $a'(y)y_{xx}DS^{-1}$, we use the estimate

$$\|a'(y)y_{xx}DS^{-1}w\|_X \leq \|a'(y)\|_{L^\infty} \|y_{xx}\|_X \|DS^{-1}w\|_{L^\infty} \leq cR\,\alpha_1(R)\,\|w\|_X,$$

which implies that $\|a'(y)y_{xx}DS^{-1}\|_X \leq cR\,\alpha_1(R)$. Altogether we obtain

$$\|B(y)\|_X \leq c(R\,\alpha_1(R) + R^2\alpha_2(R)) \equiv \lambda_1, \quad y \in W. \qquad (8.8)$$

(A3) is trivial except for (7.4), since $A(y)$ is a first-order differential operator with a smooth coefficient $a(y)$ independent of t. (7.4) follows from the estimate

$$\|(A(y)-A(z))w\|_X = \|(a(y)-a(z))w_x\|_X \leq \|a(y)-a(z)\|_X \|w_x\|_{L^\infty}$$

$$\leq c\;\alpha_1(R)\,\|y-z\|_X \|w\|_Y\,,$$

which implies that

$$\|A(y)-A(z)\|_{Y,X} \leq c\;\alpha_1(R)\,\|y-z\|_X \equiv \mu_1\|y-z\|_X, \quad y,z \in W. \qquad (8.9)$$

(A4) is trivial since $y_0 = 0$.

To verify (7.6) in (f1), we note that $\|f(y)\|_X = \|yb(y)\|_X \leq \|y\|_X\beta_0\,(R)$ and

$$\|D^2yb(y)\|_X = \|y_{xx}b(y) + 2y_x Db(y) + yD^2b(y)\|_X$$

$$\leq c(R\beta_0(R) + R^2\beta_1(R) + R^3\beta_2(R))$$

by a computation similar to the one used above. (7.7) follows from

$$\|yb(y) - zb(z)\|_X \leq \|(y-z)b(y)\|_X + \|z(b(y)-b(z))\|_X$$

$$\leq \|y-z\|_X \beta_0(R) + \|y-z\|_X R\beta_1(R).$$

We can verify (A5) and f(2) in the same way, using $\alpha_3(R)$ and $\beta_3(R)$. The detail may be omitted.

It follows that Theorem 6 holds for (8.1) if a, b have continuous derivatives up to the order 2 and Theorem 7 holds if a, b have continuous derivatives up to the order 3.

REMARK 8.1. 1. There are other choices for the spaces X, Y. In particular, we may choose the same X as above but take $Y = H^s(-\infty, \infty)$ with any integer $s \geq 2$. Then we take $S = (1 - D^2)^{s/2}$ or $S = (1 - D)^s$. All the conditions (A1) etc. can be verified as above if the existence of higher derivatives of a, b are assumed. The advantage of using a large s is that we obtain the <u>regularity theorem</u> at one stroke; if the initial value $u(0) = \phi$ is in H^s, then the solution $u(t)$ is also in H^s and depends on ϕ continuously in the H^s-norm.

One may also use non-integer values $s > 3/2$. In this case $S = (1-D^2)^{s/2}$ involves a fractional power and the computation of the commutator (8.7) becomes nontrivial, but we can use Lemma A2 of Appendix (which is prepered for this purpose).

2. It is well known that the solution of (8.1) is in $C_b^1(-\infty, \infty)$ if the initial value ϕ is in $C_b^1(-\infty, \infty)$. (We denote by C_b^1 the set of functions which is bounded and uniformly continuous together with its derivative). Unfortunately our theorems are not strong enough to recapture this result,

since the space $C_b(-\infty,\infty)$ is not reflexive. (The proof of Theorem 6 given below fails since E is not complete if we choose $X = C_b$ and $Y = C_b^1$.)

9. Proof of Theorem 6.

Since W is an open ball in Y containing $\emptyset$, we can choose $R > 0$ so that $\|\emptyset - y_0\|_Y < R$ and that $\|y - y_0\|_Y \leq R$ implies $y \in W$. Let E be the set of all functions v on $[0,T']$ to Y such that

$$\|v(t) - y_0\|_Y \leq R \qquad \text{(so that } v(t) \in W), \tag{9.1}$$

$$v \text{ is continuous from } [0,T'] \text{ to } X. \tag{9.2}$$

Here T' is a positive number $\leq T$, to be determined later.

For $v \in E$ set

$$A^v(t) = A(t,v(t)), \qquad t \in [0,T']. \tag{9.3}$$

According to (A1), $A^v(t)$ belongs to $G(X,1,\beta)$. Hence the family $\{A^v(t)\}$ is stable (see the end of §1), with stability index $1, \beta$.

LEMMA 9.1. $t \to A^v(t) \in B(Y,X)$ is continuous in norm.

Proof. (A3) says that $A^v(t) \in B(Y,X)$. Thus the lemma follows from

$$\begin{aligned}\|A^v(t')-A^v(t)\|_{Y,X} &\leq \|A(t',v(t'))-A(t',v(t))\|_{Y,X} \\ &\quad + \|A(t',v(t))-A(t,v(t))\|_{Y,X} \\ &\leq \mu_1 \|v(t')-v(t)\|_X + \|A(t',v(t))-A(t,v(t))\|_{Y,X}\end{aligned}$$

by virtue of (A3) and (9.2). ∎

By condition (A2) we have

$$SA^v(t)S^{-1} = A^v(t) + B^v(t), \tag{9.4}$$

$$B^{v}(t) = B(t,v(t)) \in B(X), \qquad \|B^{v}(t)\|_X \leq \lambda_1. \tag{9.5}$$

LEMMA 9.2. $t \to B^{v}(t) \in B(X)$ is weakly continuous (hence strongly measurable).

Proof. If $y \in Y$, we have by (9.4)

$$S^{-1}B^{v}(t)y = A^{v}(t)S^{-1}y - S^{-1}A^{v}(t)y. \tag{9.6}$$

Since $S^{-1}y \in Y$, it follows from Lemma 9.1 that the right member of (9.6) is continuous in t in the X-norm. Hence the same is true of the left member. Since $\|S^{-1}B^{v}(t)\|_X \leq \|S^{-1}\|_X \lambda_1$ by (9.5) and since Y is dense in X, it follows that $t \to S^{-1}B^{v}(t)x$ is continuous for every $x \in X$. Since $\|S^{-1}B^{v}(t)x\|_Y = \|B^{v}(t)x\|_X \leq \lambda_1\|x\|$, it follows that $t \to S^{-1}B^{v}(t)x \in Y$ is weakly continuous (see Lemma 7.4). This is equivalent to saying that $t \to B^{v}(t)x \in X$ is weakly continuous. ∎

Lemmas 9.1 and 9.2 show that the assumptions of Theorem 1 are satisfied by the family $\{A^{v}(t)\}$. Hence there exists a unique evolution operator $U^{v} = \{U^{v}(t,s)\}$ defined on $\Delta' : T' \geq t \geq s \geq 0$ with the properties described in Theorem 1.

For $v \in E$ set

$$f^{v}(t) = f(t,v(t)) \in Y, \quad t \in [0,T']. \tag{9.7}$$

LEMMA 9.3 $\|f^{v}(t)\|_Y \leq \lambda_3$, $t \to f^{v}(t)$ is continuous in the X-norm and weakly continuous (hence strongly measurable) in the Y-norm.

Proof. The first inequality follows from (f1). The continuity of f^{v} in the X-norm follows from

$$\|f^V(t')-f^V(t)\|_X \leq \|f(t',v(t'))-f(t',v(t))\|_X$$

$$+ \|f(t'v(t))-f(t,v(t))\|_X$$

$$\leq \mu_2\|v(t')-v(t)\|_X + \|f(t',v(t))-f(t,v(t))\|_X$$

by virtue of (f1) and (9.2). The weak continuity of f^V in the Y-norm follows from Lemma 7.4. ■

In view of Lemma 9.3, we can apply Theorem 2 to the linear equation of evolution

$$(L^V) \qquad du/dt + A^V(t)u = f^V(t), \qquad 0 \leq t \leq t', \quad u(0) = \phi \in W \subset Y.$$

The solution is given by

$$u(t) = U^V(t,0)\phi + \int_0^t U^V(t,s)f^V(s)ds. \tag{9.8}$$

Since $\phi \in Y$ and $f^V \in L^\infty(0,T';Y) \cap C[0,t';X]$, we have

$$u \in C[0,T';Y] \cap C^1[0,Y';X] \tag{9.9}$$

by Theorem 2, (c).

On the other hand we have the estimates (see (3.2), (3.3))

$$\|U^V\|_X \leq e^{\beta T'}, \qquad \|U^V\|_Y \leq e^{(\beta+\lambda_1)T'}, \tag{9.10}$$

since $\|S\|_{Y,X} = \|S^{-1}\|_{X,Y} = 1$; note that $\|B^V\|_{1,X} \leq \lambda_1 T'$ because $\|B^V(t)\|_X \leq \lambda_1$.

To estimate u, it is convenient to set $u' = u - y_0$ and note that u' satisfies the equation

$$du'/dt + A^V(t)u' = f^V(t) - A^V(t)y_0, \quad u'(0) = \phi - y_0. \tag{9.11}$$

Application of (9.8) to (9.11) gives

$$u(t)-y_0 = U^v(t,0)(\phi-y_0) + \int_0^t U^v(t,s)(f^v(s)-A^v(s)y_0)ds. \tag{9.12}$$

Since $\|A^v(s)y_0\|_Y \leq \lambda_2$ by (7.5) and $\|f^v(s)\|_Y \leq \lambda_3$ by Lemma 9.3, we obtain from (9.11) using (9.10)

$$\|u(t)-y_0\|_Y \leq e^{(\beta+\lambda_1)T'}(\|\phi-y_0\|_Y + (\lambda_2 + \lambda_3)T'). \tag{9.13}$$

We want u to be in E. This will be the case if the right member of (9.13) is not larger than R, since it is obvious that u is continuous in the X-norm. Since $\|\phi-y_0\|_Y < R$, this is possible if $T' > 0$ is chosen sufficiently small.

With such a choice of T', the map $v \to u \equiv \Phi v$ sends E into E.

We now make E into a metric space by the distance function

$$d(v,w) = \sup_{0\leq t\leq T'} \|v(t)-w(t)\|_X. \tag{9.14}$$

Then E is a complete metric space, since a closed ball in Y is a closed subset of X (see Lemma 7.3). Note that we did not require $v \in E$ to be continuous in the Y-norm.

We shall now show that $\Phi : E \to E$ is a strict contraction map if T' is chosen sufficiently small. To this end we use the perturbation theorem (Theorem 3), which gives

$$d(\Phi w, \Phi v) \leq e^{\beta T'}(|||f^w-f^v|||_{1,X} + |||(A^w-A^v)\Phi v|||_{1,X}).$$

Since $\|f^w(t)-f^v(t)\|_X \leq \mu_2\|w(t)-v(t)\|_X$ by (f1), we have

$$|||f^w-f^v|||_{1,X} \leq \mu_2 T' d(w,v).$$

Since $\|(A^w(t)-A^v(t))\Phi v(t)\|_X \leq \|A^w(t)-A^v(t)\|_{Y,X}\|\Phi v(t)\|_Y \leq$ $\mu_1 \|w(t)-v(t)\|_X(\|y_0\|+R)$ by (A3), (9.1) and $\Phi v \in E$, we have

$$\|(A^w-A^v)\Phi v\|_{1,X} \leq \mu_1(\|y_0\|+R)T'd(w,v).$$

Thus we obtain

$$d(\Phi w, \Phi v) \leq T'e^{\beta T'}(\mu_2 + \mu_1\|y_0\| + \mu_1 R)d(w,v), \tag{9.15}$$

which shows that Φ is a contraction map if T' is sufficiently small.

It follows from the contraction mapping theorem that Φ has a fixed point in E, which is automatically a unique solution of (Q) with the properties stated in Theorem 6.

10. Proof of Theorem 7.

We construct the solutions u^n to (Q^n) as in the proof of Theorem 6 using the same function space E. For each $v \in E$, we solve the linear equation

$$(L^{n,v}) \qquad du^n/dt + A^{n,v}(t)u^n = f^{n,v}(t), \quad 0 \leq t \leq T', \quad u^n(0) = \phi^n,$$

where

$$A^{n,v}(t) = A^n(t,v(t)), \quad f^{n,v}(t) = f^n(t,v(t)). \tag{10.1}$$

As in the previous section, one can show that the solution u^n to $(L^{n,v})$ exists and belongs to E if T' is sufficiently small, since A^n and f^n satisfy conditions (A1) to (A4) and (f1) <u>uniformly</u> and since $\|\phi^n-\phi\|_Y \to 0$. This defines a map Φ^n on E to itself: $\Phi^n v = u^n$. Again, it can be shown that Φ^n is a contraction if T' is chosen sufficiently small. The unique fixed point u^n of Φ^n gives the desired solution to (Q^n). We may also assume that with the same T', the map Φ associated with (Q) is also a contraction on E to E and its fixed point u solves (Q).

LEMMA 10.1. $\|u^n(t)-u(t)\|_X \to 0$ uniformly for $t \in [0,T']$.

Proof. Since $u^n = \Phi^n u^n$ and $u = \Phi u$, we have

$$d(u^n,u) = d(\Phi^n u^n, \Phi u) \leq d(\Phi^n u^n, \Phi^n u) + d(\Phi^n u, \Phi u)$$

$$\leq \gamma\, d(u^n,u) + d(\Phi^n u, \Phi u),$$

where $\gamma < 1$ is the common contraction factor for the Φ^n, which exists by the uniformity of conditions (A1) etc. for the (Q^n). Thus the assertion $d(u^n,u) \to 0$ follows if we show that

$$d(\Phi^n v, \Phi v) \to 0, \quad n \to \infty, \quad \text{for } v \in E. \tag{10.2}$$

But (10.2) expresses simply the continuous dependence in the X-norm of the solution of the linear equation (L^v) when the coefficients $A^v(t)$, the right member $f^v(t)$, and the initial value ϕ are varied slightly. In view of the uniformity of (A1) etc. mentioned above, (10.2) thus follows from the convergence theorem (Theorem 5). Indeed, it suffices to note that $\phi^n \to \phi$ in X,

$$A^{n,v}(t)-A^v(t) = A^n(t,v(t))-A(t,v(t)) \to 0 \quad \text{strongly in } B(Y,X)$$

by (7.10), and that

$$f^{n,v}(t)-f^v(t) = f^n(t,v(t))-f(t,v(t)) \to 0 \quad \text{strongly in } B(X)$$

by (7.12). Note that (6.2) is satisfied since $\|A^n(t,y)\|_{Y,X}$ is bounded uniformly in t, y, and n. ∎

To complete the proof of Theorem 7, we use Theorem 4. Since u^n is the solution of the linear equation $(L^{n,u^n}) \equiv (L^n)$, we have from (5.2)

$$|||u^n-u|||_{\infty,Y} \leq K'[\|\phi^n-\phi\|_Y + |||f^n-f|||_{1,Y}$$

$$+ |||(B^n-B)Su|||_{1,X} + |||h^n|||_{\infty,X}] \tag{10.3}$$

Here $f^n(t) = f^{n,u^n}(t)$, $f(t) = f^u(t)$, $B^n(t) = B^{n,u^n}(t) = B^n(t,u^n(t))$, $B(t) = B^u(t) = B(t,u(t))$, and h^n is given by

$$h^n(t) = (U^n(t,0)-U(t,0))S\phi + \int_0^t (U^n(t,s)-U(t,s))g(s)ds,$$

$$g(s) = Sf(s) - B(s)Su(s), \tag{10.4}$$

where U^n and U are the evolution operators for (L^n) and $(L) = (L^u)$ respectively. K' is a constant depending on R and T' but not on n.

The second term on the right of (10.3) is estimated by

$$\|f^n-f\|_{1,Y} = \int_0^{T'} \|f^n(t,u^n(t))-f(t,u(t))\|_Y dt$$

$$\leq \mu_4 T' \|u^n-u\|_{\infty,Y} + \int_0^{T'} \|(f^n-f)(t,u(t))\|_Y dt \tag{10.5}$$

by virtue of (f2) assumed for f^n. The third term on the right of (10.3) is estimated by

$$\|(B^n-B)S^u\|_{1,X} \leq \int_0^{T'} \|(B^n(t,u^n(t))-B(t,u(t))Su(t)\|_X dt$$

$$\leq \mu_3 \int_0^{T'} \|u^n(t)-u(t)\|_Y \|Su(t)\|_X dt + \int_0^{T'} \|(B^n-B)(t,u(t))Su(t)\|_X dt$$

$$\leq \mu_3 T' \|u^n-u\|_{\infty,Y}(\|y_0\|+R) + \int_0^{T'} \|(B^n-B)(t,u(t))Su(t)\|_X dt, \tag{10.6}$$

where we used (A5) for B^n and $\|Su(t)\|_X = \|u(t)\|_Y \leq \|y_0\|+R$.

If T' is sufficiently small that $T'(\mu_4 + \mu_3(\|y_0\|+R)) < 1$, the two terms involving $\|u^n-u\|_{\infty,Y}$ in (10.5) and (10.6), when substituted into the right member of (10.3), can be absorbed into the left member. Since $\|\phi^n-\phi\|_Y \to 0$,

$$\int_0^{T'} \|(f^n-f)(t,u(t))\|_Y dt \to 0 \quad \text{by} \quad (7.12),$$

$$\int_0^{T'} \|(B^n-B)(t,u(t))Su(t)\|_X dt \to 0 \quad \text{by} \quad (7.11),$$

(bounded convergence theorem), we can thus conclude $\|u^n-u\|_{\infty,Y} \to 0$ provided we can show that $\|h^n\|_{\infty,X} \to 0$.

In view of (10.4), this will be the case if

$$U_n(t,s)-U(t,s) \to 0 \text{ strongly in } B(X), \text{ uniformly in } t, s; \tag{10.7}$$

note that $S\phi \in X$, $Sf(t) \in X$, $Su(t) \in X$ so that $g \in L^\infty(0,T';X)$. But again (10.7) is a consequence of Theorem 5, since

$$A^n(t,u^n(t)) - A(t,u(t)) = [A^n(t,u^n(t)) - A^n(t,u(t))]$$
$$+ [A^n(t,u(t)) - A(t,u(t))] \to 0 \text{ strongly}$$

in $B(Y,X)$, by (7.4) (assumed for A^n), Lemma 10.1 and (7.10). This completes the proof of Theorem 7.

11. Remarks.

1. Theorems 6, 7 are intended to be the simplest of their kinds, while covering a considerable field of application (see Part 3). We have made a rather restrictive assumption (condition (A1)) that $A(t,y)$ is quasi-accretive (is in $G(X,1,\beta)$), in order to satisfy the stability condition for the family $\{A^v(t)\}$ rather trivially.

On the other hand, it is in general not so easy to show that a given family $\{A(t)\}$ is stable. The only practical way to do so would be to show that $A(t)$ is quasi-accretive with respect to some norm, which may be different from the original one but should be equivalent to it. Furthermore, the new norm may depend on t if the dependence is smooth. In this way we may generalize condition (A1) in the following direction.

Consider the space $N = \{\| \|_\nu\}$ of all equivalent norms in X, metrized in an obvious way. For example, we may define the distance between two norms $\| \|_\nu$ and $\| \|_\mu$ by

$$\log \max\{\sup_x \|x\|_\nu / \|x\|_\mu , \sup_x \|x\|_\mu / \|x\|_\nu\}.$$

We shall denote by X_ν the space X renormed by $\| \|_\nu$.

We may now assume that there are maps $\nu : [0,T] \times W \to N$ and $\beta : [0,T] \times W \to (-\infty, \infty)$ such that $A(t,y) \in G(X_{\nu(t,y)}, 1, \beta(t,y))$. We should also assume that ν and β are smooth in a certain, possibly complicated sense.

Here we shall not go into such a generalization, which would make the theorems rather complicated. But we note that there are problems for which the generalized theorems are necessary. As an example, we mention the symmetric hyperbolic systems of PDE in which the time derivative of the unknown has a nonlinear coefficient (see (12.13)). This problem was considered in [12] as a separate problem, but the method used there is essentially equivalent to the generalization stated above. Other examples requiring the generalized theorems include certain nonlinear wave equations, and the magnetohydrodynamic equations.

2. We have also made a rather restrictive assumption that the spaces X, Y are reflexive. An essential use of this assumption is made in the proof that the function space E (used in the proof of Theorem 6) is complete. We do not know whether or not the reflexivity can be eliminated, possibly, by adding other assumptions.

3. We have assumed that the isomorphism S of Y onto X is an isometry. This is harmless since otherwise we can renorm Y. If a different norm is used in Y, however, some change is necessary in the statement of Theorems 6, 7. Since a ball in Y does not remain a ball after renorming, the condition $\emptyset \in W$ must be replaced by a condition $\emptyset \in W_0$, where W_0 is another ball contained in W, the size of which is determined by $\|S\|_{Y,X}$ and $\|S^{-1}\|_{X,Y}$.

This remark is important when variable norms are used as noted in Remark 1 above, since it is impossible to fix S as an isometry for all norms in X.

Part 3. Applications.

12. Quasi-linear symmetric hyperbolic systems.

Consider the system of differential equations

$$\partial u/\partial t + \sum_{j=1}^{m} a_j(t,x,u)\,\partial u/\partial x_j = f(t,x,u), \quad 0 \leq t \leq T, \tag{12.1}$$

in the whole space $x \in R^m$. The unkown $u = u(t,x) = (u_1(t,x), \ldots\ldots, u_N(t,x))$ is an N vector-valued function of t and x; $a_j(t,x,p)$ are symmetric $N \times N$ matrix-valued functions and $f(t,x,p)$ is an N vector-valued function, defined for $t \in [0,T]$, $x \in R^m$ amd p in a ball $\sum \subset R^N$ containing the origin. We want to find a solution of (12.1) satisfying the initial condition

$$u(0,x) = \phi(x), \tag{12.2}$$

where the values $\phi(x)$ are in $\sum$ so that $a_j(0,x,u(0,x))$ and $f(0,x,u(0,x))$ make sense.

We assume that the a_j and f are sufficiently smooth, with the required derivatives bounded (for $p \in \sum$, of course).

We shall show that the problem (12.1-2) can be handled as an application of the abstract theorems given in Part 2. To this end, we choose

$$X = H^0(R^m;R^N), \quad Y = H^s(R^m;R^N), \quad s > m/2 + 1, \tag{12.3}$$

where H^s denotes the Sobolev space of order s of the L^2-type. s need not be an integer. As the isometric isomorphism of Y onto X, we take

$$S = (1 - \Delta)^{s/2}. \tag{12.4}$$

W is chosen as the ball in Y with center 0 and radius R_0, where R_0 s such that $\|y\|_Y < R_0$ implies $y(x) \in \sum$, which is possible because $s > m/2$. e set

$$A(t,y) = \sum_{j=1}^{m} a_j(t,s,y(x))\,\partial/\partial x_j, \quad y \in W. \tag{12.5}$$

his is a system of first order differential operators of the symmetric type (in he sense of Friedrichs [13]). For simplicity we shall call it a symmetric system. s is well known (see [13]), $A(t,y)$ is a quasi-accretive operator in X : $(t,y) \in G(X,1,\beta)$ with

$$\beta \leq (1/2) \sup_x \sum_{j=1}^{m} |\partial a_j(t,x,y(x))/\partial x_j| \leq \text{const}, \quad y \in W;$$

note that $y \in W$ implies that $\partial y/\partial x_j$ is a bounded function.

Verification of conditions (A2) to (A5) is now straightforward, being essentially he same as in §8. The crucial point is the proof that

$$B(t,y) \subset [S,A(t,y)]S^{-1} = \sum_j [S,a_j(t,x,y(x))]D_j S^{-1} \tag{12.6}$$

is a bounded operator in X, where $D_j = \partial/\partial x_j$. Since D_j is dominated by $(1-\Delta)^{1/2}$, it follows from Lemma A2 (Appendix) that (12.6) is bounded if

$$\text{grad}_x a_j(t,x,y(x)) \in H^{s-1}, \quad j = 1,\ldots,m. \tag{12.7}$$

To prove (12.7), we assume for simplicity that $a_j = a_j(p)$ does not depend on t, x explicitly (the general case is not essentially different). Then

$$\text{grad}_x a_j(y(x)) = (\text{grad}_p a_j)(y(x)).\text{grad}\, y(x) \tag{12.8}$$

Since grad $y(x) \in H^{s-1}$ with $s - 1 > m/2$, (12.8) will be in H^{s-1} if

$$\text{grad}_p a_j(y(x)) - \text{grad}_p a_j(0) \in H^{s-1}. \tag{12.9}$$

The function in (12.9) is in H^0, since a_j is smooth. To prove (12.9), therefore, it suffices to show that all its derivatives of order $[s]$ are in H^0, where $[s]$ is the Gaussian symbol. A general $[s]$-th derivative of that function is given by

$$\sum_{k=1}^{[s]} (D_p^{k+1} a_j(y(x))\cdot(D_x^{r_1}y(x))\cdot \ldots \cdot (D_x^{r_k}y(x)), \tag{12.10}$$

in a somewhat symbolic expression, where all possible terms with $r_i \geq 1$, $r_1 + \ldots + r_k = [s]$ should be added. Since $D^{r_i}y \in H^{s-r_i}$, (12.10) will be in H^0 if

$$H^{s-r_1} \cdot \ldots \cdot H^{s-r_k} \subset H^0, \quad r_1+\ldots+r_k = [s], \quad r_i \geq 1.$$

But this is easily verified using Lemma A1, noting that $s > m/2 + 1$ and $[s] \leq s$. This establishes (7.3).

A similar computation is used to verify (A5).

Other conditions (A2)-(A4) can be verified in the same way as in §8.

To verify conditions (f1) and (f2), we need the following additional assumption

$$f(\cdot,\cdot,0) \in C[0,T;H^s(R^m;R^N)]. \tag{12.11}$$

Then (f1), (f2) can be verified in the same way as above.

Under the assumptions stated above, Theorems 6 and 7 are applicable. It follows that <u>there exists a unique solution</u> u <u>to</u> (12.1-2) <u>such that</u>

$$u \in C[0,T';W] \cap C^1[0,T';H^{s-1}(R^m;R^N)], \tag{12.12}$$

<u>and</u> $u(t)$ <u>depends on</u> ϕ <u>continuously in the</u> H^s<u>-norm</u>. It should be noted that Theorem 6 gives (12.12) with only H^0 in the place of H^{s-1}, but the equation (12.1) itself shows that $du/dt \in H^{s-1}$.

REMARK. 1. The method sketched in Lax [14] is basically similar to ours.

2. Theorems 6, 7 are not applicable to a more general system

$$a_0(t,x,u)\,\partial u/\partial t + \sum_j a_j(t,x,u)\,\partial u/\partial x_j = f(t,x,u), \qquad (12.13)$$

in which a_0 is also symmetric and positive-definite. But a generalized theorem indicated in §11, 1, will be applicable. Actually (12.13) was studied in detail in [12] as a separate problem, using the same idea. (12.13) was also considered by Fischer and Marsden [15] under more restrictive assumptions.

13. Korteweg-de Vries equation.

We consider this equation in a slightly generalized form

$$u_t + u_{xxx} + a(u)u_x = 0, \quad t \geq 0, \quad -\infty < x < \infty, \qquad (13.1)$$

where a is assumed to be a smooth function.

(13.1) can be handled in the same way as the example in §8, despite the appearance of u_{xxx}; it suffices to take

$$X = H^0(-\infty,\infty), \quad Y = H^s(-\infty,\infty), \quad s \geq 3,$$

$$S = (1-D^2)^{s/2}, \quad A(t,y) = A(y) = D^3 + a(y)D, \quad f(t,y) = 0, \qquad (13.2)$$

where s need not be an integer.

To verify (A1), we note that the leading term D^3 in $A(y)$ is the generator of a contraction semi-group in X; indeed it is skew-adjoint with $D(D^3) = H^3$. The perturbing term $a(y)D$ is quasi-accretive as in §8. Since this term is relatively bounded with respect to D^3 with relative bound 0, it follows from a perturbation theorem in semigroup theory (see Kato [11]) that $A(y) \in G(X,1,\beta)$ with $\beta \leq cR\,\alpha_1(R)$ if $y \in W = \{y \in Y;\ \|y\|_Y < R\}$. (Here we used the notation (8.2).)

Verification of the other conditions is exactly the same as in §8 (see Remark 8.1, 1). The presence of D^3 in $A(y)$ does not introduce any trouble, since it commutes with S and is independent of y.

In this way we are led to the following results. If $\emptyset \in W$, (13.1) has a unique solution

$$u \in C[0,T';W] \cap C^1[0,T';H^{s-3}], \quad u(0,x) = \emptyset(x), \tag{13.3}$$

and $u(t)$ depends on $\emptyset$ continuously in the H^s-norm.

REMARK. 1. The argument given above shows clearly that the form of the term u_{xxx} in (13.1) has no special meaning. It may be replaced by $P(D)u$ with any polynomial P, of degree $m \geq 2$, with the following restriction: if m is an even number, the sign of the coefficient of the leading term must be $(-1)^{m/2}$. We may choose $Y = H^s$ with $s \geq m$.

2. It appears that so far the existence of the solution to (13.1) has been proved either by the viscosity method or by the difference method, see Sjoberg [16], Mukasa and Iino [17], Kametaka [18], Tsutsumi and Mukasa [19], Dushane [20].

3. It should be noted that (13.1) has global solutions, at least under certain conditions on the function a, but the construction of a local solution forms an essential step in the proof of the existence of global solutions.

14. The Navier-Stokes and Euler equations in R^3.

We may write the Navier-Stokes equation for an incompressible fluid in R^3 in the form

$$du/dt - \nu\Delta u + P(u\cdot \text{grad})u = f(t) \tag{14.1}$$

(see e.g. [21]). The unknown u takes values in $H^0_\sigma(R^3)$, the subspace of $L^2(R^3;R^3)$ consisting of functions v with $\text{div } v = 0$, with the associated projection operator P. ν is a nonnegative constant (kinematic viscosity).

In what follows we denote by $H^s_\sigma(R^3)$ the subspace of $H^s(R^3;R^3)$ consisting of functions v with $\text{div } v = 0$. The associated projection operator is identical with P restricted on H^s.

We shall show that Theorems 6, 7 can be applied to (14.1) with

$$X = H^0_\sigma(R^3;R^3), \quad Y = H^s_\sigma(R^3;R^3), \quad s > 5/2,$$

$$S = (1 - \Delta)^{s/2}, \tag{14.2}$$

$$A(t,y) = A(y) = - \nu \Delta + P(y \cdot \text{grad}).$$

To verify (A1), it suffices to show that if $y \in Y$, $-A(y)$ generates a contraction semi-group on X. Obviously the leading term $\nu \Delta$ is such a generator if $\nu > 0$, with domain H^2_σ. The perturbing term $- P(y \cdot \text{grad})$ is also dissipative, since $(P(y \cdot \text{grad})w,w)_X = ((y \cdot \text{grad})w,w) = 0$ due to div $y = 0$. But the perturbing term is relatively bounded with respect to $-\Delta$, with relative bound 0, since $y(x)$ is bounded. It follows that $A(y) \in G(X,1,0)$ with $D(A(y)) = H^2_\sigma$ provided $\nu > 0$.

If $\nu = 0$, a more careful definition of $A(y)$ is necessary. In this case $P(y \cdot \text{grad})$ is an operator analogous to the first-order symmetric system considered in §12. As in the latter case, one may apply $P(y \cdot \text{grad})$ first to smooth functions in X and then take the closure. Again one can prove the basic property: strong closure = weak closure (c.f. Friedrichs [22], which ensures that the closure is the generator of a contraction semigroup.

Now it is easy to verify conditions (A2) to (A5) as in §12. Here it is essential that Δ and P commute with S. The necessary computation of the commutator $[A(y),S]$ is similar to the previous case and may be omitted.

It follows from Theorems 6 and 7 that _if_ $s > 5/2$, $\phi \in H^s_\sigma$, and if $f \in C[0,T;H^0_\sigma] \cap L^1[0,T;H^s_\sigma]$, _there is a unique solution_ u _to_ (14.1) _with_ $u(0) = \phi$ _such that_

$$u \in C[0,T';H^s_\sigma(R^3)] \cap C^1[0,T';H^{s-2}_\sigma].$$

Furthermore, $u(t)$ _depends continuously on_ ϕ _and_ ν, _in the_ H^s_-topology._ This follows from Theorem 7, in which not only ϕ but also $A(t,y)$ is allowed to vary. Indeed, conditions (A1) to (A5) are satisfied uniformly for $0 \leq \nu \leq \nu_0$.

Thus we have been able to prove <u>the convergence of the viscous flow to the ideal flow as</u> $\nu \to 0$, <u>and indeed in the</u> H^s<u>-topology if</u> $\phi \in H^s_\sigma$. In a previous paper [21], this was proved only in the weak topology of H^s_σ.

For equation (14.1) in R^3 or R^2, see also Ladyzhenskaya [23], Golovkin [24], McGrath [25], Ebin and Marsden [26], Swann [27].

15. The coupled Maxwell-Dirac equations in R^3.

The equations may be written

$$\begin{cases} \partial^2 V/\partial t^2 - \Delta V = e\,\psi^*\psi \\ \partial^2 a/\partial t^2 - \Delta a = e\,\psi^*\alpha\psi, \\ \partial\psi/\partial t + \alpha\cdot(\text{grad} - ia)\psi + i(m\beta + V)\psi = 0. \end{cases} \tag{15.1}$$

Here V and a are the scalar and vector potentials of the electromagnetic field, and ψ is the spinor wave function (with 4 components) of the electron; $\alpha = (\alpha_1, \alpha_2, \alpha_3)$ and β are certain hermitian 4×4 matrices; e and m are constants.

The system (15.1) must be supplemented by the conditions

$$\begin{cases} \partial V/\partial t + \text{div}\, a = 0, \\ \text{div}(-\partial a/\partial t - \text{grad}\, V) = e\,\psi^*\psi. \end{cases} \tag{15.2}$$

But it can be shown that these are satisfied for solutions of (15.1) if they are satisfied at $t = 0$ (Gross [28]). Thus it suffices to consider (15.1) as a closed system of differential equations.

It is convenient to convert (15.1) into a first-order (in t) system by introducing $\dot{V} = \partial V/\partial t$ and $\dot{a} = \partial a/\partial t$ as additional unknowns, obtaining

$$\begin{cases} \partial V/\partial t - \dot{V} = 0, & \partial a/\partial t - \mathring{a} = 0 \\ \partial \dot{V}/\partial t - \Delta V = e\,\psi^*\psi, & \partial \mathring{a}/\partial t - \Delta a = e\,\psi^*\alpha\,\psi, \\ \partial \psi/\partial t + \alpha\cdot(\mathrm{grad} - ia)\psi + i(m\,\beta + V)\psi = 0. \end{cases} \tag{15.3}$$

The system (15.3) is rather simple as a system of differential equations, since it is semi-linear rather than quasi-linear. From the experience of the preceding examples, it is clear that one can easily obtain a local existence theorem in a Sobolev space $H^s(R^3)$ with sufficiently large s. The interest in this problem lies rather in the question of how small s can be. In what follows we shall try to answer this question.

We shall regard (15.3) as an equation of evolution in the Hilbert space

$$\begin{aligned} X &= X_M \oplus X_D, \\ X_M &= H^{\{1/2\}}(R^3;R^4) \oplus H^{-1/2}(R^3;R^4), \\ X_D &= H^0(R^3;C^4). \end{aligned} \tag{15.4}$$

Here H^s denotes the usual Sobolev space with the norm denoted by $\| \ \|_s$; $H^{\{s\}}$ denotes the space slightly larger than H^s, consisting of all (equivalence classes of) tempered distributions v such that grad $v \in H^{s-1}$, with the norm

$$\|v\|_{\{s\}} = \|\mathrm{grad}\ v\|_{s-1}. \tag{15.5}$$

In (15.4), the space $H^{\{1/2\}}$ is for the pair (V,a), $H^{-1/2}$ is for $(\dot{V},\mathring{a})$, and X_D is for ψ.

It should be noted that an element of $H^{\{1/2\}}(R^3)$ is actually a <u>function</u> v determined up to an additive constant. We agree to fix this constant by the condition $v \in L^3(R^3) + L^6(R^3)$, which is possible by the Sobolev inequality.

Another remark is in order. According to the general theory, we regard X as a <u>real</u> Hilbert space. Obviously X_M is a real space. $X_D = H^0(R^3;C^4)$ is also regarded as a real Hilbert space although it consists of complex-valued spinor functions, with the inner product $\mathrm{Re}(f,g)$ where (f,g) is the complex inner product. Accordingly, the imaginary number i that appears in (15.3) should be regarded as a linear operator in X rather than a scalar. A similar remark applies to the matrix elements of α, β.

With these remarks, (15.3) becomes an equation of evolution of the form (Q), where $u = (V, a, \dot{V}, \dot{a}, \psi)$ takes values in X and $A(t,u) = A(u)$ is a linear operator depending on u. $f(t,u) = f(u)$ comes from the right members of (15.3). The linear operator $A(y)$ may be written

$$A(y) = A_M \oplus A_D(y), \tag{15.6}$$

where

$$A_M = \begin{bmatrix} 0 & 0 & -1 & 0 \\ 0 & 0 & 0 & -1 \\ -\Delta & 0 & 0 & 0 \\ 0 & -\Delta & 0 & 0 \end{bmatrix} \tag{15.7}$$

acts in X_M and is independent of y, being a system of differential operators with constant coefficients, while $A_D(y)$ is the Dirac operator appearing in the last line of (15.3), depending on y through the potentials a and V.

It is well known that A_M is skew-adjoint in X_M; the metric in X_M was chosen exactly for this purpose. $A_D(y)$ is also skew-adjoint in X_D provided $y \in X$, which implies that a and V are in $H^{\{1/2\}}$. Indeed, the free-particle part A_D^0 of $A_D(y)$ is skew-adjoint (see e.g. [11]), while the perturbing term $A_D^1(y)$ involving a and V are relatively bounded with respect to A_D^0 with relative bound 0, because a and V are in $L^3 + L^\infty$. Note that the domain of A_D^0 is $H^1 \subset L^6$, while any function v in L^3 may be written as $v = v' + v''$ where $v'' \in L^\infty$ and $\|v'\|_{L^3}$ is arbitrarily small. Note also that A_D <u>is skew-adjoint in the real space</u> X_D <u>because it is skew-adjoint in the complex space</u> X_D.

To apply the results of Part 2, we choose

$$\begin{aligned} Y &= Y_M \oplus Y_D \\ Y_M &= H^{\{3/2\}}(R^3;R^4) \oplus H^{1/2}(R^3;R^4), \\ Y_D &= H^1(R^3;C^4), \\ S &= (1-\Delta)^{1/2}, \end{aligned} \tag{15.8}$$

with Y_D again regarded as a real Hilbert space.

We can now verify conditions (A1) to (A5) and (f1), (f2) by choosing as W any ball in Y with center 0.

(A1) is satisfied since $A(y)$ is skew-adjoint for any $y \in W$ (actually any $y \in X$) as shown above. To verify (A2), we note that A_M and A_D^0 commute with S; hence it suffices to consider $A_D^1(y)$. But

$$[S,A_D^1(y)]S^{-1} = -i\,\alpha\cdot[S,a]S^{-1} + i[S,V]S^{-1},$$

where $S = (1-\Delta)^{1/2}$ actually means its component in X_D, and a and V (which are components of y) are regarded as operators of multiplication in X_D. Since $y \in W$ implies $a, V \in H^{\{3/2\}}$, it follows from Lemma A3 (Appendix) that $[S,a]S^{-1}$ and $[S,V]S^{-1}$ are bounded operators in $X_D = H^0$, with the norm majorized by $\|a\|_{\{3/2\}}$ and $\|V\|_{\{3/2\}}$, respectively. Since these norms are majorized by $\|y\|_Y$ and since $A(y)$ is linear in y, this verifies (A2) as well as (A5). As usual, (A4) is trivially satisfied.

The first part of (A3) is satisfied since the domain of A_M contains Y_M and the domain of $A_D(y)$ is Y_D. Since $A(y)$ is linear in y, the Lipschitz-continuity (7.4) is implied by

$$\|A_D^1(y)\|_{Y_D,X_D} \leq \text{const}\,\|y\|_X,$$

which follows from

$$\|A_D'(y)\psi\|_{X_D} = \|(\alpha\cdot a - V)\psi\|_0 \leq \text{const}(\|a\|_{\{1/2\}} + \|V\|_{\{1/2\}})\|\psi\|_1$$

$$\leq \text{const}\,\|y\|_X \|\psi\|_{Y_D} ;$$

note that $\|fg\|_0 \leq \text{const}\,\|f\|_{\{1/2\}}\|g\|_1$, because $f = f' + f''$ with $\|f'\|_{L^3}$ and $\|f''\|_{L^6}$ dominated by $\|f\|_{\{1/2\}}$ while $\|g\|_{L^3}$ and $\|g\|_{L^6}$ are dominated by $\|g\|_1$.

To verify (f1) and (f2), we note that $f(y) = e(0,0,\psi^*\psi, \psi^*\alpha\psi, 0)$ if $y = (V,a,\dot{V},\dot{a},\psi)$. Thus these conditions are satisfied by

$$\|\psi^*\varphi\|_{1/2} \leq c\,\|\psi\|_1\,\|\varphi\|_1, \qquad \|\psi^*\varphi\|_{-1/2} \leq c\,\|\psi\|_1\,\|\varphi\|_0$$

and similarly for $\psi^*\alpha\psi$ (see Lemma A1).

With all the conditions verified, Theorems 6 and 7 are applicable. _The system_ (15.3) _has a unique solution such that_

$$(V,a) \in C[0,T';H^{\{3/2\}}] \cap C^1[0,T';H^{\{1/2\}}],$$

$$(\dot{V},\dot{a}) \in C[0,T';H^{1/2}] \cap C^1[0,T;H^{-1/2}],$$

$$\psi \in C[0,T';H^1] \cap C^1[0,T';H^0],$$

if the initial values satisfy the conditions

$$(V(0),a(0)) \in H^{\{3/2\}}, \quad (\dot{V}(0),\dot{a}(0)) \in H^{1/2}, \quad \psi(0) \in H^1 ,$$

and the solution depends on the initial value continuously in the related topology. These results coincide with those of Gross [28], in which a more specific method is used (the Poisson formula for the solution of the wave equation, for example). [For the Maxwell-Dirac equations in R^1 see Chadam [29], where global (in time) solutions are obtained.]

REMARK. We have aimed at obtaining the solution with the minimum smoothness assumptions on the initial data. On the other hand, it is rather easy to obtain smoother solutions by assuming smoother initial data. To this end it suffices, for example, to choose

$$Y_M = H^{\{s+1/2\}} \oplus H^{s-1/2}, \quad Y_D = H^s, \quad S = (1-\Delta)^{s/2},$$

with $s > 1$. Then the verification of the conditions is quite straightforward and much simpler than the case $s = 1$ considered above. In this case one may include all the nonlinear terms into $f(u)$, so that $A(u) = A$ becomes constant.

16. Quasi-linear wave equations in R^m.

Consider the wave equation in R^m:

$$w_{tt} - \sum_{j=1}^{m} a_j(w) w_{tx_j} - \sum_{j,k}^{1,m} a_{jk}(w) w_{x_j x_k} = f(w, w_t, \text{grad } w), \tag{16.1}$$

where w is a scalar unknown, $(a_{jk}(q))$ is a positive-definite symmetric matrix depending on the real variable q smoothly (so that it is uniformly positive-definite for q in a bounded set). f is also assumed to be smooth.

Let $(b_{jk}(q))$ be the positive square root of $(a_{jk}(q))$; $(b_{jk}(q))$ is also smooth in q. (16.1) may then be written

$$w_{tt} - \sum_j a_j(w) w_{tx_j} - \sum_{j,h,k} b_{jh}(w)(b_{hk}(w) w_{x_k})_{x_j}$$
$$= f_1(w, w_t, \text{grad } w), \tag{16.2}$$

where f_1 is another smooth function of its variables. Introducing the unknowns $w_0 = w_t$ and

$$w_j = \sum_k b_{jk}(w) w_{x_k}, \quad j = 1, 2, \ldots, m, \tag{16.3}$$

we obtain from (16.2) a first-order system in $m+2$ unknowns:

$$\begin{cases} \partial w/\partial t = w_0, \\ \partial w_0/\partial t - \sum_j a_j(w)\, \partial w_0/\partial x_j - \sum_{j,k} b_{jk}(w)\, \partial w_k/\partial x_j \\ \qquad\qquad = f_2(w, w_0, w_1, \ldots, w_m), \\ \partial w_k/\partial t - \sum_j b_{kj}(w)\, \partial w_0/\partial x_j = \sum_j c_{kj}(w) w_0 w_j, \quad k = 1, \ldots, m, \end{cases} \tag{16.4}$$

where $(c_{jk}) = (b'_{jk})(b_{jk})^{-1}$. Conversely, if $(w, w_0, \ldots, w_m)$ is a solution of (16.4) for which (16.3) is satisfied at $t = 0$, it can be shown that (16.3) is satisfied for all $t \geq 0$, so that w is a solution of (16.1).

(16.4) is a special case of the symmetric hyperbolic system considered in §12. Hence there is a unique solution $u = (w, w_0, \ldots, w_m)$ such that

$$u \in C[0,T';H^s(R^m;R^{m+2})] \cap C^1[0,T';H^{s-1}(R^m;R^{m+2})] \tag{16.5}$$

if the initial value $u(0)$ is in $H^s(R^m;R^{m+2})$ for some $s > m/2 + 1$. Actually the first component $w(t)$ of $u(t)$ is in H^{s+1} if (16.3) is satisfied at $t = 0$ (so that it is satisfied for all t), since (16.3) is an elliptic system for w with C^1-coefficients. Also $dw/dt = w_0$ is in $C^1[0,T';H^{s-1}]$. In this way we have obtained a solution of (16.1) such that

$$w \in C[0,T';H^{s+1}(R^m)] \cap C^2[0,T';H^{s-1}(R^m)] \tag{16.6}$$

if $\quad w(0) \in H^{s+1}, \quad w_t(0) \in H^s \quad$ and $\quad s > m/2 + 1$.

REMARK. The restriction $s > m/2 + 1$ is somewhat unsatisfactory; one would expect that $s > m/2$ suffices, since $w \in H^{s+1}$ with $s > m/2$ implies $w \in C^1(R^m)$. It seems that we could not obtain this result because (16.3) was an auxiliary condition not included in the main system.

A natural way to obtain the stronger result would be to write (16.1) as an evolution system in two unknowns:

$$\begin{cases} dw/dt - \dot{w} = 0, \\ d\dot{w}/dt - \sum_{j,k} a_{jk}(w) D_j D_k(w) - \sum_j a_j(w) D_j \dot{w} = f_0(w,\dot{w}), \end{cases} \tag{16.7}$$

where $u = (w,\dot{w})$ is regarded as an element of

$$X = H^{\{1\}}(R^m) \oplus H^0(R^m) \tag{16.8}$$

(For the notation $H^{\{1\}}$ see §15.)

To make the system (16.7) quasi-accretive, however, we have to choose the squared norm in $H^{\{1\}}(R^m)$ as equal to

$$\sum_{j,k} (a_{jk}(w)D_j w, D_k w)_0 .$$

This means that <u>we have to choose a norm in X depending on the unknown</u> w. The theory given in this paper is not applicable to such a situation, but a generalized theory would lead to a satisfactory result (see §11, remark 1).

17. <u>Magnetohydrodynamics</u> (including compressible fluids as a special case).

The equations considered here are examples of the symmetric hyperbolic system. Actually they are not of the simple form (12.1), so that our Theorems 6,7 are not applicable to them. But they do belong to a more general case (12.13), so that the results of [12] are applicable.

The magnetohydrodynamic equations for a homogeneous adiabatic system in R^3 may be written (see e.g. Courant and Hilbert [30])

$$\begin{cases} \partial v/\partial t + (v\cdot \mathrm{grad})v + \rho^{-1}\mathrm{grad}\ p + (\mu\rho)^{-1}(B \times \mathrm{rot}\ B) = f, \\ \partial\rho/\partial t + \mathrm{div}(\rho v) = 0, \\ \partial S/\partial t + v\cdot \mathrm{grad}\ S = 0, \\ \partial B/\partial t - \mathrm{rot}(v \times B) = 0, \\ \mathrm{div}\ B = 0, \end{cases} \tag{17.1}$$

where v is the velocity, ρ the density, p the pressure, S the entropy, B the magnetic induction, f the external force, and $\mu > 0$ is a constant (magnetic permeability). We may assume $\mu = 1$ without loss of generality; otherwise it suffices to introduce a new variable $\mu^{-1/2}B$ instead of B. It is assumed that p is a function of ρ and S only.

(17.1) can be converted into the following symmetric system for 8 unknowns (v,p,S,B) :

$$\begin{aligned}
&\rho\, \partial v/\partial t + \rho(v\cdot \mathrm{grad})v + \mathrm{grad}\ p + B \times \mathrm{rot}\ B = \rho f,\\
&(\rho c^2)^{-1}\partial p/\partial t + (\rho c^2)^{-1}\ v\cdot \mathrm{grad}\ p + \mathrm{div}\ v = 0,\\
&\partial S/\partial t + v\cdot \mathrm{grad}\ S = 0,\\
&\partial B/\partial t + (v\cdot \mathrm{grad})B - (B\cdot \mathrm{grad})v + B\ \mathrm{div}\ v = 0,
\end{aligned} \tag{17.2}$$

where ρ is regarded as a function of p and S, and where $c^2 = (\partial\rho/\partial p)^{-1}$.

Actually (17.2) is not exactly equivalent to (17.1). (17.1) implies (17.2), but the latter implies the former if and only if $\mathrm{div}\ B = 0$. It is not difficult to show, however, that if (17.2) is satisfied with $\mathrm{div}\ B = 0$ at $t = 0$, then $\mathrm{div}\ B = 0$ is true for all $t \geq 0$ so that it is equivalent to (17.1).

(17.2) is a symmetric system of the form (12.13). Indeed, $a_0(u)$ is the diagonal matrix with diagonal elements $\rho,\rho,\rho,(\rho c^2)^{-1},1,1,1,1$; and

$$a_1(u) = \begin{pmatrix}
\rho v_1 & 0 & 0 & 1 & 0 & 0 & B_2 & B_3\\
0 & \rho v_1 & 0 & 0 & 0 & 0 & -B_1 & 0\\
0 & 0 & \rho v_1 & 0 & 0 & 0 & 0 & -B_1\\
1 & 0 & 0 & v_1/\rho c^2 & 0 & 0 & 0 & 0\\
0 & 0 & 0 & 0 & v_1 & 0 & 0 & 0\\
0 & 0 & 0 & 0 & 0 & v_1 & 0 & 0\\
B_2 & -B_1 & 0 & 0 & 0 & 0 & v_1 & 0\\
B_3 & 0 & -B_1 & 0 & 0 & 0 & 0 & v_1
\end{pmatrix} \tag{17.3}$$

Thus the results of [12] are applicable to (17.2), leading to the same results as in Theorems 6, 7 with the choice of $Y = H^s(R^3;R^8)$ with $s > 5/2$. More precisely, there exists a unique solution

$$(v,p,S,B) \in C[0,T';W] \cap C^1[0,T';H^{s-1}(R^3;R^8)], \tag{17.4}$$

where W is a ball in $H^s(R^3;R^8)$ for $s > 5/2$, if the initial value is in W with $\operatorname{div} B = 0$ and if f is smooth in an appropriate sense.

REMARK. 1. p and S should be normalized to have the limiting values 0 at infinity. Since the possible ranges of these variables may not be the whole of $(-\infty, \infty)$, the set W must be restricted so as not to violate these ranges. The theorems given in [12] are general enough to satisfy these requirements.

2. If we want to consider viscosity, heat and electric conductivity, the problem becomes quite complicated and the results for symmetric systems are not applicable. Still, it is hoped that a generalized theory can be developed to deal with this case, in which the parabolicity of the equations is taken into account. (The general case (but not including the electromagnetic field) was considered by Nash [31] using the Lagrangian coordinates.)

3. For quasi-linear parabolic equations, see also Edmunds and Peletier [32].

Appendix

We prove some lemmas on estimating commutators between powers of $\Lambda = (1 - \Delta)^{1/2}$ and multiplication operators M_f, which are used frequently in the text. Here Δ is the Laplacian in R^m, and all functions $f, g, \dots$ are assumed to be smooth (say in the Schwartz space $S(R^m)$). As usual, however, the results can be extended by continuity to more general functions. $\| \ \|_s$ denotes the Sobolev $H^s(R^m)$-norm.

LEMMA A1. Let s, t be real numbers such that $-s < t \leq s$. Then

$$c\|f\|_s \|g\|_t \geq \begin{cases} \|fg\|_t & \text{if} \quad s > m/2, \\ \|fg\|_{s+t-m/2} & \text{if} \quad s < m/2, \end{cases} \tag{1}$$

where c is a positive constant depending on s, t, m.

A proof (somewhat sketchy) may be found in Palais [33].

LEMMA A2. If $s > m/2 + 1$ (s need not be an integer),

$$\|[\Lambda^s, M_f]\,\Lambda^{1-s}\| \leq c\,\|\text{grad } f\|_{s-1}, \tag{2}$$

where $\|\ \|$ on the left denotes the operator norm in $L^2(R^m)$.

Proof. Write $T = [\Lambda^s, M_f]\Lambda^{1-s}$. The Fourier transform of T is an integral operator with the kernel

$$k(x,y) = [(1+|x|^2)^{s/2} - (1+|y|^2)^{s/2}]\hat{f}(x-y)(1+|y|^2)^{(1-s)/2},$$

where $\hat{f}$ is the Fourier transform of f. Since the expression in [ ] is majorized by

$$s\,|x-y|[(1+|x|^2)^{(s-1)/2} + (1+|y|^2)^{(s-1)/2}],$$

we have $|k(x,y)| \leq k_1 + k_2$ where

$$k_1(x,y) = s(1+|x|^2)^{(s-1)/2}\,\hat{h}(x-y)(1+|y|^2)^{(1-s)/2},$$

$$k_2(x,y) = s\hat{h}(x-y), \qquad \hat{h}(x) = |x|\,|\hat{f}(x)|\,.$$

To show that T is bounded, it suffices to show that the integral operators with kernels k_1, k_2 are bounded.

To this end we return to the inverse Fourier transforms of these integral operators, which are given, apart from the constant factor s, by

$$T_1 = \Lambda^{s-1}M_h\Lambda^{1-s}, \qquad T_2 = M_h,$$

where h is the inverse Fourier transform of $\hat{h}$.

T_2 is easy to estimate:

$$\|T_2\| = \|h\|_{L^\infty} \leq c\,\|h\|_{s-1} = c\,\|\text{grad } f\|_{s-1} \tag{3}$$

because $s-1 > m/2$ and

$$\|h\|_{s-1}^2 = \int (1+|x|^2)^{s-1}\,|x|^2\,|\hat{f}(x)|^2 dx = \|\text{grad } f\|_{s-1}^2 \,. \tag{4}$$

T_1 can be estimated by

$$\|T_1 u\|_0 = \|h\,\Lambda^{1-s}u\|_{s-1} \leq c\|h\|_{s-1}\,\|\Lambda^{1-s}u\|_{s-1} = c\|\text{grad } f\|_{s-1}\,\|u\|_0 \,, \tag{5}$$

where we used Lemma A1 and (4). (3) and (5) lead to the desired result (2).

LEMMA A3. Let $m \geq 3$. Then

$$\|[\Lambda^s, M_f]\Lambda^{-s}\| \leq \begin{cases} c\|\text{grad } f\|_{m/2-1}\,, & 0 \leq s < m/2 \\ c\|\text{grad } f\|_{s-1}\,, & s > m/2. \end{cases} \tag{6}$$

Proof. Proceeding as in the proof of Lemma A2, we see that it suffices to estimate the following operators.

$$T_3 = \Lambda^{s-1}M_h\Lambda^{-s}, \quad T_4 = M_h\Lambda^{-1},$$

with h defined as before.

(i) Case $s > m/2$. T_3 can be estimated as T_1,

$$\|T_3 u\|_0 = \|h\,\Lambda^{-s}u\|_{s-1} \leq c\|h\|_{s-1}\,\|\Lambda^{-s}u\|_s = c\|h\|_{s-1}\,\|u\|_0$$

by Lemma A1 (note that $-s < s-1 < s$). For T_4, we have

$$\|T_4 u\|_0 = \|h\,\Lambda^{-1}u\|_0 \leq c\|h\|_{m/2-1}\,\|\Lambda^{-1}u\|_1 \leq c\|h\|_{s-1}\,\|u\|_0 \tag{7}$$

by Lemma A1 (applied with $s = \max\{m/2-1, 1\}$, $t = \min\{m/2-1, 1\}$). Thus we obtain $\|T_3\|$, $\|T_4\| \leq c\|h\|_{s-1} = c\|\text{grad } f\|_{s-1}$.

(ii) Case $0 \leq s < m/2$. $\|T_4\| \leq c\|h\|_{m/2-1}$ is still true (see (7)). For T_3, we have

$$\|T_3 u\|_0 = \|h\Lambda^{-s}u\|_{s-1} \leq c\|h\|_{m/2-1}\,\|\Lambda^{-s}u\|_s = c\|h\|_{m/2-1}\,\|u\|_0$$

by Lemma A1 (note that $m/2-1+s > 0$ by $m \geq 3$). Thus we obtain $\|T_3\|$, $\|T_4\| \leq c\|h\|_{m/2-1} = c\|\text{grad } f\|_{m/2-1}$.

REFERENCES

[1] J. L. Lions, Équations différentielles opérationnelles et problèmes aux limites, Springer 1961.

[2] K. Yosida, Functional Analysis, Springer 1971 (Third Edition).

[3] S. G. Krein, Linear differential equations in a Banach space, Izdat Nauka, Moscow 1967.

[4] A. Friedman, Partial differential equations, Holt, Rinehart and Winston 1969.

[5] R. W. Carroll, Abstract methods in partial differential equations, Harper and Row 1969.

[6] A. Pazy, Semi-groups of linear operators and applications to partial differential equations, Lecture Note 10, University of Maryland 1974.

[7] T. Kato, Integration of the equation of evolution in a Banach space, J. Math. Soc. Japan 5 (1953), 208-234.

[8] T. Kato, Linear evolution equations of "hyperbolic" type, J. Fac. Sci. Univ. Tokyo, Sec. I, Vol. 17 (1970), 241-258.

[9] T. Kato, Linear evolution equations of "hyperbolic" type, II, J. Math. Soc. Japan 25 (1973), 648-666.

[10] E. Hille and R. S. Phillips, Functional analysis and semi-groups, Revised Edition, Amer. Math. Soc. 1957.

[11] T. Kato, Perturbation theory for linear operators, Springer 1966.

[12] T. Kato, The Cauchy problem for quasi-linear symmetric hyperbolic systems, to appear.

[13] K. O. Friedrichs, Symmetric positive linear differential equations, Comm. Pure Appl. Math. 11 (1958), 333-418.

[14] P. D. Lax, Hyperbolic systems of conservation laws and the mathematical theory of shock waves, to appear.

[15] A. E. Fischer and J. E. Marsden, The Einstein evolution equations as a first-order quasi-linear symmetric hyperbolic systems, I, Comm. Math. Phys. 28 (1972), 1-38.

[16] A. Sjöberg, On the Korteweg-de Vries equation: existence and uniqueness, J. Math. Anal. Appl. 29 (1970), 569-579.

[17] T. Mukasa and R. Iino, On the global solution for the simplest generalized Korteweg-de Vries equation, Math. Japonicae 14 (1969), 75-83.

[18] Y. Kametaka, Korteweg-de Vries equation, Proc. Japan Acad. 45 (1969), 552-555; 556-558; 656-660; 661-665.

[19] M. Tsutsumi and T. Mukasa, Parabolic regularizations for the generalized Korteweg-de Vries equation, Funkcial. Ekvac. 14 (1971), 89-110.

[20] T. E. Dushane, Generalizations of the Korteweg-de Vries equation, Proc. Symp. Pure Math. Vol. 23, Amer. Math. Soc. 1973, pp. 303-307.

[21] T. Kato, Nonstationary flows of viscous and ideal fluids in R^3, J. Functional Anal. 9 (1972), 296-305.

[22] K. O. Friedrichs, The identity of weak and strong extensions of differential operators, Trans. Amer. Math. Soc, 55 (1944), 132-151.

[23] O. A. Ladyzhenskaya, The mathematical theory of viscous incompressible flow, Second English Edition, Gordon and Breach 1969.

[24] K. K. Golovkin, Vanishing viscosity in Cauchy's problem for hydromechanics equations, Trudy Mat. Inst. Steklov. 92 (1966), 31-49; English Translation 33-53.

[25] F. J. McGrath, Nonstationary plane flow of viscous and ideal fluids, Arch. Rational Mech. Anal. 27 (1968), 329-348.

[26] D. G. Ebin and J. Marsden, Groups of diffeomorphisms and the motion of an incompressible fluid, Ann. Math. 92 (1970), 102-163.

[27] H. Swann, The convergence with vanishing viscosity of non-stationary Navier-Stokes flow to ideal flow in R_3, Trans. Amer. Math. Soc. 157 (1971), 373-397.

[28] L. Gross, The Cauchy problem for the coupled Maxwell and Dirac equations, Comm. Pure Appl. Math. 19 (1966), 1-15.

[29] J. M. Chadam, Global solutions of the Cauchy problem for the (classical) coupled Maxwell-Dirac equations in one space dimension, J. Functional Anal. 13 (1973), 173-184.

[30] R. Courant and D. Hilbert, Mathods of Mathematical Physics, Vol. II, Interscience Publishers 1962.

[31] J. F. Nash, Le problèm de Cauchy pour les équations differentielles d'un fluide général, Bull. Soc. Math. France 90 (1962), 487-497.

[32] D. E. Edmunds and L. A. Peletier, Quasilinear parabolic equations, Ann. Scuola Norm. Sup. Pisa 25 (1971), 397-421.

[33] R. S. Palais, Foundations of global non-linear analysis, Benjamin 1968.

[This work was partly supported by NSF Grant GP-37780X.]

SPECTRAL THEORY OF PARTIAL DIFFERENTIAL OPERATORS

Joachim Weidmann

Introduction

In this paper we review some methods of perturbation theory which are especially useful for the study of spectral properties of self-adjoint differential operators.

First we describe self-adjoint differential operators with constant and not constant coefficients in $L_2(\mathbb{R}^m)$ or $L_2(G)$, where G is an open subset of $\mathbb{R}^m$ (sections 1 and 2). In section 3 we give some abstract perturbation results concerning perturbations which are relatively compact with respect to the unperturbed operator or with respect to the square of the unperturbed operator. Roughly speaking the results say that the essential spectrum (and frequently the singular sequences) are preserved. In section 4 we show how these results are applicable to differential operators, especially to operators of the Schrödinger or Dirac type. We study perturbations which are small at infinity. It turns out that these perturbations are very well adapted to this situation. In section 5 finally we show that similar conditions are also useful in scattering theory (and therefore in the perturbation theory of the absolutely continuous spectrum).

We introduce some notations. With $L_2(G)$ we denote the Hilbert space of square integrable complex valued functions. By $L_{2,M}(G) := [L_2(G)]^M = L_2(G, \mathbb{C}^M)$ we denote the corresponding space of $\mathbb{C}^M$-valued functions. The inner products and norms are in both cases denoted by $\langle .,. \rangle$ and $\|.\|$, respectively. The additional index M is used in the same sense in other cases: $C^{\infty}_{o,M}$, $W^{r}_{2,M}$ etc.

The Fourier transform in $L_2(\mathbb{R}^m)$ (and in the same way in $L_{2,M}(\mathbb{R}^m)$) is defined by

$$Ff(x) = (2\pi)^{-m/2} \operatorname*{l.i.m.}_{N \to \infty} \int_{|y| \leq N} e^{-ixy} f(y)dy;$$

here xy stands for the inner product of $x,y \in \mathbb{R}^m$. With $\partial_j = i \frac{\partial}{\partial x_j}$ we then have

$$F\partial_j f = -x_j Ff, \quad Fx_j f = \partial_j Ff,$$

where x_j is the j-th component of x.

1. Differential operators with constant coefficients.

The simplest differential operators are of course those with constant coefficients. Such an operator T can be written in the form

$$T = P(\partial) = P(\partial_1, \ldots, \partial_m),$$

where P is a polynomial in m variables,

$$\partial = (\partial_1, \ldots, \partial_m), \quad \partial_j = i \frac{\partial}{\partial x_j}$$

(the imaginary factor i is included to make ∂_j formally self-adjoint; this means, for $u,v \in C^{\infty}_{o}(\mathbb{R}^m)$ we have $\langle \partial_j u, v\rangle = \langle u, \partial_j v\rangle$). A quite natural domain of definition for such a differential operator is $C^{\infty}_{o}(\mathbb{R}^m)$, which is dense in $L_2(\mathbb{R}^m)$.

For $u,v \in C^{\infty}_{o}(\mathbb{R}^m)$ we have

$$\langle P(\partial)u, v\rangle = \langle u, P^*(\partial)v\rangle,$$

where P^* is the polynomial with coefficients which are conjugate to the coefficients of P. Therefore T_o, the restriction of T to $D(T_o) = C^{\infty}_{o}(\mathbb{R}^m)$, is symmetric if the coefficients of P are real.

By means of the Fourier transform F the operator T_o is transformed to

$$M_{P,O} = F^{-1} T_o F,$$

where $M_{P,O}$ is the restriction of the multiplication by P to the domain $D(M_{P,O}) = F^{-1} C_o^\infty(\mathbb{R}^m)$, which is also dense in $L_2(\mathbb{R}^m)$. From this it is clear that $M_{P,O}$, and therefore T_o, are closable. $D(\overline{M_{P,O}})$, and therefore $D(\overline{T_o})$, obviously contain the Schwartz' space $S(\mathbb{R}^m)$; this immediately implies that $\overline{M_{P,O}} = M_P$, the maximal operator of multiplication by P, i.e.

$$D(M_P) = \{f \in L_2(\mathbb{R}^m): Pf \in L_2(\mathbb{R}^m)\},$$
$$M_P f = Pf \quad \text{for } f \in D(M_P).$$

Since M_P is self-adjoint if and only if P is real valued (i.e. the coefficients of P are real), the operator $T = \overline{T_o}$ is self-adjoint if and only if P has real coefficients. This implies that T_o is essentially self-adjoint in this case, and

$$T = T_o^* = F M_P F^{-1}.$$

In general T is a normal operator, i.e.

$$D(T) = D(T^*), \ \|T^* f\| = \|Tf\| \quad \text{for } f \in D(T).$$

T^* is the operator generated by P^* in the same sense as T was generated by P. (We could say: T_o is essentially normal, since $\overline{T_o}$ is normal.)

The domain of T apparently is given by

$$D(T) = FD(M_P) = \{f \in L_2(\mathbb{R}^m) : PF^{-1} f \in L_2(\mathbb{R}^m)\}.$$

From $T = T_o^*$ and the definition of the derivatives for distributions it follows that $f \in D(T)$ if and only if $P(\partial)f \in L_2(\mathbb{R}^m)$, $Tf = P(\partial)f$. (Here the derivatives are taken in the sense of distributions. This means: take the regular distribution D_f which is generated by f, calculate $P(\partial)D_f$ in the sense of distributions; this is again a regular distribution generated by an L_2-function g; define $P(\partial)f = g$.) This is another useful description of D(T) and T.

Let P_1 and P_2 be polynomials with

$$|P_1(\xi)| \leq C_1 + C_2|P_2(\xi)|, \quad \xi \in \mathbb{R}^m$$

for suitable constants C_1 and C_2. Let further T_1 and T_2 be the operators generated by P_1 and P_2, respectively. It follows that $D(T_1) \subset D(T_2)$ and for a suitable constant C

$$\|T_1 f\| \leq C(\|f\| + \|T_2 f\|).$$

For example, if

$$T := -\Delta = \sum_{j=1}^{m} \partial_j^2 ,$$

then every derivative (in the sense of distributions) of $f \in D(T)$ up to order 2 lies in $L_2(\mathbb{R}^m)$. If, on the other hand, $T = \partial_1 \partial_2$, then $\partial_1 \partial_2 f$ is (in general) the only derivative of f which lies in $L_2(\mathbb{R}^m)$ for $f \in D(T)$.

If T is an <u>elliptic</u> operator of order r, i.e. P is a polynomial of order r and

$$C_1 + |P(\xi)| \geq C_2|\xi|^r , \quad \xi \in \mathbb{R}^m$$

for some constants C_1 and C_2, then D(T) is equal to the well known <u>Sobolev space</u>

$$W_2^r(\mathbb{R}^m) := \{f \in L_2(\mathbb{R}^m): |x|^r F^{-1} f \in L_2(\mathbb{R}^m)\}$$

$$= \{f \in L_2(\mathbb{R}^m): |x|^r Ff \in L_2(\mathbb{R}^m)\}.$$

This definition is meaningful for all $r \in \mathbb{R}$, and we shall use it later for all positive r. For r = 0 we have $W_2^r(\mathbb{R}^m) = L_2(\mathbb{R}^m)$. From the above considerations it follows that $C_o^\infty(\mathbb{R}^m)$ is <u>dense</u> in $W_2^r(\mathbb{R}^m)$ with respect to the <u>Sobolev norm</u>

$$\|f\|_r := \|f\|_{W_2^r} = \|(1+|\cdot|)^r Ff\|.$$

With this norm $W_2^r(\mathbb{R}^m)$ becomes a Hilbert space. One of the most important facts about these Sobolev spaces is given by the following theorem.

<u>1.1.Theorem</u> Let $0 \leq s < r$. Then $W_2^r(\mathbb{R}^m) \subset W_2^s(\mathbb{R}^m)$ and the imbedding is continuous. For every $\varepsilon > 0$ there is a $C \geq 0$ such that

$$\|f\|_s \leq \varepsilon\|f\|_r + C\|f\| \quad \text{for every } f \in W_2^r(\mathbb{R}^m).$$

For every $\varphi \in C_o^\infty(\mathbb{R}^m)$ and $f \in W_2^r(\mathbb{R}^m)$ we have $\varphi f \in W_2^r(\mathbb{R}^m)$ and the map

$$J_\varphi : W_2^r(\mathbb{R}^m) \to W_2^s(\mathbb{R}^m),\ f \mapsto \varphi f$$

is compact.

The first and the second statement are immediate consequences of the definition of the norms in $W_2^\tau(\mathbb{R}^m)$. The remaining results follow easily from the fact that the multiplication by φ is represented by a convolution with the rapidly decreasing function $(2\pi)^{-m/2}F^{-1}\varphi$ in the Fourier representation.

Sometimes <u>local Sobolev spaces</u> are used. We define

$$W_{2,loc}^r(\mathbb{R}^m) := \{f \in L_{2,loc}(\mathbb{R}^m) : \varphi f \in W_2^r(\mathbb{R}^m) \text{ for every } \varphi \in C_o^\infty(\mathbb{R}^m)\}.$$

It is possible to define an inductive limit topology such that $W_{2,loc}^r(\mathbb{R}^m)$ becomes a (locally convex) Fréchet space; but we do not need such a topological structure.

If P does not satisfy the ellipticity condition, then one could define a "generalized Sobolev space" by

$$W_2(\mathbb{R}^m, P) := \{f \in L_2(\mathbb{R}^m) : PF^{-1}f \in L_2(\mathbb{R}^m)\}.$$

It is easy to prove some quite similar results for such spaces, but we do not go into details.

All questions concerning *spectral properties* of the operators considered above are now almost trivial, since they are *unitarily equivalent* to multiplication operators (in fact to multiplication by polynomials). By means of this unitary equivalence we see that the spectrum of T is equal to the range of P (remember that the range of a polynomial is always closed). Therefore, the spectrum of a self-adjoint differential operator T is either a half line (if T is semi-bounded), or the entire real line (if T is not semi-bounded). Since λ is an eigenvalue of the operator of multiplication by a function t if and only if $t^{-1}(\lambda)$ has positive measure, it follows that our operators T have *no eigenvalues.*

The *spectral resolution* $\hat{E}$ of the self-adjoint operator M_P is given by

$$(\hat{E}(b)-\hat{E}(a))f = \chi_{\{x\in\mathbb{R}^m: a\leq P(x)\leq b\}} f.$$

From this we can see that the spectra of M_P and T are *absolutely continuous*, which means that the measure on $\mathbb{R}$, which is generated by

$$\mu([a,b)) = \langle f, (\hat{E}(b)-\hat{E}(a))f\rangle$$

is absolutely continuous with respect to the Lebesgue measure for every $f \in L_2(\mathbb{R}^m)$. The spectral projection $E(b)-E(a)$ of T is given by the convolution with

$$(2\pi)^{-m/2} F\,\chi_{\{x\in\mathbb{R}^m: a\leq P(x)\leq b\}}$$

(which lies in $L_2(\mathbb{R}^m)$).

The *resolvent* of M_P is the multiplication operator

$$(\lambda-M_P)^{-1}f(x) = (\lambda-P(x))^{-1}f(x)$$

for λ not in the range of P. Therefore the resolvent of T can be given as a convolution operator, but in general this is possible only in the sense of distributions.

For the very important operator $T = -\Delta$ in $\mathbb{R}^3$ we have $\sigma(T)=[0,\infty)$ and

$$(\lambda-T)^{-1}f(x) = \frac{-1}{4\pi}\int|x-y|^{-1}\exp[i\sqrt{\lambda}\,|x-y|]f(y)dy,$$

where $\operatorname{Im}\sqrt{\lambda} > 0$ for $\lambda \notin [0,\infty)$. It is also possible to calculate the unitary group $\exp(-itT)$ by means of the Fourier transform, since the unitary group of M_P is multiplication by $\exp(-itP)$. We get for $T=-\Delta$ and $m \in \mathbb{N}$

$$(e^{-itT}f)(x) = \underset{N\to\infty}{\text{l.i.m.}}\,(4\pi it)^{-m/2}\int_{|x|\leq N} e^{-|x-y|/4it}f(y)dy.$$

These things are only a little more complicated for differential operators in the Hilbert spaces of $\mathbb{C}^M$-valued functions $L_{2,M}(\mathbb{R}^m) := L_2(\mathbb{R}^m)^M = L_2(\mathbb{R}^m,\mathbb{C}^M)$. In this case we assume that P is a matrix valued function on $\mathbb{R}^m$, where all the entries are polynomials. Again by means of the Fourier transform on $L_{2,M}(\mathbb{R}^m)$ we get

$$T_o = FM_{P,0}F^{-1},$$

where now $M_{P,0}$ is the operator of (matrix×vector) multiplication by P. By the same technique as above we prove that T_o is closable and that T_o^* is unitarily equivalent to the maximal multiplication operator M_{P^*}, where $P^*(x)$ is the matrix which is hermitian conjugate to $P(x)$). It follows that $T := \overline{T}_o$ is essentially self-adjoint if and only if $P(x) = P^*(x)$; then $T = T_o^*$.

One of the most important examples of this kind is the Dirac operator for a free electron. In this case we have $m = 3$, $M = 4$ and

$$P(x) = c(y_1\alpha_1 + y_2\alpha_2 + y_3\alpha_3) + \mu c^2\beta,$$

μ and c are positive constants (μ the mass of the electron, c the velocity of the light),

$$\alpha_j = \begin{pmatrix} 0 & \sigma_j \\ \sigma_j & 0 \end{pmatrix} \quad (j=1,2,3), \quad \beta = \begin{pmatrix} 1 & 0 & 0 & 0 \\ 0 & 1 & 0 & 0 \\ 0 & 0 & -1 & 0 \\ 0 & 0 & 0 & -1 \end{pmatrix}$$

and

$$\sigma_1 = \begin{pmatrix} 0 & 1 \\ 1 & 0 \end{pmatrix}, \ \sigma_2 = \begin{pmatrix} 0 & -i \\ i & 0 \end{pmatrix}, \ \sigma_3 = \begin{pmatrix} 1 & 0 \\ 0 & -1 \end{pmatrix}.$$

Since $P^*(x) = P(x)$ the operator T_o is essentially self-adjoint; $T := \overline{T}_o$ is self-adjoint with the domain $W^1_{2,M}(\mathbb{R}^m) = W^1_2(\mathbb{R}^m)^M$.

The eigenvalues of the matrix $P(x)$ are given by $h_j(x)$ with

$$h_1(x) = h_2(x) = -h_3(x) = -h_4(x) = \mu c^2\left(1+\frac{|x|^2}{\mu^2 c^2}\right)^{1/2}$$

These eigenvalues and a suitable choice of corresponding eigenvectors depend smoothly on x (see [25], section 4). Therefore T is unitarily equivalent to the multiplication operator M_H, where

$$H(x) = \begin{pmatrix} h_1(x) & 0 & 0 & 0 \\ 0 & h_2(x) & 0 & 0 \\ 0 & 0 & h_3(x) & 0 \\ 0 & 0 & 0 & h_4(x) \end{pmatrix},$$

from which we can easily see that the <u>spectrum</u> is

$$\sigma(T) = (-\infty, -\mu c^2] \cup [\mu c^2, \infty).$$

It is also not difficult to show that this spectrum is <u>absolutely continuous.</u>

Finally we should mention that the differential operators are just a subclass of those operators which are <u>Fourier transforms of multiplication operators</u> (in the same sense as differential operators are Fourier transforms of multiplications by polynomials). This more general class of operators has many properties which are the same as for differential operators. If, for example, the function

$P: \mathbb{R}^m \to \mathbb{R}$ is smooth and grad $P \neq 0$ almost everywhere, then M_P (and therefore T) has an absolutely continuous spectrum and there exists at least some part of a reasonable scattering theory, e.g. [24,25]. One great difference lies in the fact that differential operators are *local operators* (i.e. $\mathrm{supp}(T\varphi) \subset \mathrm{supp}\ \varphi$), but these more general operators are not local.

2. Perturbed operators

Most operators which appear in applications do not have constant coefficients, but in many cases the terms of highest order have constant(or asymptotically constant) coefficients. Therefore it seems reasonable to consider these operators as *perturbed operators*, the *unperturbed operator* being one with constant coefficients. Therefore all the basic problems of spectral theory can be studied from the view point of *perturbation theory*: assume that we know the unperturbed operator T_o very well; what can we say about the perturbed operator. Almost everything we are going to describe in this paper is perturbation theory in some sense.

The first problem is, to give a reasonable definition of the perturbed operator, i.e. to find a domain of definition such that this operator is self-adjoint or, at least, essentially self-adjoint. One of the most usefull (abstract) results was given by Rellich and Kato (see [12],V.4).

Let A and B be operators in a Hilbert space H. We say that B is *A-bounded*, if $D(A) \subset D(B)$ and if there exist non-negative numbers a and b such that

$$\|Bx\| \leq a\|Ax\| + b\|x\|, \quad x \in D(A).$$

The infimum of all possible numbers a is called the *A-bound* of B.

<u>2.1. Theorem</u> Let A be a (essentially) self-adjoint operator in a Hilbert space H and assume that B is symmetric and A-bounded with A-bound less than 1. Then A+B is (essentially) self-adjoint and $\overline{A+B} = \overline{A}+\overline{B}$ (this implies $D(\overline{A+B}) = D(\overline{A})$).

It turns out, that very often one proves the A-boundedness with relative bound < 1 of a perturbation B by proving that it is A-bounded with A-bound 0. This is so, if one considers differential operators where the coefficients are taken from a suitable Stummel class.

For a real number ρ we say that a function $q: \mathbb{R}^m \to \mathbb{C}$ belongs to the local Stummel class $M_{\rho,loc}(\mathbb{R}^m)$, if

$$M_{q,\rho}(x) := \begin{cases} \left\{ \int_{|x-y|\leq 1} |q(y)|^2 |x-y|^{\rho-m} dy \right\}^{1/2} & \text{for } \rho < m, \\ \left\{ \int_{|x-y|\leq 1} |q(y)|^2 dy \right\}^{1/2} & \text{for } \rho \geq m \end{cases}$$

exists for every $x \in \mathbb{R}^m$ and if $M_{q,\rho}(.)$ is locally bounded. For $\rho \geq m$ we have $M_{\rho,loc}(\mathbb{R}^m) = L_{2,loc}(\mathbb{R}^m)$. We say that q belongs to the (global) Stummel class $M_\rho(\mathbb{R}^m)$, if

$$M_{q,\rho} := \sup_{x\in\mathbb{R}^m} M_{q,\rho}(x) < \infty.$$

We have obviously $M_{\rho,loc} \subset M_{\tau,loc}$ and $M_\rho \subset M_\tau$ if $\rho \leq \tau$. With this notion the following theorem holds.

<u>2.2. Theorem</u> a) Let $\rho < 2r$, $q \in M_\rho(\mathbb{R}^m)$. Then for every $u \in W_2^r(\mathbb{R}^m)$ we have

$$\|qu\| \leq CM_{q,\rho}\|u\|_r,$$

where C depends on m, r and ρ only.

b) Let $\rho < 2r$, $q \in M_\rho(\mathbb{R}^m)$. Then for every $\varepsilon > 0$ there exists a $C \geq 0$

such that

$$\|qu\| \leq \varepsilon\|u\|_r + C\|u\|$$

for every $u \in W_2^r(\mathbb{R}^m)$.

c) Let $\rho < 2r$, $q \in M_\rho(\mathbb{R}^m)$ and assume that

$$\int_{|x-y|\leq 1} |q(y)|^2 dy \to 0 \qquad \text{for } |x| \to \infty.$$

Then the map

$$W_2^r(\mathbb{R}^m) \to L_2(\mathbb{R}^m), \; u \mapsto qu$$

is compact.

Proof. a) This proof is very technical and quite long. We refer the reader to the book of Schechter [17], Chapter 6 Theorem 2.1.

b) Let $s \in \mathbb{R}$ be such that $\rho < 2s < 2r$. Then by Theorem 1.1 the embedding $W_2^r(\mathbb{R}^m) \to W_2^s(\mathbb{R}^m)$ has the property: for every $\varepsilon > 0$ there is a $C \geq 0$ such that

$$\|u\|_s \leq \varepsilon\|u\|_r + C\|u\|, \qquad u \in W_2^r(\mathbb{R}^m).$$

Therefiore b) follows by applying part a) of the theorem for s instead of r.

c) If $\rho \geq m$, then our condition and the definition of $M_{q,\rho}(.)$ (for $\rho \geq m$) imply

$$M_{q,\rho}(x) = \Big\{ \int_{|x-y|\leq 1} |q(y)|^2 dy \Big\}^{1/2} \to 0 \quad \text{for } |x| \to \infty.$$

In this case we choose $\tau = \rho$ for later use.

If $\rho < m$, then we choose $\tau \in \mathbb{R}$ such that $\rho < \tau < \min\{m, 2r\}$. Then by means of Hölder's inequality $\left(s = \frac{m-\rho}{\tau-\rho}, \; s' = \frac{m-\rho}{m-\tau}\right)$

$$M_{q,\tau}(x)^2 = \int_{|x-y|\leq 1} \Big\{|q(y)|^{2/s}\Big\}\Big\{|q(y)|^{2/s'}|x-y|^{\tau-m}\Big\}dy$$

$$\leq \Big\{ \int_{|x-y|\leq 1} |q(y)|^2 dy\Big\}^{1/s} \Big\{ \int_{|x-y|\leq 1} |q(y)|^2 |x-y|^{\rho-m} dy\Big\}^{1/s'}.$$

Therefore we have in both cases

$$M_{q,\tau}(x)^2 \to 0 \quad \text{for } |x| \to \infty$$

for some $\tau < 2r$.

Let $\varphi \in C^\infty(\mathbb{R})$ with $0 \leq \varphi(t) \leq 1$,

$$\varphi(t) = \begin{cases} 1 & \text{for } t \leq 1 \\ 0 & \text{for } t \geq 2. \end{cases}$$

Define $\varphi_n : \mathbb{R}^m \to \mathbb{R}$ by

$$\varphi_n(x) = \varphi(n^{-1}|x|),\ x \in \mathbb{R}^m.$$

Then, for every $s \in \mathbb{R}$ with $\tau < 2s < 2r$ and every $n \in \mathbb{N}$, the map

$$W_2^r(\mathbb{R}^m) \to W_2^s(\mathbb{R}^m),\ u \mapsto \varphi_n u$$

is compact by Theorem 1.1. Applying part a of this theorem for r replaced by s we find that the map

$$W_2^r(\mathbb{R}^m) \to L_2(\mathbb{R}^m),\ u \mapsto q\varphi_n u$$

is compact for every $n \in \mathbb{N}$. Furthermore the maps

$$W_2^r(\mathbb{R}^m) \to L_2(\mathbb{R}^m),\ u \mapsto q(1-\varphi_n)u$$

converge to zero for $n \to \infty$. This implies that the map

$$W_2^r(\mathbb{R}^m) \to L_2(\mathbb{R}^m),\ u \mapsto qu$$

is compact. Q.E.D.

A simple consequence of this theorem is

2.3. Theorem If T_o is a closed (e.g. self-adjoint) operator with $D(T_o) = W_2^r(\mathbb{R}^m)$ and V a differential operator of the form

$$Vu = \sum_{|\alpha|<r} q_\alpha \partial^\alpha u,$$

where $q_\alpha \in M_{\rho_\alpha}$ with $\rho_\alpha < 2(r-|\alpha|)$. Then V is T_o-bounded with relative bound 0. If, in addition,

$$\int_{|x-y|\leq 1} |q_\alpha(y)|^2 dy \to 0 \quad \text{for } |x| \to \infty,$$

then V is T_o-compact (for the definition of T_o-compact, see section 3).

<u>2.4. Example</u> Let $T_o = -\Delta$ with $D(T_o) = W_2^2(\mathbb{R}^m)$, and let V be defined by $D(V) = D(T)$ and

$$Vu = 2\sum_{j=1}^m b_j \partial_j u + (q + \sum_{j=1}^m b_j^2)u$$

with real valued functions b_j and q,

$$b_j^2 \in M_\rho(\mathbb{R}^m),\ q \in M_\rho(\mathbb{R}^m) \quad \text{for some } \rho < 4,$$

$$\operatorname{div} b = \sum_{j=1}^m \partial_j b_j = 0 \text{ in the sense of distributions.}$$

Then V is T_o-bounded with T_o-bound 0 and therefore the Schrödinger operator T,

$$Tu = T_o u + Vu = \sum_{j=1}^m (\partial_j + b_j)^2 u + qu$$

$$= -\Delta u + 2\sum_{j=1}^m b_j \partial_j u + (q + \sum_{j=1}^m b_j^2)u$$

is self-adjoint on $D(T) = W_2^2(\mathbb{R}^m)$. If in addition

$$\int_{|x-y|\leq 1} \left\{q(y)^2 + \sum_{j=1}^m b_j^2(y)\right\} dy \to 0 \quad \text{for } |x| \to \infty,$$

then V is T_o-compact.

<u>2.5. Example</u> Let T_o be the Dirac operator of the free electron in $L_{2,4}(\mathbb{R}^3)$ (see section 1) and V the operator of multiplication by a 4×4 hermitian matrix valued function q, where all entries of q lie in $M_\rho(\mathbb{R}^m)$ for some $\rho < 2$. Then V is T_o-bounded with T_o-bound 0 and $T = T_o + V$ is self-adjoint on $D(T) = W_{2,4}^1(\mathbb{R}^3)$. If in addition for every entry q_{jk} of q

$$\int_{|x-y|\leq 1} |q_{jk}(y)|^2 dy \to 0 \quad \text{for } |x| \to \infty,$$

then V is T_o-compact.

By means of very specific considerations it can be shown (see Kato [12],V.5) that, in order to have self-adjointness of the perturbed Dirac operator on the same domain as T_o, q may also contain sufficiently small Coulomb terms; but in this case the perturbation is not relatively compact with respect to T_o (relative bound > 0).

Much more general conditions are possible if we do only wish that $T = T_o + V$ is essentially self-adjoint on $C_o^\infty(\mathbb{R}^m)$ (but not, that the self-adjoint extension of T has the same domain as that of T_o). For the Schrödinger operator (Example 2.4) the first general result of this kind was given by Ikebe and Kato [8] (notice that they considered operators with variable highest order coefficients). Recently this result has been generalized by Simon [21,22,23] and Kato [13]. The latest, and probably most general result was recently given by Simader [20]. The following theorem combines the results of Ikebe-Kato and Simader.

<u>2.6. Theorem</u> Let q and $b_j (j=1,2,\dots,m)$ be real valued,

$$b_j \in C^1(\mathbb{R}^m), \ \operatorname{div} b = 0,$$

$q = q_1 + q_2$ with

$$q_1 \in L_{2,loc}(\mathbb{R}^m), \ q_1(x) \geq -c|x|^2,$$

$$q_2 \in M_\rho(\mathbb{R}^m) \quad \text{for some } \rho < 4.$$

Then the operator T defined in 2.4 is essentially self-adjoint on $C_o^\infty(\mathbb{R}^m)$. If in addition $q \in M_{\rho,loc}(\mathbb{R}^m)$ for some $\rho < 4$, then the domain of the self-adjoint extension is given by

$$D(\bar{T}) = \left\{u \in W_{2,loc}^2(\mathbb{R}^m) \cap L_2(\mathbb{R}^m) : Tu \in L_2(\mathbb{R}^m)\right\}.$$

In the original proof Ikebe and Kato first calculated T^*

(i.e. they determined $D(T^*)$ explicitly) and then proved that T^* is symmetric. Simader just shows that $R(\pm i-T)$ is dense. It seems that the adjoint T^* cannot be calculated explicitly in the situation considered by Simader; this means that we do not know explicitly the self-adjoint extension of T in this case.

One can prove similar results for the Dirac operator. One of the most general results in this direction was given by Jörgens [10].

2.7. Theorem Let T_o be the Dirac operator of the free electron, V the multiplication by a 4×4 hermitian matrix valued function $q = q_1 + q_2 + q_3$ with

a) $|q_1(x)| \leq \sum_{j=1}^{n} \gamma_j |x-x_j|^{-1}$ with $|\gamma_j| < 1/2$, $x_j \neq x_k$,

b) $|q_2| \in L_{3,loc}(\mathbb{R}^3)$

c) $|q_3| \in M_{\rho,loc}(\mathbb{R}^3)$ with $\rho < 2$.

Then $T = T_o + V$ is essentially self-adjoint on $C^{\infty}_{o,4}(\mathbb{R}^3)$. The domain of the self-adjoint extension is given by

$$D(\overline{T}) = \{u \in W^1_{2,4,loc}(\mathbb{R}^3) \cap L_{2,4}(\mathbb{R}^3) : Tu \in L_{2,4}(\mathbb{R}^3)\}.$$

For many applications it is also important to study differential operators in $L_2(G)$, where G is an open subset of $\mathbb{R}^m$. Conditions for the essential self-adjointness on $C^{\infty}_o(\mathbb{R}^m)$ of operators of the form considered in 2.4 (the operators of Schrödinger type and even for a more general class of operators) have been given by Jörgens [9]. In this case, similar to the Ikebe-Kato result, the domain of the self-adjoint extension consists of all functions $u \in L_2(G) \cap W^2_{2,loc}(G)$ with $Tu \in L_2(G)$; here Tu is calculated in the sense of distributions; the local Sobolev space $W^r_{2,loc}(G)$ is the set of all $u \in L_{2,loc}(G)$ such that $\varphi u \in W^r_2(\mathbb{R}^m)$ for every $\varphi \in C^{\infty}_o(G)$. No result of this kind seems to be known for a more general class of differential operators; actually it seems very difficult to guess how such a result might look!

3. The invariance of the essential spectrum; abstract results.

The essential spectrum $\sigma_e(A)$ of a self-adjoint operator A in a Hilbert space H is defined to be the set consisting of all limit points of the spectrum and the eigenvalues of infinite multiplicity. It is known that a real number λ belongs to $\sigma_e(A)$ if and only if one of the following conditions is satisfied:

(3.1) For every $\varepsilon > 0$ the projection $E(\lambda+\varepsilon) - E(\lambda-\varepsilon)$ is infinite dimensional, or

(3.2) there exists a sequence (u_n) in D(A) such that $u_n \rightharpoonup 0$, $u_n \not\to 0$ and $(A-\lambda)u_n \to 0$; such a sequence is called a singular sequence for A and λ.

An important problem is the invariance of the essential spectrum under perturbations. In this connection relative compactness plays an important role.

Let A and B be operators in H. B is said to be A-compact if $D(A) \subset D(B)$ and the map $B: D(A) \to H$ is compact (here D(A) is equiped with the graph norm $\|x\|_A := (\|x\|^2 + \|Ax\|^2)^{1/2}$). An equivalent condition is: from $x_n \rightharpoonup 0$ and $Ax_n \rightharpoonup 0$ it follows that $Bx_n \to 0$ (for details see [11]).

For a self-adjoint operator A with the spectral resolution E it follows easily from the von Neumann spectral theorem: B is A-compact if and only if B is A-bounded with A-bound 0 and the restriction of B to $R(E(N) - E(-N))$ is compact for every $N \geq 0$.

The following result is well known. It follows immediately from the above characterizations of the essential spectrum and of the relative compactness.

3.3. Theorem Let A be self-adjoint and let B be symmetric and A-compact. Then $\sigma_e(A+B) = \sigma_e(A)$.

For many interesting applications to differential operators the requirement of A-compactness is too restrictive. Recently Schechter [18] proved that in certain cases A^2-compactness is also sufficient. His result holds also for non self-adjoint operators in Banach spaces; the proof makes strong use of Fredholm theory. For this reason we will give here a simpler proof for self-adjoint operators.

3.4. Theorem Let A be a self-adjoint operator in H, B symmetric and A-bounded such that A+B is self-adjoint. If B is A^2-compact, then $\sigma_e(A+B) = \sigma_e(A)$.

3.5. Remarks Gustafson and Weidmann [6] proved that for a self-adjoint operator A the A^n-compactness of B (for some n) and the A-boundedness is sufficient to guarantee the A^2-compactness of B. An important difference to theorem 3.3 lies in the fact that in 3.3 the operators A and A+B have the same singular sequences, but we are not able to prove this in 3.4 (see also Theorem 4.3). The above characterization of A-compactness for self-adjoint A gives us the following result: B is A^2-compact if and only if B is A^2-bounded with A^2-bound 0 (this holds if B is A-bounded) and the restriction of B to $R(E(N)-E(-N))$ is compact for every $N \geq 0$ (E the spectral resolution of A).

Proof of Theorem 3.4. (i) $\sigma_e(A) \subset \sigma_e(A+B)$: We actually prove: every singular sequence of A and λ is a singular sequence of A + B and λ. Let (u_n) be a singular sequence of A and λ. Then

$$(I-E(\lambda+1)+E(\lambda-1))u_n \to 0 \quad \text{and} \quad A(I-E(\lambda+1)+E(\lambda-1))u_n \to 0;$$

hence by the A-boundedness of B

$$B(I-E(\lambda+1)+E(\lambda-1))u_n \to 0.$$

From the A^2-compactness of B it follows that $B(E(\lambda+1)-E(\lambda-1))u_n \to 0$ and therefore $[(A+B)-\lambda]u_n \to 0$.

(ii) $\sigma_e(A+B) \subset \sigma_e(A)$: Let $\lambda \in \sigma_e(A+B)$. Then there exists a singular sequence (u_n) for A+B and λ

$$(u_n) \subset D(A),\ u_n \rightharpoonup 0,\ u_n \not\to 0,\ (A+B-\lambda)u_n \to 0.$$

Since A is (A+B)-bounded we have also $(A-i)u_n \rightharpoonup 0$. Let now $v_n := (A-i)^{-1}u_n$; then

$$(v_n) \subset D(A^2),\ v_n \rightharpoonup 0, (A-i)^2 v_n = (A-i)u_n \rightharpoonup 0.$$

Now the A^2-compactness of B implies $Bv_n \to 0$ and therefore

$$\begin{aligned}(A+B-i)(A-\lambda)v_n &= [(A+B-\lambda)(A-i) - (\lambda-i)B]v_n \\ &= (A+B-\lambda)u_n - (\lambda-i)Bv_n \to 0.\end{aligned}$$

Since $(A+B-i)$ is continuously invertible, this implies $(A-\lambda)v_n \to 0$. In order to prove that v_n is a singular sequence of A and λ, it remains to show $v_n \not\to 0$.

Assume $v_n \to 0$. Without restriction we may assume that $\|u_n\| \geq \varepsilon > 0$ for every $n \in \mathbb{N}$. Then from

$$\|v_n\|^2 = \int_{-\infty}^{\infty} |\lambda-i|^{-2}\, d\|E(\lambda)u_n\|^2 \to 0,\ n \to \infty$$

it follows for every $N \geq 0$

$$(E(N)-E(-N))u_n \to 0,\ n \to \infty,$$

$$\liminf_{n \to \infty} \|(I - E(N) + E(-N))u_n\| \geq \varepsilon,$$

and therefore (for every N and sufficiently large n)

$$\|(A-i)u_n\|^2 = \int_{-\infty}^{\infty} |\lambda-i|^2\, d\|E(\lambda)u_n\|^2 \geq N^2 \varepsilon$$

From this it follows that $\|(A-i)u_n\|^2 \to \infty$ for $n \to \infty$.

This is a contradiction to the boundedness of the sequence $((A-i)u_n)$.
Q.E.D.

So far we have quite general conditions which guarantee the invariance of the essential spectrum; in the following section we shall give simple conditions which are sufficient for A^2-compactness if A is a self-adjoint differential operator.

We should note that the conditions of Theorem 3.4 are, in general, not sufficient to guarantee that A+B is semi-bounded if A is semi-bounded. As an example consider A in $L_2(\mathbb{R})$ defined by

$$D(A) = \{u \in L_2(\mathbb{R}) : u \text{ continuous, } u(-1) = u(1) = 0;\ u \text{ and } u' \text{ locally absolutely continuous in } (-\infty,-1)\cup(-1,1)\cup(1,\infty),\ u'' \in L_2(\mathbb{R})\},$$

$$Au = -u''.$$

For B we choose the operator defined by

$$D(B) = D(A),$$

$$Bu(x) = \begin{cases} +2u''(x) & \text{in } (-1,1) \\ 0 & \text{in } (-\infty,-1) \cup (1,\infty). \end{cases}$$

Then it is easy to show that B is A-bounded and A^2-compact, and A+B is self-adjoint. But $-\infty$ is a cluster point of eigenvalues of A+B.

A simpler example is as follows: Let A be a positive self-adjoint operator with a discrete spectrum and B = -2A. - Such a situation can of course not occur if B is A-bounded with A-bound less than 1.

It can be expected that sharper results hold, if the perturbation is semi definite. For example, if A has the spectrum $[\mu,\infty)$ and B non-negative, then the spectrum of A+B is contained in $[\mu,\infty)$. But what happens, if there are holes in the spectrum (or in the essential spectrum) of A, as it occurs, for example, in the case of the Dirac operator? A first result of this type is given by Glazman [5],

Theorem 7bis; this theorem is the same as our Corollary 3.9 b.

3.6. Theorem Let A be a self-adjoint operator in the Hilbert space H,

$$H = H_1 \oplus H_2 \oplus H_3, \quad \dim H_3 = m < \infty.$$

For the orthogonal projections P_j onto H_j we assume

$$P_j D(A) \subset D(A) \quad (j=1,2,3).$$

If

$$\langle Af, f\rangle \begin{cases} \leq a\|f\|^2 & \text{for } f \in P_1 D(A), \\ \geq b\|f\|^2 & \text{for } f \in P_2 D(A), \end{cases}$$

then we have

$$\dim (E(b-) - E(a)) \leq m.$$

Proof. Assume that $\dim (E(b-) - E(a)) \geq m + 1$. Then there is an $\varepsilon > 0$ such that

$$\dim (E(b-\varepsilon) - E(a+\varepsilon)) \geq m + 1.$$

We define $B \in B(H)$ by

$$B = \int_{a+\varepsilon}^{b-\varepsilon} \left(\frac{a+b}{2} - t\right) dE(t), \quad \|B\| \leq \frac{b-a}{2} - \varepsilon.$$

The operator $\tilde{A} = A+B$ has the following properties: $\tilde{A}$ is self-adjoint, $D(\tilde{A}) = D(A)$,

$$\langle \tilde{A}f, f\rangle \begin{cases} \leq \left(\frac{b+a}{2} - \varepsilon\right) \|f\|^2 & \text{for } f \in P_1 D(\tilde{A}), \\ \geq \left(\frac{b+a}{2} + \varepsilon\right) \|f\|^2 & \text{for } f \in P_2 D(\tilde{A}), \end{cases}$$

and $\frac{a+b}{2}$ is an eigenvalue of $\tilde{A}$ with multiplicity not smaller than m+1. Therefore there exists an eigenelement f of $\tilde{A}$ for the eigenvalue $\frac{a+b}{2}$ with $f \neq 0$, $f \perp H_3$. This f can be written in the form

$$f = P_1 f + P_2 f$$

and a simple calculation shows (remember $\tilde{A}f = \frac{a+b}{2} f$)

$$(\frac{a+b}{2}+\varepsilon)\|P_2 f\|^2 \leq \langle \tilde{A}P_2 f, P_2 f\rangle = \langle \tilde{A}(I-P_1)f,(I-P_1)f\rangle$$

$$= \langle \tilde{A}f,f\rangle - 2\mathrm{Re}\,\langle \tilde{A}f,P_1 f\rangle + \langle \tilde{A}P_1 f,P_1 f\rangle$$

$$\leq \frac{a+b}{2}\,(\|f\|^2 - 2\mathrm{Re}\langle f,P_1 f\rangle + \|P_1 f\|^2) - \varepsilon\|P_1 f\|^2$$

$$= \frac{a+b}{2}\,\|P_2 f\|^2 - \varepsilon\,\|P_1 f\|^2.$$

This is a contradiction. Q.E.D.

There are some simple corollaries of Theorem 3.7. The proofs are easy and will be left to the reader.

3.7. Corollary If in Theorem 3.7 $H_3 = \{0\}$, then $(a,b) \subset \rho(A)$.

3.8. Corollary a) Let A be a self-adjoint operator in H, $(0,b) \subset \rho(A)$, $B \geq 0$, $D(B) \supset D(A)$ such that A+B is also self-adjoint. Assume that for every $x \in R(E(0))$ we have

$$\langle Bx,x\rangle \leq -\langle Ax,x\rangle + \eta\|x\|^2 \quad \text{for some } \eta < b.$$

Then $(\eta,b) \subset \rho(A+B)$.

b) If $(0,b) \cap \sigma_e(A) = \emptyset$, then $(\eta,b) \cap \sigma_e(A+B) = \emptyset$.

3.9. Corollary a) Let A be a self-adjoint operator in H, $(a,b) \subset \rho(A)$, B symmetric with $0 \leq B \leq \eta < b-a$. Then $(a+\eta,b) \subset \rho(A+B)$.

b) If $(a,b) \cap \sigma_e(A) = \emptyset$, then $(a+\eta,b) \cap \sigma_e(A+B) = \emptyset$.

3.10. Theorem Let A be a self-adjoint operator in H, $\sigma_e(A) \cap (a,b) = \emptyset$ and assume that b is not a limit point of eigenvalues of A which are contained in (a,b). If B is compact and $B \geq 0$, then $\sigma_e(A+B) \cap (a,b) = \emptyset$ and b is not a limit point of eigenvalues of A+B which are contained in (a,b).

Proof. Only the last statement needs a proof. Since there is an $\varepsilon > 0$ such that $(b-\varepsilon,b) \subset \rho(A)$, we may assume that $(a,b) \subset \rho(A)$.

Let $\mu_1 \geq \mu_2 \geq \ldots$ be the eigenvalues of B, u_j the corresponding eigenelements. Then there is an $m \in \mathbb{N}$ such that $\mu_n \leq \mu_m = b-a-\varepsilon < b-a$ for $n \geq m$ and some $\varepsilon > 0$. Let

$$M := L\{u_1,u_2,\ldots,u_m\},$$
$$H_3 := E(a)M,$$
$$H_1 := R(E(a)) \ominus H_3$$
$$H_2 := R(I-E(b-)),$$

where E is the spectral resolution of A. Then it is clear that $H = H_1 \oplus H_2 \oplus H_3$ holds and

$$\langle (A+B)f,f\rangle \geq \langle Af,f\rangle \geq b\|f\|^2 \quad \text{for } f \in P_2D(A).$$

From $H_1 \perp M$ it follows that $\|Bf\| \leq (b-a-\varepsilon)\|f\|$ for $f \in H_1$, hence

$$\langle (A+B)f,f\rangle \leq a\|f\|^2 + (b-a-\varepsilon)\|f\|^2$$
$$\leq (b-\varepsilon)\|f\|^2 \quad \text{for } f \in P_1D(A).$$

Since $\dim H_3 \leq m < \infty$ it follows from Theorem 3.6 that there are at most m points of $\sigma(A+B)$ contained in $(b-\varepsilon,b)$. Therefore b cannot be a limit point of eigenvalues contained in (a,b). Q.E.D.

It seems remarkable that this result can be extended not only to A-compact perturbations, but also to some A^2-compact perturbations.

3.11. Theorem Let A be a self-adjoint operator in H, $\sigma_e(A) \cap (a,b) = \emptyset$ and assume that b is not a limit point of eigenvalues of A which are contained in (a,b). If B is non-negative, A^2-compact and A-bounded with relative bound less than 1, then $\sigma_e(A+B) \cap (a,b) = \emptyset$ and b is not limit point of eigenvalues of A+B which are contained in (a,b).

Proof. Again we have to prove only the last statement. Without restriction we may assume that $(0,b) \subset \rho(A)$. First we note that there exists a non-increasing function

$$\eta\colon [1,\infty) \to \mathbb{R}^+ \quad \text{with} \lim_{t\to\infty} \eta(t) < 1$$

such that for every $x \in D(A) \cap R(E(t) - E(-t))^{\perp}$ we have

$$\|Bx\| \leq \eta(t)\|Ax\|$$

(this follows easily from the fact that B has an A-bound less than 1).

Let now $P_t = E(t) - E(-t)$ for $t > 0$. Then we can write

$$B = B_t + K_t$$

with

$$B_t = (I-P_t)B(I-P_t),$$

$$K_t = (I-P_t)BP_t + \overline{P_tB(I-P_t)} + P_tBP_t.$$

It is clear from our assumption that P_tBP_t and $(I-P_t)BP_t$ are compact; from $[P_tB(I-P_t)]^* = (I-P_t)BP_t$ it follows that $\overline{P_tB(I-P_t)}$ is also compact. Hence K_t is compact.

Choose now $t \geq 0$ such that $\eta(t) \leq 1$ (this is possible since $\eta(t) < 1$ for t sufficiently large). Then with $|A| = (A^2)^{1/2}$ it follows immediately for every $x \in D(A)$

$$\|B_tx\| \leq \|B(I-P_t)x\| \leq \|A(I-P_t)x\| \leq \||A|x\|.$$

The theorem of Heinz (see [12], Theorem V.4.12) implies

$$\langle B_tx,x\rangle \leq \langle |A|x,x\rangle \quad \text{for } x \in D(A).$$

Hence for every $x \in E(0)D(A)$

$$\langle (A+B_t)x,x\rangle = \langle (-|A|+B_t)x,x\rangle \leq 0.$$

On the other hand we have for $x \in (I-E(b-))D(A)$

$$\langle (A+B_t)x,x\rangle \geq \langle Ax,x\rangle \geq b\|x\|^2.$$

From Corollary 3.7 we now get

$$(0,b) \subset \rho(A+B_t).$$

The remaining part follows from Theorem 3.10 since K_t is compact.

4. Perturbations which are small at infinity

In [11] Jörgens and the author introduced a general class of perturbations which are "relatively small at infinity". This class contains the class of relatively compact perturbations (see Theorem 4.1) and seems to be a very natural class of perturbations (perturbations usually represent some kind of interactions and it is reasonable that these should be small, in some sense, near infinity). In [11] it seems that the invariance of the essential spectrum under a perturbation which is small at infinity depends heavily on the nature of the unperturbed operator. But recently Böcker [3] showed that in the case considered in [11] the perturbation is relatively compact with respect to the square of the unperturbed operator. Then the invariance of the essential spectra follows from the results of section 3. In [28] these considerations have been extended to operators in $L_2(G)$, G an open subset of $\mathbb{R}^m$. The notion of "relative smallness at A" has been introduced, where A is usually some closed subset of the boundary of G, including infinity (if G is unbounded). We give here the most important results without proofs and then turn to some simple applications.

Let T and V be operators in $L_2(G)$ with $D(T) \subset D(V)$ and let A be a closed subset of $\overline{G}$. We say that V is *T-A-small*, if for every $\varepsilon > 0$ there exists a compact subset K of $\overline{G} \setminus A$, such that for every $u \in D(T)$ with $\operatorname{supp} u \subset \overline{G} \setminus K$ we have

$$\|Vu\| \leq \varepsilon(\|u\| + \|Tu\|)$$

(supp always means "essential support"). For $G = \mathbb{R}^m$ and $A = \emptyset$ this reduces to the definition of T-smallness at infinity. Notice that we do not require that A is a subset of the boundary of G; but, if A contains interior points, then the T-A-smallness of V implies $Vu = 0$ for $u \in D(T)$ with $\text{supp } u \subset A$. It is clear that the T-A-smallness of V implies the (T+V)-A-smallness of V (see [28], Lemma 2.6). Since the requirement of T-A-smallness is stronger for larger A, we should always choose A as small as possible.

Our first theorem gives some characterization of T-compact operators by means of a T-A-smallness condition.

4.1. Theorem Let T and V be closable operators in $L_2(G)$.
a) If V is T-compact, then V is T-bounded with T-bound 0 and V is T-A-small for every closed subset A of $\overline{G}$ with measure zero.
b) Let A be a closed subset of $\overline{G}$ and assume that for every $\varphi \in C_o^\infty(\mathbb{R}^m \setminus A)$

$$\varphi D(T) \subset D(T)$$

and the mappings

$$u \mapsto \varphi u, \ u \mapsto \varphi Tu - T\varphi u$$

are T-compact. If V is T-bounded with relative bound 0 and T-A-small, then V is T-compact. (For the proof see [28], Theorem 2.7.)

The next theorem gives a similar result concerning the T^2-compactness for self-adjoint T.

4.2. Theorem Let T be a self-adjoint operator in $L_2(G)$ and A a closed subset of $\overline{G}$, such that for every $\varphi \in C_o^\infty(\mathbb{R}^m \setminus A)$

$$\varphi D(T) \subset D(T)$$

and the mappings

$$u \mapsto \varphi u \ , \ u \mapsto \varphi T u - T \varphi u$$

are T-compact. If V is an operator in $L_2(G)$ which is T-bounded and T-A-small, then V is T^2-compact. (For the proof see [28] Theorem 2.9.)

If we combine Theorem 4.2 with Theorem 3.4, then this gives a convenient condition for the invariance of the essential spectrum. But under a little more restrictive conditions we can even show that T and T+V have the same singular sequences (see [28], Theorem 3.2).

4.3. Theorem Let T and V be as in Theorem 4.2 and assume that T+V is also self-adjoint.

a) Every singular sequence for T and λ is also a singular sequence for T+V and λ.

b) If in addition for every $\varphi \in C_0^\infty(\mathbb{R}^m \setminus A)$ the mapping

$$u \mapsto \varphi V u - V \varphi u$$

is T-compact, then T and T+V have the same singular sequences.

Our first example is trivial, it only serves as an illustration.

4.4. Example (for a similar example see [28]). Let $G = (0,1)$ and the operator T defined by

$$D(T) = \{u \in L_2(0,1): u \text{ and } u' \text{ are locally absolutely continuous in } (0,1], (x^2u')' \in L_2(0,1), u(1) = 0\},$$

$$Tu(x) = -(x^2u')' \quad \text{for } u \in D(T).$$

A fundamental system of the differential equation $-(x^2u')' = \lambda u$ is given by

$$u_j(x) = x^{\rho_j}, \ \rho_{1,2} = -\frac{1}{2} \pm (\frac{1}{4} - \lambda)^{1/2} .$$

These solutions are non-oscillatory for $\lambda < \frac{1}{4}$ and no solution of the differential equation lies in $L_2(0,1)$ for $\lambda \geq \frac{1}{4}$. Applying Satz 1.1.b and Satz 3.1.b of [27] we find

$$\sigma_e(T) = [\tfrac{1}{4},\infty).$$

For $\lambda < \frac{1}{4}$ the only L_2-solution is $u_\lambda(x) = cx^\rho$, where $\rho = -\frac{1}{2}+(\frac{1}{4}-\lambda)^{1/2}$. Since $u_\lambda(1) = 0$ implies $u_\lambda = 0$ it is clear that T has no eigenvalues, i.e. $\sigma(T) = \sigma_e(T)$.

Let $\varphi \in C^\infty(\mathbb{R})$, $\varphi(x) = 1$ in a neighbourhood of 0 and $0 \leq \varphi(x) \leq 1$; let $\psi = 1-\varphi$.

Now define for $a,b \in \mathbb{R}$ the operator V_1 by

$$D(V_1) = D(T),\ V_1u = -a(x^2\varphi u')' + b\varphi u$$

Then as above we can show that for arbitrary $a > -1$ and $b \in \mathbb{R}$

$$\sigma_e(T+V_1) = [\tfrac{1}{4},\infty) = \sigma_e(T)$$

On the other hand define V_o by

$$D(V_o) = D(T),\ V_ou = -a(x^2\psi u')' + b\psi u.$$

Then $T + V_o$ is self-adjoint for arbitrary $a > -1$ and $b \in \mathbb{R}$. Again the same calculation as above shows

$$\sigma_e(T+V_o) = [\tfrac{1}{4}(1+a)+b,\infty).$$

V_o is T-$\{1\}$-small, but not T-$\{0\}$-small and the essential spectrum is not preserved. V_1 is T-$\{0\}$-small, but not T-$\{1\}$-small and it does preserve the essential spectrum. Therefore we might guess that for this operator T the T-$\{0\}$-smallness of the perturbation suffices for the invariance of the essential spectrum. Actually it is not difficult to show that the conditions of Theorem 4.2 and 4.3 are satisfied with $A = \{0\}$.

<u>4.5. Example</u> Let T be an elliptic differential operator with constant coefficients of order r in $L_2(\mathbb{R}^m)$, $D(T) = W_2^r(\mathbb{R}^m)$. Then T

satisfies the assumptions of Theorem 4.2. Let

$$V = \sum_{|\alpha|\le m} q_\alpha \partial^\alpha$$

where

$$q_\alpha \in L_\infty(\mathbb{R}^m),\ q_\alpha(x) \to 0 \quad \text{for } |x| \to \infty \quad (|\alpha| = m),$$

$$\left.\begin{aligned} &q_\alpha \in M_\rho(\mathbb{R}^m) \quad \text{for some } \rho < 2(r-|\alpha|) \\ &\int_{|x-y|\le 1} |q_\alpha(y)|^2 dy \to 0 \quad \text{for } |x| \to \infty \end{aligned}\right\} (|\alpha| < m).$$

Then V is T-bounded and T-∅-small (i.e. T-small at infinity in the sense of [11]). We omit the simple proofs of these facts. It is also clear that the T-bound of V is less than 1 if $\sum_{|\alpha|=m} \sup\{|q_\alpha(x)| : x\in\mathbb{R}^m\} < 1$. An immediate consequence of these facts is that T+V is self-adjoint on $W_2^r(\mathbb{R}^m)$ and that the essential spectrum of T+V is equal to the range of the polynomial which generates T. Actually T and T+V have the same singular sequences.

<u>4.6. Example</u> Let $T = -\Delta$ with $D(T) = W_2^2(\mathbb{R}^m)$. Let V be a symmetric perturbation of the form

$$Vu = \sum_{j,k=1}^{m} a_{jk}\partial_j\partial_k u + \sum_{j=1}^{m} d_j \partial_j u + qu$$

with

$$a_{jk} \in L_\infty(\mathbb{R}^m),\ a_{jk}(x) \to 0 \quad \text{for } |x| \to \infty,$$

$$d_j \in M_\rho(\mathbb{R}^m) \text{ for some } \rho < 2,\ \int_{|x-y|\le 1} |d_j(y)|^2 dy \to 0 \text{ for } |x| \to \infty,$$

$$q \in M_\rho(\mathbb{R}^m) \text{ for some } \rho < 4,\ \int_{|x-y|\le 1} |q(y)|^2\, dy \to 0 \text{ for } |x| \to \infty.$$

Then V is T^2-compact. If in addition T+V is self-adjoint, then T and T+V have the same singular sequences; T and T+V have the essential spectrum $[0,\infty)$.

<u>4.7. Example</u> Let T be the Dirac operator of the free electron in $L_{2,4}(\mathbb{R}^3)$, V a symmetric differential operator of first order in

$_{2,4}(\mathbb{R}^3)$,

$$Vu = \sum_{j=1}^{3} q_j \partial_j u + qu$$

where all entries of q_j and q satisfy

$$q_j^{kl} \in L_\infty(\mathbb{R}^3),\ q_j^{kl}(x) \to 0 \quad \text{for } |x| \to \infty,$$

$$q^{kl} \in M_\rho(\mathbb{R}^3) \quad \text{for some } \rho < 2,$$

$$\int_{|x-y| \leq 1} |q^{kl}(y)|^2 dy \to 0 \quad \text{for} \quad |x| \to \infty.$$

Then V is T^2-compact. If in addition $T+V$ is self-adjoint, then T and $T+V$ have the same singular sequences; T and $T+V$ have the essential spectrum $(-\infty, -\mu c^2] \cup [\mu c^2, \infty)$.

For the discussion of general operators of the Schrödinger type on an open subset G of $\mathbb{R}^m$ we first extend our definition of $M_{\rho,loc}(\mathbb{R}^m)$ to open subsets $G \subset \mathbb{R}^m$. We say that q belongs to the local Stummel class $M_{\rho,loc}(G)$ if for every compact subset K of G we have

$$\sup_{x \in K} \int_{\substack{|x-y| \leq 1 \\ y \in K}} |q(y)|^2 |x-y|^{\rho - m} dy < \infty \quad \text{if } \rho < m,$$

$$\sup_{x \in K} \int_{\substack{|x-y| \leq 1 \\ y \in K}} |q(y)|^2 dy < \infty \quad \text{if } \rho \geq m.$$

Again we have $M_{\rho,loc}(G) = L_{2,loc}(G)$ for $\rho \geq m$. The local Sobolev space $W_{2,loc}^r(G)$ has been defined in section 2.

4.8. Lemma Let G be an open subset of $\mathbb{R}^m$, T a differential operator on G of the form

$$Tu = -\Delta u + 2 \sum_{j=1}^{m} b_j \partial_j u + (q + \sum_{j=1}^{m} b_j^2) u$$

with $q \in M_{\rho,loc}(G)$ and $b_j^2 \in M_{\rho,loc}(G)$ for some $\rho < 4$, div $b = 0$ (in the sense of distributions). Then for every relatively compact open subset Ω of G and every $\tau < \text{dist}(\Omega, \complement G)$ there exists a constant $C(\Omega, \tau)$ such that for every $u \in W_{2,loc}^2(G)$

$$\|\Delta u\|_\Omega + \|\text{grad } u\|_\Omega \leq C(\Omega,\rho)\left\{\|u\|_{\Omega_\tau} + \|Tu\|_{\Omega_\tau}\right\}$$

where $\Omega_\tau = \{x \in \mathbb{R}^m : \text{dist}(x,\Omega) < \tau\}$ and $\|\cdot\|_\Omega$ is the norm in $L_2(\Omega)$.

The proof for $G = \mathbb{R}^m$ is given in [11], Lemma 3.5. But this proof works also in our more general situation, since only the restrictions to Ω_ρ of all functions play a role in the proof.

4.9. Example Let T be an operator of Schrödinger type in $L_2(G)$ as considered in Lemma 4.8. If T is self-adjoint with $D(T) \subset W^2_{2,loc}(G)$ then the assumptions of Theorem 4.2 are satisfied with $A = \partial G$.

Proof Assume that $\Omega \subset G$ is relatively compact, open and contains supp φ. We have to show that every bounded sequence (u_n), for which (Tu_n) is bounded, contains a subsequence (u_{n_k}) such that (φu_{n_k}) and $(\varphi Tu_{n_k} - T\varphi u_{n_k})$ are convergent. From Lemma 4.8 it follows that the sequences $(\nabla(\psi u_n))$ and $(\Delta(\psi u_n))$ are bounded for $\psi \in C_o^\infty(\mathbb{R}^m)$ with $\psi(x) = 1$ in $\overline{\Omega}$, $\psi(x) = 0$ for $x \notin \Omega_\rho$ $(\rho < \text{dist}(\Omega, G))$. Now the result follows from Theorem 2.2 (notice that $\varphi T - T\varphi$ is a differential operator of first order with coefficients lying in $M_{\rho,loc}(G)$ and having compact support.

4.10. Example Let $T = T_o + B$, where T_o is the Dirac operator of the free electron in $L_{2,4}(\mathbb{R}^3)$ and B is the multiplication operator by 4×4 hermitian matrix valued function, where all the entries of B belong to $M_{\rho,loc}(\mathbb{R}^3)$ for some $\rho < 2$. Then T is essentially self-adjoint on $C^\infty_{o,4}(\mathbb{R}^3)$ by Theorem 2.6. The self-adjoint extension of T satisfies the assumptions of Theorem 4.2 with $A = \partial G$. This is easily proved, since $u \mapsto \varphi u$ and $u \mapsto T\varphi u - \varphi Tu$ are multiplication operators by C_o^∞-functions.

One should expect similar results (as in examples 4.9 and 4.10) for general elliptic operators with constant highest order coefficients. So far the author did not succeed in proving such a result. The proof given in [11] depends heavily on the special structure of Schrödinger operators.

There are, of course, also perturbations which are not differential operators (and therefore not local): e.g. integro-differential operators and some kinds of pseudo differential operators. Simple examples are given in [11], section 4 and [25], section 4.

5. The absolutely continuous spectrum and scattering theory.

From examples (see for example Aronszajn [1]) it is well known that we cannot expect a reasonable perturbation theory for the continuous spectrum, neither for differential operators nor in an abstract setting.

From scattering theory we know that the situation is much better with the *absolutely continuous spectrum.* But, let us first give some definitions. If T is a self-adjoint operator in a Hilbert space H, then the *absolutely continuous subspace* H_{ac} of T is defined by

$$H_{ac} = \{u \in H: \langle E(.)u,u\rangle \text{ is absolutely continuous}\}.$$

The restriction T_{ac} of T to H_{ac} is called the *absolutely continuous part* of T; the spectrum of T_{ac} is the *absolutely continuous spectrum* of T, $\sigma_{ac}(T)$. In this section we consider operators T_o and T_1; to all symbols H_{ac}, σ_{ac}, T_{ac} we add the index 0 or 1 if they refer to T_o or T_1 respectively.

Let us now assume that T_o has a purely absolutely continuous spectrum (i.e. $H_{ac,o}= H$); this is always the case if T_o is a differential operator with constant coefficients, $T_o \neq 0$. Then the wave operators $W_\pm(T_1,T_o)$ are defined by

$$W_\pm(T_1,T_o) = \underset{t\to\pm\infty}{\text{s-lim}}\ \exp(itT_1)\exp(-itT_o)$$

if they exist ("s-lim" means "strong limit").

If the wave operators $W_\pm = W_\pm(T_1,T_o)$ exist, then these are isometries and the ranges $R(W_\pm)$ are subspaces of $H_{ac,1}$ and T_o is unitarily equivalent to $T_1|_{R(W_\pm)}$, the restriction of T_1 to $D(T_1)\cap R(W_\pm)$. The unitary equivalence is given by $W_\pm$. In general this is not very much information, since we do not know anything about the restriction of T_1 to $R(W_\pm)^\perp$. But there are results which guarantee that the ranges of $W_\pm$ are equal to all of $H_{ac,1}$. One of these results is given by

5.1. Theorem ([12],note the end of X.4.5). If $(\lambda-T_1)^{-n}-(\lambda-T_o)^{-n}$ is in the trace class for every non-real λ, then the wave operators $W_\pm(T_1,T_o)$ exist and $R(W_+) = R(W_-) = H_{ac,1}$. (The simplest characterization of an operator of the trace class is, that it is a product of two Hilbert-Schmidt operators.)

This result has been applied in [26] to the 1-dimensional case. There we have proven essentially the following theorem.

5.2. Theorem Let T_o be a self-adjoint (elliptic) differential operator with constant coefficients in $L_2(\mathbb{R})$ of order ≥ 2 and assume that V is a symmetric operator in $L_2(\mathbb{R})$ such that: $D(T_o) \subset D(V)$, $T_1 = T_o + V$ is self-adjoint and there exist constants $C \geq 0$ and $\Theta < -1$ such that for every $r \geq 0$

$$\|Vu\| \leq C(1+r)^\Theta\{\|u\|+\|T_ou\|\},$$

if $u \in D(T_o)$ and $|\text{supp}\ u| \geq r$. Then the wave operators $W_\pm(T_1,T_o)$

exist and we have $R(W_+) = R(W_-) = H_{ac,1}$.

So far we are not able to give such a general result for higher dimensions or for $m = 1$ and an operator T_o of order 1. But it is not difficult to prove the following extension which covers "smooth and rapidly decreasing" perturbations.

5.3. Theorem Let T_o be a self-adjoint elliptic differential operator with constant coefficients in $L_2(\mathbb{R}^m)$ of order $r \geq 1$. Let V be a symmetric operator in $L_2(\mathbb{R}^m)$ such that: $D(T_o) \subset D(V)$, $T_1 = T_o + V$ is self-adjoint and there exist constants $C \geq 0$ and $\Theta < -m$ such that for every $r \geq 0$

$$\|Vu\| \leq C(1+r)^{\Theta} \{\|u\| + \|T_o u\|\}$$

if $u \in D(T_o)$ and $|\text{supp } u| \geq r$. Assume further that $D(T_o^k) = D(T_1^k)$ for $k \in \mathbb{N}$, $k \leq p$, $p > \frac{m}{r} + 1$. Then the wave operators $W_{\pm}(T_1, T_o)$ exist and $R(W_+) = R(W_-) = H_{ac,1}$.

For the proof we use the identity (see (5.1) of [26])

$$R_o^{2^n} - R_1^{2^n} = R_o^{2^n} VR_1 + R_o^{2^n-1} VR_1^2 + \ldots + R_o^2 VR_1^{2^n-1} + R_o VR_1^{2^n},$$

where $R_j = (\lambda - T_j)^{-1}$. If n is sufficiently large, then every term on the right hand side contains either a factor $R_o^p V$ or a factor VR_1^p, where $p-1 > \frac{m}{r}$. Now, as in [26] we write

$$VR_o^p = \sum_{n \in \mathbb{N}} V\varphi_n R_o^p$$

where φ_n has its support in $\{x \in \mathbb{R}^m : n-1 \leq |x| \leq n+1\}$. The maps

$$\Phi_n : W_2^{rp}(\mathbb{R}^m) \to W_2^r(\{x \in \mathbb{R}^m : n-1 \leq |x| \leq n+1\}), \quad u \mapsto \varphi_n u$$

are in the trace class and the trace norms can be estimated by Mn^{m-1}, where M is some constant (indepent of n). This can be easily deduced from the results in Yosida [29], X.2. Since the operators

$$V_n:\ W_2^r\Big(\{x \in \mathbb{R}^m : n-1 \le |x| \le n+1\}\Big) \to L_2(\mathbb{R}^m),\ u \mapsto Vu$$

are bounded by $M'n^{\Theta}$ with some constant M' by our assumption, we see that $\sum_{n\in\mathbb{N}} V\varphi_n R_o^p = \sum_{n\in\mathbb{N}} V_n \Phi_n R_o^p$ converges with respect to the trace norm. Hence VR_o^p is in the trace class. Taking the adjoint we see that also $R_o^p V$ is in the trace class. Since $(\lambda - T_o)^p R_1^p$ is bounded (this follows from $D(T_1^p) = D(T_o^p)$ by means of the closed graph theorem), the operator

$$VR_1^p = VR_o^p(\lambda - T_o)^p R_1^p$$

is also in the trace class. Therefore every term in the above identity for $R_o^{2^n} - R_1^{2^n}$ is in the trace class. This proves the theorem.

<u>5.4. Example</u> Let $T_o = -\Delta$ in $L_2(\mathbb{R}^m)$. If V is a differential operator of order less or equal 2 with C^∞-coefficients, such that all derivatives are bounded, the second order coefficients are sufficiently small and all coefficients tend to 0 like $|x|^{-m-\varepsilon}$ for $|x| \to \infty$ and some $\varepsilon > 0$. Then the above theorem is applicable, i.e. the wave operators $W_\pm(T_1, T_o)$ exist for $T_1 = T_o + V$ and $R(W_+) = R(W_-) = H_{ac,1}$.

<u>5.5. Example</u> If T_o is the Dirac operator of the free electron in $L_{2,4}(\mathbb{R}^m)$, then the corresponding result holds for a differential operator V of order less or equal 1, if the coefficients tend to zero like $|x|^{-3-\varepsilon}$ for some $\varepsilon > 0$.

<u>5.6. Remark</u> Using Birman's <u>local theory of wave operators</u> (see [2] for further references) and similar techniques as in the proof of Theorem 5.3 we can prove the result of Theorem 5.3 without the assumption that $D(T_o^k) = D(T_1^k)$ for $1 < k \le p$. This implies the results of Example 5.4 and Example 5.5 without the smoothness assumptions. As a special case we get the existence of $W_\pm(T_1, T_o)$ and $R(W_+) = R(W_-) = H_{ac,1}$ if T_o is as in Theorem 5.3 and V is T_o-bounded with relative bound less than 1 and has compact support (i.e. $Vu = 0$ for $u \in D(T_o)$

with $|\text{supp } u| \geq r_o$ for some $r_o \geq 0$).

There are many related results concerning the examples 5.4 and 5.5; some of them are, at least partially, much stronger (see for example [2,4,8,13,16,19]). But it should be noted that Theorem 5.3 also covers perturbations which are not differential operators. These results seem to indicate that strong scattering results are possible with such smallness conditions near infinity.

Much weaker conditions are possible if one wants only to prove the existence of wave operators (e.g. [15,24,25]). There are even sufficient conditions [15,25] which do not require the relative smallness at infinity. A very old result of this kind is due to Kuroda [15] and says that for $T_o = -\Delta$ and $T_1 = T_o + q$ with

$$(5.7) \qquad q(1+|\cdot|)^{1+\varepsilon-m/2} \in L_2(\mathbb{R}^m) \text{ for some } \varepsilon > 0$$

the wave operators $W_\pm(T_1,T_o)$ exist. This result has been extended to more general operators in [25].

For $m \geq 3$ the multiplication by such a q is (i.g.) not T_o-small at infinity in general. For an example choose $q(x) = -1$ in the union of infinitely many balls B_n of radii r_n with centers x_n in $\mathbb{R}^m$. If the sequence (x_n) tends to infinity fast enough (depending on (r_n)), then Kuroda's condition (5.7) is satisfied. But if the sequence (r_n) does not converge to 0, then this operator is not T_o-small at infinity: let $r_n \geq \rho_o$, $u \in C_o^\infty\{x \in \mathbb{R}^m : |x| < \rho_o\}$, $u \neq 0$, $u_n(x) = u(x-x_n)$, then $qu_n = -u_n \not\to 0$.

Actually such perturbations do not even preserve the essential spectrum. In order to see this we choose $\rho_o > 0$ so large that there is a $u \in C_o^\infty(\{x \in \mathbb{R}^m : |x| < \rho_o\})$ such that

$$<-\Delta u - u, u> \leq -\frac{1}{2} \|u\|^2.$$

Then we have for every n (with the above notation)

$$<-\Delta u_n + qu_n, u_n> \leq -\frac{1}{2} \|u_n\|^2$$

and, since the u_n have mutually disjoint support, the same inequality holds for every linear combination. Therefore there is an infinite dimensional subspace M such that this inequality holds for every $u \in M$. By Theorem 3.6 we then have $\dim E(-\frac{1}{2}) = \infty$. Since $-\Delta + q \geq -1$ holds, there must be at least one point of $\sigma_e(T)$ in $[-1, -\frac{1}{2}]$.

It is also easy to show that such conditions do not in general imply $R(W_+) = R(W_-) = H_{ac,1}$. We cannot prove this for $m \leq 3$. For $m \geq 4$ we give the following counter example. Choose

$$q(x) = \begin{cases} -1 & \text{if } \sum_{j=2}^{m} x_j^2 \leq \rho \\ 0 & \text{elsewhere.} \end{cases}$$

This function obviously satisfies Kuroda's condition (5.7). But by means of a separation of variables it is easy to show that, if ρ is sufficiently large, $\sigma(T_o + q) = [\mu, \infty)$ with some $\mu < 0$; furthermore this operator is purely absolutely continuous. Therefore the relation $R(W_+) = R(W_-) = H_{ac,1}$ cannot hold.

This indicates that such weak conditions are not sufficient for a reasonable scattering theory. The best possible result which one might expect is the following: "Let T_o be a self-adjoint differential operator with constant coefficients. Assume (i) V is T_o-small at infinity, (ii) $T_1 = T_o + V$ is self-adjoint and (iii) the wave operators $W_\pm(T_1, T_o)$ exist. Then $R(W_+) = R(W_-) = H_{ac,1}$."

6. Appendix (added after the end of the symposium)

In the course of this symposium the author was able to conclude from part ii) of the proof of Theorem 3.4 that every singular sequence of $A + B$ and λ is also a singular sequence of A and λ; together with part i) this means that under the assumptions of Theorem 3.4 the operators A and $A + B$ have the same singular sequences. This simplifies several later results.

The symmetry in this result lead the author to the conjecture that "B is also $(A+B)^2$-compact" or more general: "If A_1 and A_2 are self-adjoint with the same domain D and $D \subset D(B)$, then B is A_1^2-compact if and only if B is A_2^2-compact". The proof of the first conjecture is contained in the following theorem which the author owes to T.Kato. So far we could neither prove nor disprove the second conjecture. It is (in general) not true if the operators A_j are not self-adjoint.

Theorem (T.Kato, private communication). Let A_1 and A_2 be self-adjoint with $D(A_1) = D(A_2)$. Assume that $B := A_2 - A_1$ is A_1^2-compact. Then

a) B is also A_2^2-compact,

b) $\sigma_e(A_1) = \sigma_e(A_2)$,

c) an operator V is A_1-bounded and A_1^2-compact if and only if V is A_2-bounded and A_2^2-compact.

Proof. Let $R_j = (i-A_j)^{-1}$ for $j=1,2$. Then BR_j are bounded operators in H for $j=1,2$ and BR_1^2 is a compact operator in H. We also have

$$R_2 - R_1 = R_1BR_2 = R_2BR_1 \tag{1}$$

and therefore

$$R_2^2 = (I+R_2B)\, R_1^2(I+BR_2). \tag{2}$$

a) From (2) we get

$$BR_2^2 = (I+BR_2)\ BR_1^2(I+BR_2).$$

Hence BR_2^2 is compact because BR_1^2 is compact. Hence B is A_2^2-compact.

b) From (2) we get also

$$R_2^2 - R_1^2 = (BR_1^{*2})^* R_2 + R_2 BR_1^2 (I+BR_2)$$

which is compact since BR_1^{*2} and BR_1^2 are compact. Hence $\sigma_e(R_2^2) = \sigma_e(R_1^2)$ and therefore $\sigma_e(A_2) = \sigma_e(A_1)$. (Here a simple spectral mapping theorem for the essential spectrum is used).

c) Let V be A_1-bounded and A_1^2-compact. Then by (1)

$$VR_2 = VR_1(I-BR_2)$$

is a bounded operator in H. Therefore V is B-bounded. By (2)

$$VR_2^2 = (VR_1^2 + VR_2BR_1^2)(I+BR_2)$$

is compact since VR_1^2 and BR_1^2 are compact. Hence V is B^2-compact. The converse is proved in the same way since the assumptions are completely symmetric with respect to A_1 and A_2 by part a) of this theorem. Q.E.D.

Part b) of the above theorem gives a very simple prove of the invariance of the essential spectrum. If we combine part a) of this theorem with part i) of the proof of Theorem 3.4, then we get a simple proof of the fact that also the singular sequences are preserved. The complicated part ii) of the proof of Theorem 3.4 is not needed.

Note that part a) of the theorem holds also if A_j are arbitrary closed operators in a Banach space and $\rho(A_1) \cap \rho(A_2) \neq \emptyset$.

References

1. Aronszajn, N.: On a problem of Weyl in the theory of singular Sturm Liouville equations. Amer.J.Math. 79, 597-610 (1951).

2. Birman, M.Sh.: Scattering problems for differential operators with constant coefficients. Functional Anal.Appl. 3, 167-180 (1969)

3. Böcker, U.: Invarianz des wesentlichen Spektrums bei Schrödinger-operatoren. Math.Ann. 207, 349-352 (1974).

4. Eckardt, K.-J.: Scattering Theory for Dirac Operators. Math.Z. (to appear).

5. Glazman, I.M.: Direct Methods of Qualitative Spectral Analysis of Singular Differential Operators. Israel Program for Scientific Translations, Jerusalem 1965.

6. Gustafson, K. and J.Weidmann: On the essential spectrum. J.Math. Anal.Appl. 25, 121-127 (1969).

7. Ikebe, T. and T.Kato: Uniqueness of the self-adjoint extension of singular elliptic differential operators. Arch.Rational Mech. Anal. 9, 77-92 (1962).

8. Ikebe, T. and T.Tayoshi: Wave and Scattering Operators for Second Order Elliptic Operators in $\mathbb{R}^3$. Publ.Res.Inst.Math.Sci., Kyoto University, Ser.A, 4, No.2, 483-496 (1968).

9. Jörgens, K.: Wesentliche Selbstadjungiertheit singulärer elliptischer Differentialoperatoren zweiter Ordnung in $C_o^\infty(G)$. Math. Scand. 15, 5-17 (1964).

10. ___: Perturbations of the Dirac Operator. Conf.Ordinary and Partial Diff.Equ., Dundee/Scotland 1972, Lecture Notes in Math. 280, 87-102 (1972).

11. Jörgens, K. and J.Weidmann: Spectral Properties of Hamiltonian operators. Lecture Notes in Math. 313, 1973.

12. Kato, T.: Perturbation theory for linear operators. Berlin-Heidelberg-New York: Springer 1966

13. ___: Some results on potential scattering. Proceedings of the International Conference on Functional Analysis and Related

Topics, 206-215, Tokyo: University of Tokyo Press 1969.

14. ___: Schrödinger operators with singular potentials. Israel J. Math. 13, 135-148 (1972).

15. Kuroda, S.T.: On the existence and unitary property of the scattering operators. Nuovo Cimento 12, 431-454 (1959).

16. ___: Scattering theory for differential operators. I operator theory, II self-adjoint elliptic operators. J.Math.Soc. Japan 25, 75-104 and 222-234 (1973).

17. Schechter, M.: Spectra of partial differential operators. Amsterdam-London: North-Holland Publ.Comp. 1971.

18. ___: On the essential spectrum of an arbitrary operator I. J.Math. Anal.Appl. 13, 205-215 (1966).

19. ___: Scattering theory for second order elliptic operators. Preprint 1974.

20. Simader, Ch.: Bemerkungen über Schrödinger-Operatoren mit stark singulären Potentialen. Preprint 1974 (to appear in Math. Zeitschr.)

21. Simon, B.: Essential self-adjointness of Schrödinger operators with positive potentials. Math.Ann. 201, 211-220 (1973).

22. ___: Schrödinger operators with singular magnetic vector potentials. Math.Z. 131, 361-370 (1973).

23. ___: Essential self-adjointness of Schrödinger operators with singular potentials. Arch.Rational Mech.Anal. 52, 44-48 (1973).

24. Veselič, K. and J. Weidmann: Existenz der Wellenoperatoren für eine allgemeine Klasse von Operatoren. Math.Z. 134, 255-274 (1973).

25. ___: Asymptotic estimates of wave functions and the existence of wave operators. J.Functional Analysis (to appear).

26. ___: Scattering theory for a general class of differential operators. Proc.Conf.Spectral Theory and Asymptotics of Differential Equations, Scheveningen, September 1973.

27. Weidmann, J.: Zur Spektraltheorie von Sturm-Liouville-Operatoren. Math.Z. 98, 268-302 (1967).

28. ___: Perturbations of self-adjoint operators in $L_2(G)$ with applications to differential operators. Conf.Ordinary and Partial Diff.Equ. Dundee/Scotland 1974 (to appear in Lecture Notes in Math.)

29. Yosida, K.: Functional Analysis. Berlin-Heidelberg-New York: Springer 1966.

DIFFERENT APPLICATIONS OF CONVEX AND NONCONVEX OPTIMIZATION, ESPECIALLY TO DIFFERENTIAL EQUATIONS

L. Collatz

Summary

Some applications of nonlinear optimization, especially of convex, pseudoconvex, quasiconvex and other types of optimization, to analysis and differential equations are described.

Applications of linear optimization to many fields of mathematics and sciences are well known and often used. Here we are looking on some nonlinear types of optimization. Especially we give a collection of examples of the variants of convexity which occur in many different fields of applications and therefore should be considered more than hitherto. One gets often exact inclusions for desired quantities by using the described methods.

This paper does not intend to state theorems but to describe roughly the situation of applicability of quasiconvex optimization a.o. to stimulate for further research.

1. Variants of convexity and concavity

We repeat some well known definitions (compare, for instance, Mangasarian [69], Stoer Witzgall [7o]).

Let B be a convex subset of the n-dimensional space R^n of points $x = \{x_1,\ldots,x_n\}$. We consider real valued functions $\varphi(x)$, defined in B.

If	for all x,y in B and		then $\varphi(x)$ is called
$\varphi(\alpha x+(1-\alpha)y) \leq \alpha\varphi(x)+(1-\alpha)\varphi(y)$		all α with $0\leq\alpha\leq1$	convex
" $\geq$ "		all α with $0\leq\alpha\leq1$	concave
" $\leq \varphi(y)$	$\varphi(x)\leq\varphi(y)$	all α with $0\leq\alpha\leq1$	quasiconvex
" $\geq$	$\varphi(x)\geq\varphi(y)$	all α with $0\leq\alpha\leq1$	quasiconcave
$\varphi(x) \geq \varphi(y)$	$(x-y)^T \text{grad } \varphi(y)\geq0$		pseudoconvex
$\varphi(x) \leq \varphi(y)$	" ≤0		pseudoconcave

In this case of pseudoconvexity and pseudoconcavity the differentiability of $\varphi(x)$ in B is assumed.

The reason for introducing these generalizations of convexity is the fact, that the classical idea of convexity and concavity is not invariant under monotone one to one mappings $\varphi \to \phi$ of the dependent function φ (for these mappings a minimum of φ corresponds to a minimum of ϕ).

Example. The exponential function maps each convex function into a convex function, but not each concave function into a concave function, fig.1; $e^{|x|}$ is convex, but $e^{-|x|}$ is not concave, but only quasiconcave.

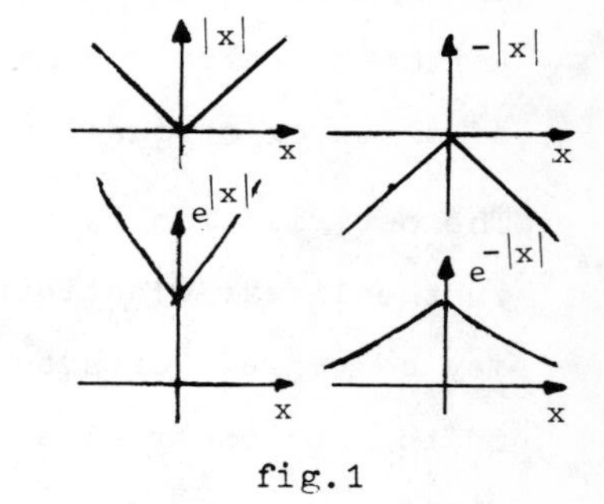

fig.1

2. Nonlinear Optimizations in Geometry and Applications

I. Convex optimization occurs frequently in geometrical problems, but there are many slightly more complicated problems with quasiconvex optimization. We select just one simple geometrical example.

Example. The cellar problem. A big cellar of given domain B with area F should be divided in n subdomains of area $\frac{1}{n}$ F in such a way, that the sum Φ of the length of all new rectilinear walls is minimal, fig.2.
(Collatz-Wetterling 71 , p.21o).

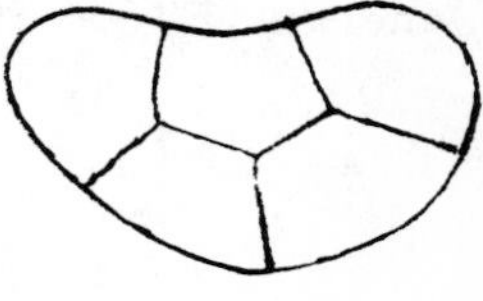
fig.2

As simple example let B consist of three squares of unit length in the x-y-plane, fig.3, and $n=3$. Taking the point $x = y = a$ as intersection point, the function $\Phi(a)$ is not convex, but strictly quasiconvex, fig.4.

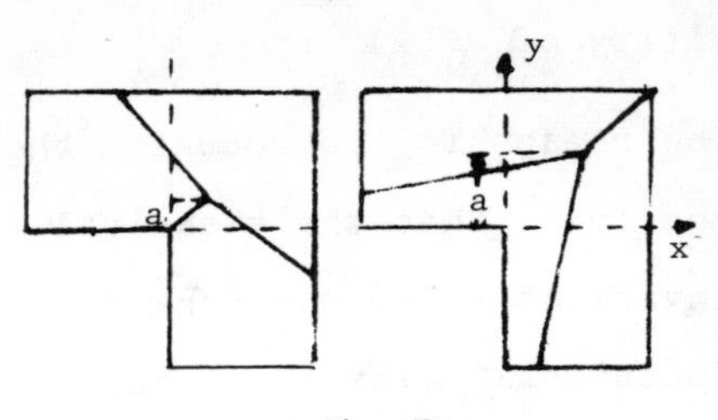

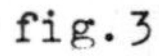
fig.3

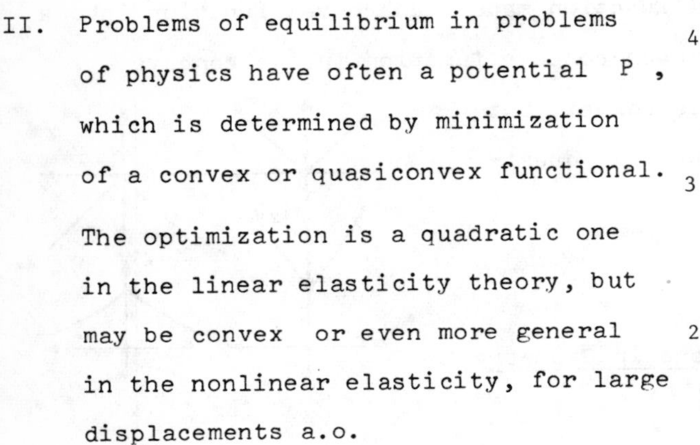

II. Problems of equilibrium in problems of physics have often a potential P, which is determined by minimization of a convex or quasiconvex functional. The optimization is a quadratic one in the linear elasticity theory, but may be convex or even more general in the nonlinear elasticity, for large displacements a.o.

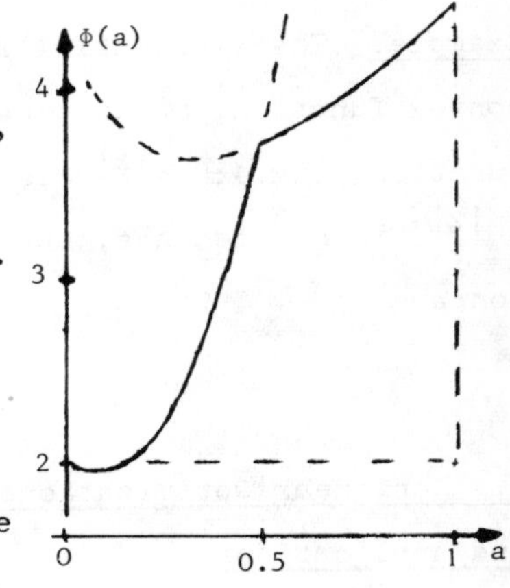

fig.4

<u>Example.</u> The weight G and the elastic forces K_1, K_2 are working at the mass m, fig.5. The forces K_j may depend nonlinearly on the prolongations z_j -for instance by a rule of the form

$K_j = \frac{a_j}{b_j + z_j}$; Fig.6 shows a typical graph for the corresponding elastic potential energy P_j as function of z_j .

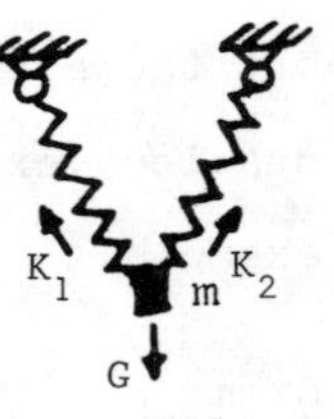

fig.5

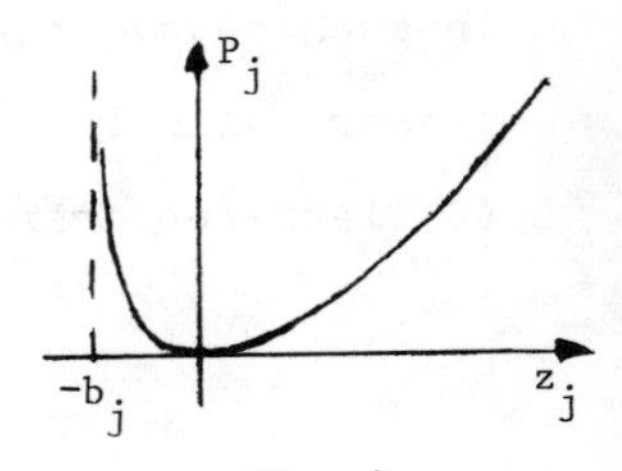

fig.6

<u>Example.</u> The classical Newtonian gravitational potential P or the classical potential P in electrostatics and magnetostatics are given by

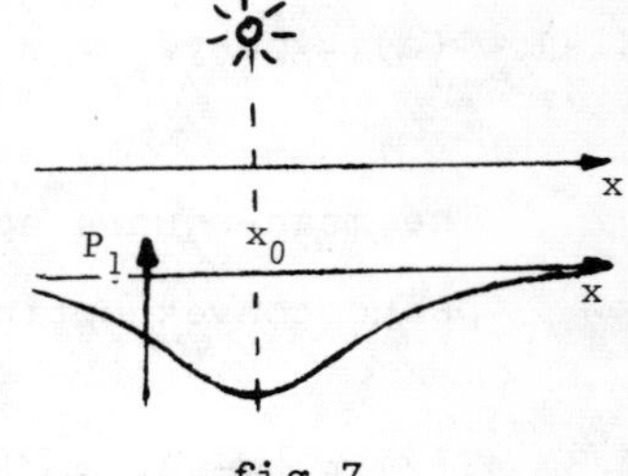

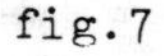

fig.7

$$(2.1) \qquad P = -\frac{c}{r}$$

where c is a constant and r is the distance between the considered point $(x,y,z) \in R^3$ and the mass-point or pole in the point (x_o,y_o,z_o). Along the x-axis, fig.7, the potential is

$$(2.2) \quad P = P_1(x) = -c\left[(x-x_o)^2 + y_o^2 + z_o^2\right]^{-1/2}$$

and along the x-y-plane:

$$(2.3) \quad P = P_2(x) = -c\left[(x-x_o)^2 + (y-y_o)^2 + z_o^2\right]^{-1/2} .$$

Both of these functions P_1,P_2 are for $z_o \neq 0$ not convex, but pseudoconvex.

III. Tschebyscheff-Approximation (abbreviated T.A.)

<u>Example.</u> The function $y(x) = \begin{cases} 1-x & \text{for } 0 \leq x \leq 1 \\ 0 & \text{for } x \geq 1 \end{cases}$

see fig.8, may be approximated by a function $w(x,a)$ of the type

$$w(x,a) = e^{-ax} \quad \text{for} \quad 0 \leq x < +\infty .$$

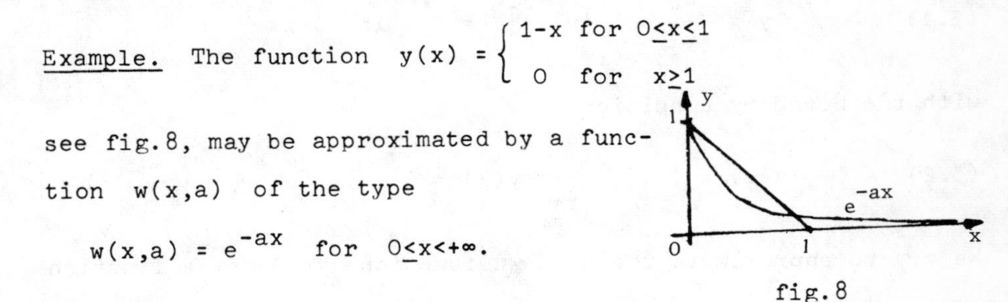

fig.8

The T.A. is the quasiconvex optimization, fig.9

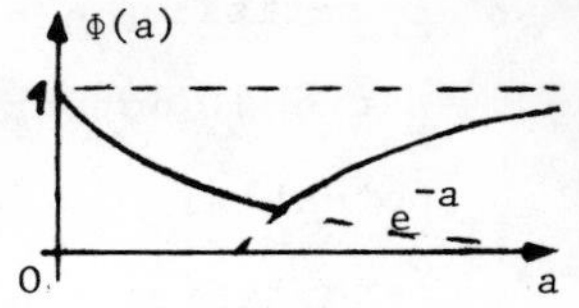

$$(2.4) \quad \Phi(a) = \| w-y \|_\infty = \sup_{x\in[0,\infty)} |w(x,a)-y(x)| = \text{Min.}$$

The mean-square approximation is the pseudoconvex optimization, fig.1o

$$(2.5) \quad \Psi(a) = \| w-y \|_2 = \left[\int_0^\infty [w(x,a)-y(x)]^2 dx\right]^{1/2} = \text{Min.}$$

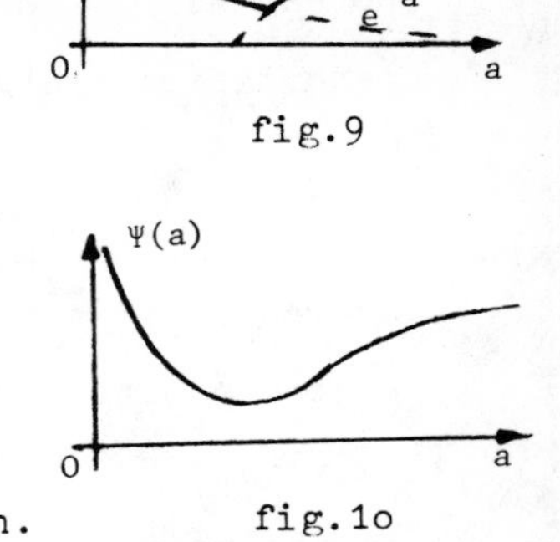

fig.9

fig.1o

3. Application to differential equations

For general description of application of optimization to differential equations compare Collatz-Wetterling [71] ,p.154-16o. Here we give special examples for illustrating, how quasiconvex optimization occur and how one can get inclusion theorems for desired quantities in many cases.

Example. Let be given the nonlinear differential equation for a function $y(x)$

$$(3.1) \qquad Ty = \frac{d^2y}{dx^2} - x\cdot[y(x)]^2 = 0 \quad \text{in} \quad 0<x<\infty$$

with the boundary conditions

$$(3.2) \qquad y(0) = 1 , \qquad \lim_{x\to\infty} y(x) = 0$$

We try to approximate the unknown function $y(x)$ by a function of the form

$$(3.3) \qquad w_1(x,a) = \frac{a}{x+a} \quad \text{or} \quad w_2(x,a) = \left(\frac{a}{x+a}\right)^3$$

which satisfies (3.2); the defect Tw should be as small as possible. The T.A.

$$\Phi_j(a) = \| Tw_j(x,a) \|_\infty = \text{Min.} \quad (j=1 \text{ or } 2)$$

is a nonconvex, but quasiconvex optimization, fig.11.

The norm is defined as in (2.4).

I thank Mr. H. Günther and Mr. D. Schütt for numerical calculations on a computer.

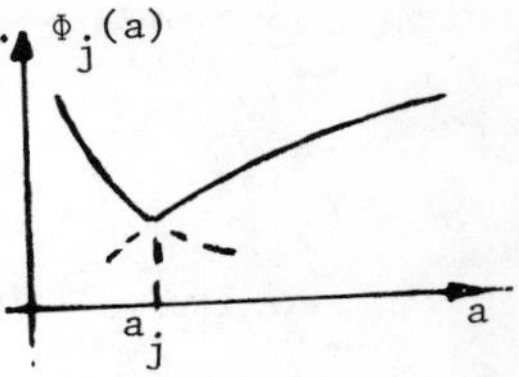

fig.11

The optimal values of a are

$$a_1 = 2.1, \quad \|Tw\|_\infty = 0.47; \quad a_2 = 6.2, \quad \|Tw\|_\infty = 0.32$$

4. Oneparametric Optimization for Differential and Integral Equations

A special type of optimization occurs frequently in differential equations, which we will call oneparametric optimization. Here the function to be minimized is one single parameter a

$$(4.1) \qquad a = \text{Min.}$$

and we have an infinite number of restrictions, often in the form

$$(4.2) \qquad Q(x,a) \leq 0 \quad \text{for all } x \in B$$

B for instance as in § 1.

More general a may be a vector $a = (a_1, \ldots, a_p)$ and we replace (4.1) by

$$(4.3) \qquad F(a) = \text{Min.},$$

where F is a given functional.

Example. Let be given the differential equation

$$(4.4) \qquad Ty = -\frac{d^2y}{dx^2} + y = r(x), \quad \text{for instance} \quad r(x) = \frac{\sin x}{x}$$

with the boundary conditions

$$(4.5) \qquad y(\pm 1) = 0 .$$

We approximate $y(x)$ by $w(x,a) = a(1-x^2)$.
Then we have the oneparametric optimization

$$(4.6) \qquad a = \text{Min.}, \qquad Tw(x,a) \geq r(x)$$

An admissable value a_1 for a which fulfils $T(w,a_1) \geq r(x)$ gives an upper bound $w(x,a_1)$ for $y(x)$; correspondingly

$$(4.7) \qquad a = \text{Max.}, \qquad Tw(x,a) \leq r(x)$$

with an admissable value $a=a_2$ gives a lower bound $w(x,a_2) \leq y(x)$. The solution is $a_1 = 0.4207$, $a_2 = \frac{1}{3}$; one gets better results with two parameters α, β and $w = \alpha(1-x^2)+\beta(1-x^4)$. This gives

$$0.3263(1-x^2) + 0.01525(1-x^4) \leq y(x) \leq 0.3283(1-x^2) + 0.0154(1-x^4) ,$$

for instance $$0.34155 \leq y(0) \leq 0.3437 .$$

The same method goes through nonlinear problems too.

Example. We consider instead of (4.4) the equation

$$(4.8) \qquad Ty = -\frac{d^2y}{dx^2} - \frac{1}{3}[y(x)]^2 = 0$$

with the boundary conditions

$$(4.9) \qquad y(\pm 1) = 1 .$$

Here we use the approximate solutions

$$w(x,a) = 1+a(1-x^2)$$

which satisfy (4.9).

One sees immediately

$$Tw(x,0) \leq 0 \leq Tw(x,1)$$

and one has the inclusion

$$w(x,0) = 1 \leq y(x) \leq 1 + (1-x^2) .$$

Using the oneparametric optimization (4.6) (4.7) one could improve these bounds: $w(x,\frac{1}{6}) \leq y(x) \leq w(x,2-\sqrt{3})$.

Another typical example for oneparametric optimization belongs to an infinite interval.

Example. For the problem

$$\hat{T}y = \frac{dy}{dx} - \frac{1}{x^2+y^2} = 0 , \quad y(1) = 0 \tag{4.10}$$

we take as approximate solution

$$w(x,a) = a(1 - \frac{1}{x}) .$$

We get $\hat{T}w(x,a) = [Tw(x,a)] \cdot \Phi$, where $\Phi > 0$ has a fixed sign, with

$$Tw(x,a) = a^3(1- \frac{1}{x})^2 + (a-1)x^2 .$$

Replacing $r(x)$ by zero the formulas (4.6) (4.7) give the corresponding optimization problem.

The solution is

$$0.945(1 - \frac{1}{x}) \leq y(x) \leq 1 - \frac{1}{x} \quad \text{for} \quad 1 \leq x < \infty .$$

The methods are also applicable to integral equations.

Example. We consider the nonlinear integral equation of Urysohn-type

$$(4.11) \qquad y(x) = Ty(x) \quad \text{with} \quad Tz(x) = \lambda \int_0^1 \frac{dt}{1+xt-[z(t)]^2}$$

The approximate solution $w(x) = a = \text{const.}$ gives

$$Ta = \frac{\lambda}{x} \ell n\left[1 + \frac{x}{1-a^2}\right]$$

and the oneparametric optimization for a lower bound $a \le y(x)$ ("chained approximation") is:

$$a = \text{Max} , \quad 0 \le \frac{\lambda}{x} \ell n(1 + \frac{x}{1-a^2}) - a$$

An upper bound can be got analogeously. For $\lambda = \frac{3}{8}$, for instance, one has immediately the inclusion $0 \le y(x) \le \frac{1}{2}$. One can get better results in a more complicated way by using the vector $a = (a_1,a_2)$ with constants a_j and the optimization (see (4.2,(4.3))

$$F(a) = \| Ta_2-Ta_1\|_\infty = \text{Min.}$$

with the restrictions

$$a_2 \ge Ta_2 \ge Ta_1 \ge a_1 \quad .$$

5. Partial Differential Equations

We consider the following problem, which occured in the theory of suction pumps. Gas is entering at one end $x=0$ of a tube. The other end $x=\ell$ is a free end with pressure $p=0$. The pressure $p(x,t)$ satisfies the nonlinear parabolic equation (Guenther [74])

(5.1) $$\frac{\partial p}{\partial t} = \frac{1}{2}\frac{\partial^2(p^2)}{\partial x^2} = p\,\frac{\partial^2 p}{\partial x^2} + \left(\frac{\partial p}{\partial x}\right)^2 \quad \text{in } B:\ 0<x<\ell,\ t>0\ ,$$

with the initial boundary conditions

(5.2) $$\begin{cases} p \cdot \dfrac{\partial p}{\partial x} = \Psi(t) & \text{for } x=0,\ t>0 \\ p(x,0) = \varphi(x) & \text{for } t=0,\ 0<x<\ell \\ p(\ell,t) = 0 & \text{for } t>0\ . \end{cases}$$

$\varphi(x)$ and $\Psi(t)$ are given functions; we take as example $\ell = 1$, $\varphi(x) = 1-x$, $\Psi(t) = -1$. By $p^2 = q$ the boundary conditions are transformed into linear ones; one gets, fig.12

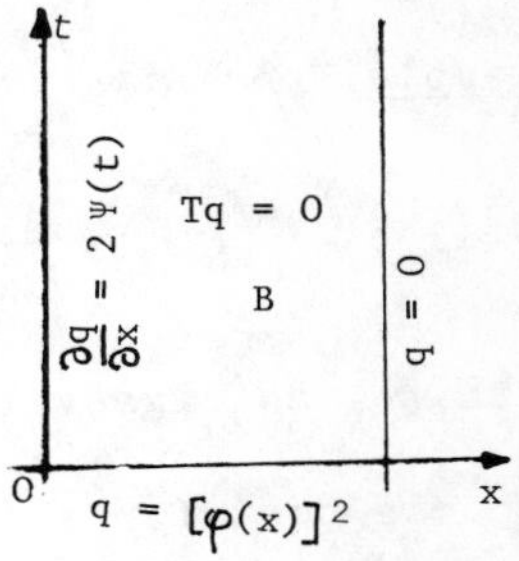

(5.3) $$Tq = \frac{\partial q}{\partial t} - \sqrt{q}\,\frac{\partial^2 q}{\partial x^2} = 0 \quad \text{in } B\ .$$

(5.4) $$\begin{cases} \dfrac{\partial q}{\partial x} = 2\Psi(t) & \text{for } x=0,\ t>0, \text{ here } \dfrac{\partial q}{\partial x}=-2 \\ q(x,0) = [\varphi(x)]^2 & \text{for } t=0,\ 0<x<\ell\ , \text{ here } q(x,0)=(1-x)^2 \\ q(\ell,t) = 0 & \text{for } t>0 \end{cases}$$

fig. 12

For a lower bound v and an upper bound w for q one has the optimization (compare for instance Collatz [66])

(5.5) $$w-v \le \delta,\ \delta = \text{Min.}$$

with the restrictions

$$Tv \le 0 \le Tw \quad \text{in } B$$

v,w are supposed to satisfy the boundary conditions.

In the simplest case one has

$$v = (1-x)^2 \leq q(x,t) \leq w = (1-x)^2+2t+t^2$$

For further examples see for instance Abadie [7o], Collatz [69], Collatz-Krabs [73].

6. Eigenvalue Problems

Other types of optimization occur in eigenvalue problems.

Example. Be given the problem

$$\frac{d^2y}{dx^2} + (x^2+\lambda)y = 0 \ , \ y(0) = y(\pi) = 0 \tag{6.1}$$

with λ as eigenvalue.

Suppose that for a function $w(x) \in C^2[0,\pi]$ the function

$$\Phi[w] = \frac{-w'' - x^2 w}{w} \tag{6.2}$$

has values in a finite interval J, then there exists an eigenvalue λ_s with

$$\underset{J}{\mathrm{Min}}\ \Phi \leq \lambda_s \leq \underset{J}{\mathrm{Max}}\ \Phi \tag{6.3}$$

If $w \geq 0$ in J, then the first eigenvalue λ_1 is included by (6.3). Here we choose $w = \sin x\, e^{-a \cos x}$, then

$$\Phi = 1 - x^2 - 3a \cos x - a^2 \sin^2 x \ . \tag{6.4}$$

Fig.13 shows $\Phi(x)$ for $a=1$ and $a=1.5$

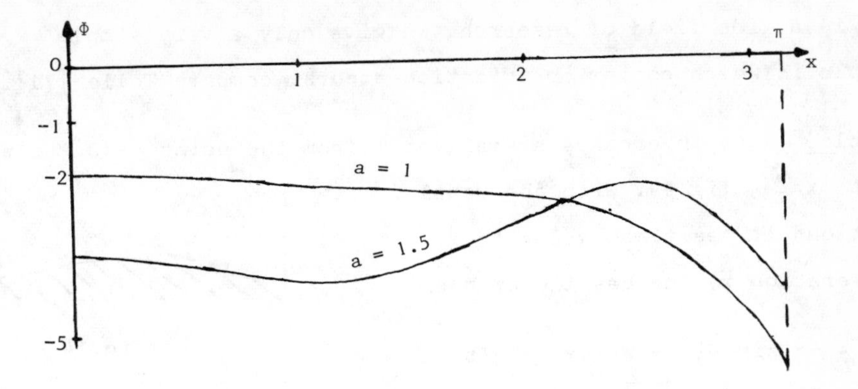

fig.13

For a lower bound k for λ_1 one has the optimization

(6.5) $$\Phi \geq k \ , \quad k = \text{Max.}$$

and for an upper bound $\hat{k}$

(6.6) $$\Phi \leq \hat{k} \ , \quad \hat{k} = \text{Min.}$$

One gets for $a=1.18$ and $a=1.55$ the inclusion $-4.3 \leq \lambda_1 \leq -2.6$.

Example. As another example we consider

(6.7) $$y'' + \lambda(1+\sin x)y = 0 \ , \quad y(0) = y(\pi) = 0$$

For $w = \sin x + a \sin 3x$ we have

(6.8) $$\Phi = -\frac{w''}{(1+\sin x)w} = \frac{1 + 9aS}{(1+\sin x)(1+aS)} \text{ with } S = 3-4\sin^2 x \ .$$

Then again one has the optimization (6.5) (6.6); in (6.4) a occurs quadratically, in (6.8) linear rationally.

7. Optimal Control

This is a wide field of research, we give only a very simple example in which convex optimization occurs, compare Tolle [71].

Example. A train crosses a small hill from the point $x=0$ to the point $x=\ell$, fig.14. With the usual notations (t as time, velocity $\dot{x}$, acceleration $\ddot{x}$) one has the problem

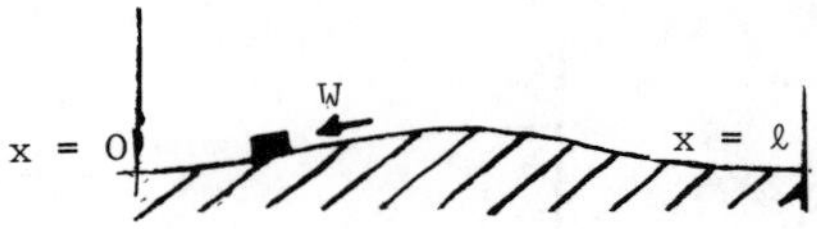

fig.14

$$\ddot{x} = -\alpha \sin \frac{2\pi x}{\ell} - W(x,\dot{x}) + u(t) ,$$

$$T = \int_0^T dt = \text{Min.},$$

$$x(0) = \dot{x}(0) = \dot{x}(T) = 0 , \quad x(T) = \ell, \quad -b \leq u(t) \leq a$$

with α, a, b as given constants, $W(x,\dot{x})$ as given resistance. Let us assume, that the train uses in the time interval $0<t<\tau$ the greatest possible force, and after this it is only braking:

$$u(t) = \begin{cases} a & \text{for } 0<t<\tau \\ \leq 0 & \text{for } \tau<t<T \end{cases}$$

Considering τ as desired parameter, $T(\tau)$ is convex for $W = 0$ and monotone decreasing if W is monotone increasing in $\dot{x}$; in the latter case the optimization may be quasiconvex.

References

Abadie, J. [7o]: Integer and nonlinear programming. Amsterdam: North Holland Publ. Comp. 197o, 544 p.

Collatz, L. [66]: Functional Analysis and Numerical Mathematics. Academic Press 1966, 473 p.

Collatz, L. [69]: Nichtlineare Approximationen bei Randwertaufgaben. Wiss. Z. Hochsch. Architektur u. Bauwesen Weimar, V. IKM 1969, 169-182.

Collatz, L. [7o]: Applications of nonlinear optimization to approximation problems. In: Integer and nonlinear programming. Amsterdam: North Holland Publ. Comp. 197o, p. 285-3o8.

Collatz, L. and W. Wetterling [71]: Optimierungsaufgaben, Springer, 2. Aufl. 1971, 222 p. English edition to appear.

Collatz, L. and W. Krabs [73]: Approximationstheorie, Teubner Stuttgart 1973, 2o8 p.

Guenther, R. [74]: Some nonlinear parabolic differential equations, to appear.

Mangasarian, O.L. [69]: Nonlinear programming. New York: Mc Graw Hill 1969, 22o p.

Stoer, J. and C. Witzgall [7o]: Convexity and optimization in finite dimensions I. Berlin-Heidelberg-New York: Springer 197o, 293 p.

Tolle, H. [71]: Optimierungsverfahren, Berlin-Heidelberg-New York: Springer 1971, 291 p.

RESULTS AND PROBLEMS IN THE SPECTRAL THEORY OF PERIODIC DIFFERENTIAL EQUATIONS

M. S. P. EASTHAM

1. Introduction

We consider the second-order equation

$$\{p(x)y'(x)\}' + \{\lambda s(x) - q(x)\}y(x) = 0 \quad (-\infty < x < \infty), \tag{1.1}$$

where p(x), s(x), and q(x) are real-valued and all have period a. Also, p'(x), s(x), and q(x) are piece-wise continuous, $p(x) \geqslant p_0$ (>0) and $s(x) \geqslant s_0$ (>0). We denote by λ_n and μ_n $(n = 0, 1, \ldots)$ the eigenvalues in the periodic and semi-periodic problems associated with (1.1) and the x-interval (0,a). It is known that the λ_n and μ_n occur in the order

$$\lambda_0 < \mu_0 \leq \mu_1 < \lambda_1 \leq \lambda_2 < \mu_2 \leq \mu_3 < \cdots.$$

Further, if λ lies in any of the open intervals $(-\infty, \lambda_0)$ and (μ_{2m}, μ_{2m+1}), $(\lambda_{2m+1}, \lambda_{2m+2})$ $(m = 0, 1, \ldots)$, then all non-trivial solutions of (1.1) are unbounded in $(-\infty, \infty)$. These intervals are called the <u>instability intervals</u> of (1.1). Apart from $(-\infty, \lambda_0)$, some or all of the instability intervals will be absent in the case of double eigenvalues. If λ lies in any of the complementary open intervals (λ_{2m}, μ_{2m}), $(\mu_{2m+1}, \lambda_{2m+1})$, then all solutions of (1.1) are bounded in $(-\infty, \infty)$, and these intervals are called the <u>stability intervals</u> of (1.1). For the theory which is summarized here, see e.g. (1), Chap. 2

We are concerned here with the lengths l_n of the instability intervals, where

$$l_{2m+1} = \mu_{2m+1} - \mu_{2m}, \quad l_{2m+2} = \lambda_{2m+2} - \lambda_{2m+1}.$$

The following estimates for l_n as $n \to \infty$ are known:

(i) $l_n = o(n^2)$;

(ii) $l_n = O(n)$ if s(x) is piecewise smooth;

(iii) $l_n = o(n)$ if s'(x) exists and is piecewise continuous;

(iv) $l_n = O(1)$ if s'(x) and p'(x) are piecewise smooth;

(v) $l_n = o(1)$ if s"(x) and p"(x) exist and are piecewise continuous.

The sequence of results continues. The general "o-result" is:

(vi) $l_n = o(n^{-r})$ if $s^{(r+2)}(x)$, $p^{(r+2)}(x)$, $q^{(r)}(x)$ all exist and are piecewise continuous.

It will be noticed that the estimates for l_n improve as the differentiability conditions on p(x), s(x), and q(x) are increased. Proofs of (i)-(vi), based on the use of the modified Prüfer transformation, are given in (1), §4.5. The earliest of these results are (v) and (vi)—see Borg (2) and Hartman and Putnam (3). An example of (ii) in which $s(x)$ is a step function was analysed by Hochstadt (4).

The best-possible nature of (ii) and (iv) can be demonstrated by examples. For (ii), $s(x)$ as a step function will do. For (iv) (as well as (ii)), $s(x)$ as a piecewise linear function will do, although here the calculations are lengthy and involve Bessel functions. More generally, the best-possible nature of (ii) and (iv) will follow from the work in §3 below, where it is shown exactly how the O-term depends on the discontinuities of $s(x)$ and $s'(x)$. In addition, (v) is known to be best-possible as an immediate consequence of the following refinement of (v):

(v') Let $p(x) = s(x) = 1$ and let c_n denote the n-th complex Fourier coefficient of $q(x)$ referred to the interval $(0,a)$. Then

$$l_n = |c_n| + O(n^{-\frac{1}{2}}). \tag{1.2}$$

For this result, see (3), (5) § 21.5, and (1) §§4.4-5. It is not known whether (i) and (iii) are best-possible, but one obviously conjectures that they are, and a proof of this would be of interest.

Lastly, following on from (vi) above, we mention the result of Simonyan (6) that, if $p(x) = 1$, $q(x) = 0$ and if $s(x)$ is the restriction of an entire or meromorphic function $s(z)$ of a certain kind, then $l_n = O(ne^{-An})$, where A is a positive constant which can be specified.

2. The Prüfer transformation method

In §1, we mentioned that (i)-(vi) can be proved by means of the modified Prüfer transformation. In this section, we introduce this method but, rather than simply reproduce material from (1) Chapter 4, we use the method to obtain an improvement of (ii) which makes the O-term more explicit.

Theorem 1 Let $s(x)$ be piecewise smooth and denote the discontinuities of $s(x)$ in $[0,a)$ by a_r $(1 \leq r \leq N)$, where $0 \leq a_1 < \ldots < a_N < a$ and $a_1 = 0$ means that $s(a - 0) \neq s(0 + 0)$. Let

$$\omega_r = \tan^{-1}\left\{ \frac{1}{2} \left| \left(\frac{s(a_r+0)}{s(a_r-0)} \right)^{\frac{1}{4}} - \left(\frac{s(a_r-0)}{s(a_r+0)} \right)^{\frac{1}{4}} \right| \right\},$$

where $0 < \omega_r < \frac{1}{2}\pi$. Then, for $\lambda = \lambda_{2m+1}$ or $\lambda = \lambda_{2m+2}$,

$$|\lambda^{1/2} - 2(m+1)\pi I^{-1}| \leq I^{-1} \sum_{r=1}^{N} \omega_r + o(1) \tag{2.1}$$

as $m \to \infty$, where

$$I = \int_0^a \{s(x)/p(x)\}^{\frac{1}{2}}\, dx.$$

The same result holds for μ_{2m} and μ_{2m+1} but with $2(m + 1)$ replaced by $2m + 1$.

<u>Proof</u> It is convenient to move the interval [0,a] slightly (if necessary) to make $a_1 > a$. Then $s(0) = s(a)$. This does not affect the λ_n and μ_n. We write $a_0 = 0$ and $a_{N+1} = a$. Also, we need only consider $q(x) = 0$ since the inclusion of $q(x)$ would only make a bounded difference to λ_n and μ_n, and this would not affect (2.1). Thus we consider here

$$\{p(x)y'(x)\}' + \lambda s(x)y(x) = 0. \tag{2.2}$$

We introduce the modified Prüfer transformation (cf (1), §4.1) relevant to (2.2):

$$\begin{aligned} \lambda^{\frac{1}{2}}\{s(x)p(x)\}^{\frac{1}{2}}y(x) &= \rho(x)\sin\theta(x) \\ p(x)y'(x) &= \rho(x)\cos\theta(x), \end{aligned} \tag{2.3}$$

where $\lambda = \lambda_{2m+1}$ or λ_{2m+2} and $y(x)$ is a corresponding periodic eigenfunction of (2.2). Of course, $\rho(x)$ and $\theta(x)$ depend on λ but we do not bring this fact into the notation. It follows from (2.3) that, at any point where $s'(x)$ exists,

$$\theta'(x) = \lambda^{\frac{1}{2}}\left(\frac{s(x)}{p(x)}\right)^{\frac{1}{2}} + \frac{1}{4}\frac{\{s(x)p(x)\}'}{s(x)p(x)}\sin 2\theta(x) \tag{2.4}$$

(1, §4.1). A further consequence of (2.3) is that, since $s(0) = s(a)$ and $p(0) = p(a)$,

$$\theta(a) - \theta(0) = 2k\pi, \tag{2.5}$$

where k is an integer. Now it is known that $y(x)$ has exactly $2(m + 1)$ zeros in [0,a) (1, Theorem 3.1.2) and it follows from this and (2.4) and (2.5) that, in the standard situation where $s'(x)$ exists throughout [0,a] (1,§§4.1-2), we have

$$k = m + 1. \tag{2.6}$$

We now check that (2.6) can still be effected even though $s(x)$ is only piecewise smooth.

If a_r is a zero of either $y(x)$ or $y'(x)$, then, by (2.3), we can take

$$\theta(a_r + 0) = \theta(a_r - 0) \quad (= N_r\pi \text{ or } (N_r + \tfrac{1}{2})\pi)$$

for some integer N_r. If a_r is not such a zero, then we see from (2.3) that $\theta(a_r \pm 0)$ lie in the same quadrant. Hence we can take it that

$$|\theta(a_r + 0) - \theta(a_r - 0)| < \tfrac{1}{2}\pi. \tag{2.7}$$

We can now say from (2.4) that $\theta(x)$ is increasing at the zeros of $y(x)$ and that, although $\theta(x)$ may have jumps at the a_r, no such jump straddles a value of $\theta(x)$ which is a multiple of π. Hence the basic property of $\theta(x)$, which is given in (1, 4.1.7) in a different notation, continues to hold, and hence (2.6) continues to hold.

Now, integration of (2.4) over (a_r, a_{r+1}) gives

$$\theta(a_{r+1} - 0) - \theta(a_r - 0) = \lambda^{\frac{1}{2}}\int_{a_r}^{a_{r+1}}\left(\frac{s(x)}{p(x)}\right)^{\frac{1}{2}}dx + o(1),$$

on using the Riemann-Lebesgue type result of (1), Lemma 4.1.1. Hence, by (2.5) and (2.6),

$$2(m+1)\pi = \theta(a) - \theta(0)$$
$$= \sum_{r=0}^{N} \{\theta(a_{r+1}-0) - \theta(a_r+0)\} + \sum_{r=1}^{N} \{\theta(a_r+0) - \theta(a_r-0)\}$$
$$= \lambda^{\frac{1}{2}} I + \sum_{r=1}^{N} \{\theta(a_r+0) - \theta(a_r-0)\} + o(1). \tag{2.8}$$

To estimate $\theta(a_r + 0) - \theta(a_r - 0)$ for a particular r, we write

$$t_\pm = \tan \theta(a_r \pm 0), \quad s_\pm = s(a_r \pm 0).$$

From (2.3),
$$t_+/t_- = (s_+/s_-)^{\frac{1}{2}}$$
and so
$$\tan\{\theta(a_r + 0) - \theta(a_r - 0)\} = (t_+ - t_-)/(1 + t_+t_-)$$
$$= \frac{(t_+t_-)^{\frac{1}{2}}}{1 + t_+t_-} \left\{\left(\frac{s_+}{s_-}\right)^{\frac{1}{4}} - \left(\frac{s_-}{s_+}\right)^{\frac{1}{4}}\right\}$$

Since $2(t_+t_-)^{\frac{1}{2}} \leq 1 + t_+t_-$, we have
$$|\tan\{\theta(a_r + 0) - \theta(a_r - 0)\}| \leq \tan \omega_r$$
and hence, by (2.7), $|\theta(a_r + 0) - \theta(a_r - 0)| \leq \omega_r$. Then (2.1) follows from (2.8). The proof for μ_{2m} and μ_{2m+1} is similar.

<u>Corollary</u> Under the conditions on s(x) in the theorem,

$$l_n \leq 4n\pi I^{-2} \sum_{1}^{N} \omega_r + o(n). \tag{2.9}$$

In particular, if s(x) is differentiable in (0,a) but (s(0) =) $s(0 + 0) \neq s(a - 0)$ (= s(a)), then

$$l_n \leq 4n\pi I^{-2}\omega + o(n), \tag{2.10}$$

where

$$\omega = \tan^{-1}\left\{\tfrac{1}{2}\left|\left(\frac{s(0)}{s(a)}\right)^{\frac{1}{4}} - \left(\frac{s(a)}{s(0)}\right)^{\frac{1}{4}}\right|\right\} \tag{2.11}$$

and $0 < \omega < \frac{1}{2}\pi$.

The corollary shows how l_n depends both on the number of discontinuities of s(x) and on the size of the jumps. It is not clear whether the multiple of n in (2.9) is best-possible for all N, but we shall see in §3 that, in the case N = 1 (which is (2.10)), it is best-possible.

We state without proof the improvement of (iv) in §1 which can be obtained by the modified Prüfer transformation method. The proof is given in the thesis (7) as is the above proof of Theorem 1 together with a comparison of the results with those obtained by means of the Liouville transformation in §3 below.

<u>Theorem 2</u> Let s'(x) and p'(x) be piecewise smooth and denote the discontinuities of $\{s(x)p(x)\}'$ by b_r $(1 \leq r \leq N)$, where $0 \leq b_1 < \ldots < b_N < a$. Then

$$l_n \leq \tfrac{1}{2}\sum_{r=1}^{N} s^{-\frac{3}{2}}(b_r)p^{-\frac{1}{2}}(b_r) \left|(ps)'(b_r + 0) - (ps)'(b_r - 0)\right| + o(1).$$

3. The Liouville transformation method

In this section, we consider (1.1) when $s(x)$ has a piecewise continuous second derivative in $(0,a)$ and $s(0)$ may be unequal to $s(a)$, where we write $s(0) = s(0 + 0)$ and $s(a) = s(a - 0)$. Thus we have the case $N = 1$ of §2 with the extra differentiability condition on $s(x)$ in $(0,a)$. For convenience we take $p(x) = 1$ but, as we wish (inter alia) to improve the $o(n)$ term in (2.10), we do not now take $q(x) = 0$ Thus we shall examine the eigenvalues λ_n for

$$y''(x) + \{\lambda s(x) - q(x)\} y(x) = 0. \tag{3.1}$$

After an application of the Liouville transformation

$$t = \int_0^x s^{\frac{1}{2}}(u)\, du, \quad z(t) = s^{\frac{1}{4}}(x) y(x),$$

(3.1) becomes

$$\ddot{z}(t) + \{\lambda - Q(t)\} z(t) = 0, \tag{3.2}$$

where

$$Q(t) = q(x)/s(x) - s^{-\frac{3}{4}}(x)\{s^{-\frac{1}{4}}(x)\}''.$$

Since

$$y'(x) = \{s^{-\frac{1}{4}}(x)\}' z(t) + s^{\frac{1}{4}}(x)\dot{z}(t),$$

the periodic boundary conditions $y(a) = y(0)$, $y'(a) = y'(0)$ become

$$\left.\begin{aligned} z(A) &= \sigma z(0) \\ \sigma \dot{z}(A) + \rho z(A) &= \dot{z}(0) + \tau z(0), \end{aligned}\right\} \tag{3.3}$$

where

$$A = \int_0^a s^{\frac{1}{2}}(u)\, du \text{ (= the I of §2)}$$

$$\left.\begin{aligned} \sigma &= \{s(a)/s(0)\}^{\frac{1}{2}} \\ \rho = s^{-\frac{1}{4}}(0)\{s^{-\frac{1}{4}}(x)\}'_{x=a},\ &\ \tau = s^{-\frac{1}{4}}(0)\{s^{-\frac{1}{4}}(x)\}'_{x=0} \end{aligned}\right\} \tag{3.4}$$

We note that $\sigma = 1$ means $s(0) = s(a)$ and that $\rho = \tau$ if also $s'(0) = s'(a)$. Now (3.3) are self-adjoint boundary conditions for (3.2) and they can be dealt with in the usual way as follows.

Let $\theta(t,\lambda)$ and $\phi(t,\lambda)$ be the solutions of (3.2) such that

$$\theta(0,\lambda) = 1,\ \dot{\theta}(0,\lambda) = 0;\ \phi(0,\lambda) = 0,\ \dot{\phi}(0,\lambda) = 1.$$

Then the condition that (3.2) has a solution satisfying (3.3) is

$$\begin{vmatrix} \theta - \sigma & \phi \\ \sigma\dot{\theta} + \rho\theta - \tau & \sigma\dot{\phi} + \rho\phi - 1 \end{vmatrix} = 0,$$

where $\theta = \theta(A,\lambda)$ etc. Since $W(\theta,\phi)(t,\lambda) = 1$, this gives

$$\theta + \sigma^2 \dot{\phi} + (\rho\sigma - \tau)\phi = 2\sigma. \tag{3.5}$$

This equation is a generalization of the usual discriminant condition $D(\lambda) = \theta + \dot{\phi} = 2$, to which it reduces when $\sigma = 1$ and $\rho = \tau$. When the usual asymptotic formulae for θ and $\dot{\phi}$ (1, 4.3.3-4) are substituted into (3.5), we obtain

$$(1 + \sigma^2)\cos A\surd\lambda = 2\sigma + O(\lambda^{-\frac{1}{2}})$$

i.e.,

$$\cos A\surd\lambda = \cos\omega + O(\lambda^{-\frac{1}{2}}),$$

where ω is the angle defined in (2.11). If $\omega \neq 0$, this gives

$$A\surd\lambda = 2M\pi \pm \omega + O(\lambda^{-\frac{1}{2}}), \tag{3.6}$$

here M is an integer. Here, + will occur for λ_{2m+2} and - for λ_{2m+1}. rom (2.1), we have M = m + 1 and hence

$$\lambda = 4(m+1)^2\pi^2A^{-2} \pm 4(m+1)\pi A^{-2}\omega + O(1).$$

hus

$$l_{2m+2} = 8(m+1)\pi A^{-2}\omega + O(1).$$

ith the similar result for l_{2m+1} we obtain:

<u>Theorem 3</u> In (3.1), let s"(x) exist and be piecewise continuous in 0,a) and let $s(0) \neq s(a)$. Then

$$l_n = 4n\pi A^{-2}\omega + O(1), \tag{3.7}$$

here ω is as in (2.11).

Note that here the inequality (2.10) has been improved to an asymptotic equality subject to the existence of s"(x). Also, the o(n) erm in (2.10) is improved to O(1) in (3.7).

It is possible to take (3.7) further if more accurate asymptotic ormulae for θ and $\dot\phi$ are used. From (1, 4.3.9 and 4.3.11), we have

$$\theta = \cos A\surd\lambda + \tfrac{1}{2}\lambda^{-\frac{1}{2}}\int_0^A\{\sin A\surd\lambda - \sin(2t - A)\surd\lambda\}\,Q(t)\,dt + O(\lambda^{-1}),$$

$$\dot\phi = \cos A\surd\lambda + \tfrac{1}{2}\lambda^{-\frac{1}{2}}\int_0^A\{\sin A\surd\lambda + \sin(2t - A)\surd\lambda\}\,Q(t)\,dt + O(\lambda^{-1}).$$

These, together with

$$\phi = \lambda^{-\frac{1}{2}}\sin A\surd\lambda + O(\lambda^{-1}),$$

substituted into (3.5) give

$$\cos A\surd\lambda + \lambda^{-\frac{1}{2}}(K + k)\sin A\surd\lambda + \tfrac{1}{2}\lambda^{-\frac{1}{2}}\frac{\sigma^2 - 1}{\sigma^2 + 1}\int_0^A \sin\{(2t - A)\surd\lambda\}\,Q(t)\,dt = \cos\omega + O(\lambda^{-1}) \tag{3.8}$$

where

$$K = \tfrac{1}{2}\int_0^A Q(t)\,dt \tag{3.9}$$

and

$$k = (\rho\sigma - \tau)/(1 + \sigma^2).$$

By (3.6), we have, if $\omega \neq 0$,

$$A\surd\lambda = 2(m+1)\pi \pm \omega + \delta, \tag{3.10}$$

where $\delta = O(\lambda^{-\frac{1}{2}})$, and substitution into (3.8) gives

$$\cos\omega \mp \delta\sin\omega \pm \lambda^{-\frac{1}{2}}(K + k)\sin\omega + \tfrac{1}{2}\lambda^{-\frac{1}{2}}(\sigma^2 - 1)/(\sigma^2 + 1)\,J_\pm = \cos\omega + O(\lambda^{-1}),$$

where

$$J_\pm = \int_0^A Q(t)\sin\left(\frac{2t}{A}\{2(m+1)\pi \pm \omega\} \mp \omega\right)dt.$$

Since $\omega \neq 0$, we obtain

$$\delta = \lambda^{-\frac{1}{2}}(K + k) \pm \tfrac{1}{2}\lambda^{-\frac{1}{2}}\frac{\sigma^2 - 1}{\sigma^2 + 1}\frac{1}{\sin\omega}J_\pm + O(\lambda^{-1}).$$

Substituting this into (3.10) and squaring, we obtain

$$\lambda A^2 = 4(m+1)^2\pi^2 \pm 4(m+1)\pi\omega + \omega^2 +$$
$$+ 2A(K+k) \pm A\frac{\sigma^2-1}{\sigma^2+1}\frac{1}{\sin\omega} J_\pm + O(m^{-1}).$$

Here again, we have + for λ_{2m+2} and - for λ_{2m+1}. Hence

$$l_{2m+2} = 8(m+1)\pi\omega A^{-2} + A^{-1}\frac{\sigma^2-1}{\sigma^2+1}\frac{1}{\sin\omega}(J_+ + J_-) + O(m^{-1}).$$

This together with the corresponding result for μ_{2m} and μ_{2m+1} gives:

Theorem 4 Under the same conditions as Theorem 3,

$$l_n = 4n\pi A^{-2}\omega +$$
$$+ 2A^{-1}\frac{\sigma^2-1}{\sigma^2+1}\frac{1}{\sin\omega}\int_0^A\left\{Q(t)\cos\left(\frac{2t}{A}-1\right)\omega\right\}\sin\frac{2n\pi t}{A}\,dt + O(n^{-1}).$$

Note that this has the form $l_n = 4n\pi A^{-2}\omega + o(1)$, which may be compared with (3.7).

To obtain (3.6), and so Theorems 3 and 4, we took $\omega \neq 0$. To examine the case $\omega = 0$, we must return to (3.5) and use even better asymptotic formulae for $\theta, \phi, \dot{\phi}$. The result here is:

Theorem 5 In (3.1), let $s''(x)$ exist and be piecewise continuous in $(0,a)$ and let $s(0) = s(a)$. Then

$$l_n = \tfrac{1}{2}A^{-1}s^{-3/2}(0)\,|s'(a) - s'(0)| + o(1). \tag{3.11}$$

In (3.5) we now have $\sigma = 1$ and so (3.5) is now

$$\theta + \dot{\phi} + (\rho - \tau)\phi = 2.$$

As in (1, 4.4.1), this gives

$$2\cos A\sqrt{\lambda} + 2\lambda^{-\frac{1}{2}}K\sin A\sqrt{\lambda} -$$
$$- \lambda^{-1}\int_0^A Q(t)\,dt\int_0^t \sin\{(t-u-A)\sqrt{\lambda}\}\sin\{(t-u)\sqrt{\lambda}\}\,Q(u)\,du$$
$$+ (\rho-\tau)\left(\lambda^{-\frac{1}{2}}\sin A\sqrt{\lambda} + \lambda^{-1}\int_0^A \sin\{(A-t)\sqrt{\lambda}\}\,Q(t)\sin t\sqrt{\lambda}\,dt\right)$$
$$= 2 + O(\lambda^{-3/2}), \tag{3.12}$$

where K is as in (3.9). When $\omega = 0$, (3.6) must be replaced by

$$A\sqrt{\lambda} = 2M\pi + \delta, \tag{3.13}$$

where $M = m + 1$ and only $\delta = O(\lambda^{-\frac{1}{4}})$. We now use (3.13) in (3.12) to improve the estimate on δ. We get

$$-\delta^2 + (2K + \rho - \tau)\lambda^{-\frac{1}{2}}\delta -$$
$$-\lambda^{-1}\int_0^A Q(t)\,dt\int_0^t \sin^2\{(t-u)\sqrt{\lambda}\}\,Q(u)\,du - (\rho-\tau)\lambda^{-1}\int_0^A Q(t)\sin^2 t\sqrt{\lambda}\,dt$$
$$= O(\delta^4) + O(\delta^3\lambda^{-\frac{1}{2}}) + O(\delta\lambda^{-1}) + O(\lambda^{-3/2}). \tag{3.14}$$

Since $\delta = O(\lambda^{-\frac{1}{4}})$, the right-hand side of (3.14) is $O(\lambda^{-1})$. Then (3.14) gives $\delta = O(\lambda^{-\frac{1}{2}})$ and hence the right-hand side of (3.14) is in fact $O(\lambda^{-3/2})$. Arguing as for (1, 4.4.4), (3.14) gives

$$- \delta^2 + (2K + \rho - \tau)\lambda^{-\frac{1}{2}}\delta -$$
$$- \tfrac{1}{4}\lambda^{-1}\left\{\left(\int_0^A Q(t)\, dt\right)^2 - \left(\int_0^A Q(t)\cos \frac{2M\pi t}{A}\, dt\right)^2 - \left(\int_0^A Q(t)\sin \frac{2M\ t}{A}\, dt\right)^2\right\}$$
$$- \tfrac{1}{2}(\rho - \tau)\lambda^{-1}\left\{\int_0^A Q(t)\, dt - \int_0^A Q(t)\cos \frac{2M\pi t}{A}\, dt\right\} = O(\lambda^{-3/2}).$$

Hence, denoting the cosine and sine integrals by c_M and s_M, we have

$$\delta^2 - (2\ K + \rho - \tau)\lambda^{-\frac{1}{2}}\delta +$$
$$+ \lambda^{-1}\{K^2 - \tfrac{1}{4}(c_M^2 + s_M^2) + (\rho - \tau)K - \tfrac{1}{2}(\rho - \tau)c_M\} = O(\lambda^{-3/2}).$$

Solving the quadratic for δ, we obtain

$$\delta = \tfrac{1}{2}\lambda^{-\frac{1}{2}}[2K + \rho - \tau \pm \{(\rho - \tau + c_M)^2 + s_M^2\}^{\frac{1}{2}}] + O(\lambda^{-\frac{3}{4}}).$$

On substituting this into (3.13), we obtain

$$l_{2m+2} = 2A^{-1}\{(\rho - \tau + c_M)^2 + s_M^2\}^{\frac{1}{2}} + O(m^{-\frac{1}{2}}). \qquad (3.15)$$

This is actually more precise than (3.11). In fact, since $c_M = o(1)$ and $s_M = o(1)$, (3.15) gives

$$l_{2m+2} = 2A^{-1}|\rho - \tau| + o(1)$$
$$= \tfrac{1}{2}A^{-1}s^{-3/2}(0)|s'(a) - s'(0)| + o(1)$$

by the definition of ρ and τ in (3.4). Note also that, when $\rho = \tau$, (3.15) reduces to (v') in §1 above. In the same way as (3.7) improves (2.10) to an asymptotic equality, Theorem 5 improves the case N = 1 of Theorem 2.

4. Two "best-possible" questions

In this section we take $p(x) = s(x) = 1$ and we suppose that

$$\int_0^a q(x)\, dx = 0. \qquad (4.1)$$

A general reference for this section is (1), §§5.4-5.

I. By (4.1), $\lambda_0 \leq 0$. The question has been raised as to what are the best-possible constants A_1 and A_2 such that

$$\lambda_0 \geqslant - A_1\left(\int_0^a |q(x)|\, dx\right)^2 \qquad (4.2)$$

and

$$\lambda_0 \geqslant - A_2 a\int_0^a q^2(x)\, dx. \qquad (4.3)$$

It is known that $\quad 1/4\pi^2 \leq A_2 \leq A_1 \leq 1/16. \qquad (4.4)$

The left-hand inequality here is due to Putnam (8) and the right-hand one to Kato (9). In (10) an example of a step-function $q(x)$ was given which shows that in fact $A_1 = 1/16$. However, this kind of example appears to give no new information on A_2 and the question remains as to what is the exact value of A_2.

Inequalities of the type (4.2-3) have found a recent application in the estimation of the least point of the spectrum associated with singular second-order differential operators - see (10).

II. In (11), Putnam proved that the length of any instability interval (α,β) satisfies

$$\beta - \alpha \leq 2\left(a^{-1}\int_0^a q^2(x)\,dx\right)^{\frac{1}{2}} \tag{4.5}$$

provided that the mid-point $\gamma = \frac{1}{2}(\beta + \alpha) \geqslant 0$, and he raised the question of whether (4.5) holds for all (α,β). Using (4.3), he showed that (4.5) does hold for all (α,β) if

$$\int_0^a q^2(x)\,dx \leqslant 256a^{-3}. \tag{4.6}$$

However, in (12), an example was given of a q(x) for which (4.5) does not hold for (μ_0,μ_1). This q(x) satisfies

$$\int_0^a q^2(x)\,dx > 2^{10}\pi^2 256a^{-3}.$$

Can the difference between this and (4.6) be reduced? A useful lower bound for $\mu_0 - \lambda_0$ in terms of q(x) is required if (4.6) is to be widened.

5. Higher-order and partial differential equations

We have referred often to (1) for information on the well- developed spectral theory of (1.1). There is the constant question of to what extent this theory can be extended (a) to higher-order self-adjoint ordinary differential equations and (b) to the periodic Schrödinger equation in N dimensions.

In the case of (a), we have of course the basic Floquet theory and it is known that the spectrum in $L^2(-\infty,\infty)$ coincides with the stability set (13, p.1491). The method of singular sequences is therefore available (1, §5.2) to estimate the lengths of the instability intervals and this method was used in (7) to extend some of the results (i) - (v) in §1 above. In (14), the case where the differential operator is a product of first-order operators is considered and some oscillation properties of the periodic and semi-periodic eigenfunctions are given. However, no analogue of the modified Prüfer transformation in §2 above is known which could take advantage of these properties and so extend the results of §2.

Although no Floquet is available for the periodic Schrödinger equation, it is possible to make progress using some general methods from spectral theory and an account is given in (1, Chapter 6). We mention also the recent paper (15) in which it is proved that the spectrum is continuous and this settles the question of the existence of eigenvalues of infinite multiplicity which was left open in (1, §6.10).

References

1. M.S.P.Eastham, The spectral theory of periodic differential equations (Scottish Academic Press, Edinburgh, 1973).
2. G.Borg, Acta Math. 78 (1946) 1-96.
3. P.Hartman and C.Putnam, Amer. J. Math. 72 (1950) 849-62.
4. H.Hochstadt, Amer. Math. Monthly 70 (1963) 18-26.
5. E.C.Titchmarsh, Eigenfunction expansions, part 2 (Oxford, 1958).
6. S.G.Simonyan, Differential equations 6 (1970) 965-71.
7. A.Ntinos, London University Ph.D. thesis (1974).
8. C.R.Putnam, Quart. Appl. Math. 9 (1951) 310-4.
9. T.Kato, ibid. 10 (1952) 292-4.
10. M.S.P.Eastham, Proc. Roy. Soc. Edinburgh (A) 72 (1974) 9-16.
11. C.R.Putnam, Quart. Appl. Math. 11 (1953) 496-8.
12. M.S.P.Eastham, Proc. Amer. Math. Soc. 21 (1969) 643-7.
13. N.Dunford and J.T.Schwartz, Linear operators (Interscience, 1963).
14. J.W.Lee, J. Diff. Equations 11 (1972) 592-606.
15. L.E.Thomas, Comm. Math. Phys. (1974).

Sobolev embeddings

W. D. Evans

1. Introduction.

Embedding theorems for Sobolev spaces are well known to have an important role to play in the theory of partial differential equations. They are a useful tool in problems of existence and regularity of solutions of elliptic equations (both linear and non-linear). Furthermore the properties of maps between the Sobolev spaces, such as the compactness of the embedding maps as expressed in the Rellich Theorem, have applications to the spectral theory of linear elliptic operators.

In recent years the properties of Sobolev spaces and the relationships between them have been much investigated, the work being largely motivated by the importance and need of such results for differential equations. The properties of mappings (especially embeddings) between Sobolev spaces in an unbounded domain have been studied ([1], [2], [3], [6], [7]). Also Orlicz-Sobolev spaces and their role in the study of elliptic equations with strong non-linear terms have been investigated (see e.g. [4], [5], [7], [8], [10]).

Before proceeding further we must introduce some notation and give a brief account of the background required.

Let Ω be a domain in $\mathbb{R}^n$ $(n \geq 1)$ and let k be a positive integer and p a real number > 1. We denote by $H_0^{k,p}(\Omega)$ (another notation is $W_0^{k,p}(\Omega)$) the completion of $C_0^k(\Omega)$, the space of k times continuously differentiable functions in Ω whose supports are compact subsets of Ω, with respect to the norm

$$\|u\|_{k,p} = \sum_{i=0}^{k} \|\mathbb{D}^i u\|_{0,p}$$

where $\|\cdot\|_{0,p}$ is the usual $L^p(\Omega)$ norm. We have used the notation

$$|D^i u|^2 = \sum_{|\alpha|=i} |D^\alpha u|^2$$

where the summation extends over all n-tuples $\alpha = (\alpha_1, \alpha_a, \ldots, \alpha_n)$ of non-negative integers with $|\alpha| = \alpha_1 + \ldots + \alpha_n = i$ and

$$D^\alpha u(x) = \frac{\partial^{|\alpha|}}{\partial x_1^{\alpha_1} \ldots \partial x_n^{\alpha_n}} u(x).$$

$C^k(\bar{\Omega})$ is the space of functions whose first k derivatives are continuous in Ω and can be continuously extended to $\bar{\Omega}$. We denote by $H^{k,p}(\Omega)$ the completion of the space of functions $u \in C^k(\bar{\Omega})$ which are such that $\|u\|_{k,p} < \infty$, with respect to the norm $\|\cdot\|_{k,p}$. It is well known that if the boundary $\partial\Omega$ is sufficiently smooth e.g. if Ω satisfies the segment condition, then $H^{k,p}(\Omega)$ coincides with the space of functions $u \in L^p(\Omega)$ whose first k distributional derivatives are also in $L^p(\Omega)$. Also, $H_0^{k,p}(\mathbb{R}^n) = H^{k,p}(\mathbb{R}^n)$ but the two spaces are not identical in general.

We also need to define an Orlicz space $L_\Phi(\Omega)$. An Orlicz function Φ is a real-valued, continuous, convex, even function on $\mathbb{R}$ which satisfies

$$\lim_{t\to 0} \frac{\Phi(t)}{t} = 0, \quad \lim_{t\to\infty} \frac{\Phi(t)}{t} = \infty .$$

By $L'_\Phi(\Omega)$ we mean the set of function u which satisfy $\int_\Omega \Phi(|u(x)|)dx < \infty$. The Orlicz space $L_\Phi(\Omega)$ is the linear hull of $L'_\Phi(\Omega)$ with the Luxemborg norm

$$\|u\|_\Phi = \inf\{k : \int_\Omega \Phi(\frac{|u(x)|}{k})dx \leq 1\}.$$

$L_\Phi(\Omega)$ is a Banach space, in general neither separable nor reflexive. The closure in $L_\Phi(\Omega)$ of the bounded functions with compact supports in $\bar{\Omega}$ is a linear subspace $E_\Phi(\Omega)$ of $L_\Phi(\Omega)$. $E_\Phi(\Omega)$ is separable and also $E_\Phi(\Omega)$ is in $L'_\Phi(\Omega)$. The two spaces E_Φ and L_Φ are identical if and only if Φ satisfies the

Δ_2-condition i.e. $\Phi(2t) \leq K\Phi(t)$ for some constant $K > 0$ and all $t \geq 0$. The L^p spaces $(p > 1)$ are clearly examples of Orlicz spaces whose Orlicz function $(\Phi(t) = |t|^p)$ satisfies the Δ_2-condition. A detailed exposition of Orlicz spaces may be found in [9]. For bounded Ω, Orlicz functions Φ, Ψ which satisfy $\Phi(t) \leq \Psi(\lambda t)$, $\Psi(t) \leq \Phi(\mu t)$ for large t (and some $\lambda, \mu > 0$) are said to be equivalent (written $\Phi \sim \Psi$) and give rise to identical Orlicz spaces. However for unbounded Ω, two Orlicz functions have to be comparable in this way for all values of t if they are to give rise to identical Orlicz spaces. This is why in such conditions as the Δ_2-condition mentioned above the inequality $\Phi(2t) \leq K\Phi(t)$ has to hold for all values of t in order to have the same implications as when Ω is bounded.

The Orlicz-Sobolev space $H^kL_\Phi(\Omega)$ can now be defined as the space of functions $u \in L_\Phi(\Omega)$ whose first k partial derivatives in the distributional sense are in $L_\Phi(\Omega)$. It is a Banach space with norm

$$\|u\|_{k,\Phi} = \sum_{i=0}^{k} \|D^i u\|_\Phi .$$

Similarly we can define $H^kE_\Phi(\Omega)$ and $H_0^kL_\Phi(\Omega)$. If Ω (bounded) satisfies the segment condition then as before $C^k(\overline{\Omega})$ (indeed $C^\infty(\overline{\Omega})$) is a dense subset of $H^kE_\Phi(\Omega)$.

2. The spaces $H_0^{1,p}(\Omega)$, $p > 1$.

We shall hereafter take Ω to be an arbitrary domain in $\mathbb{R}^n$, $n \geq 1$, unless otherwise stated. We shall always imply by the inclusion symbol $\subset$ that the natural embedding is continuous. The first result is the Sobolev Embedding Theorem for the space $H_0^{1,p}(\Omega)$.

Theorem 1.

(i) _If_ $p \leq n$ _then_ $H_0^{1,p}(\Omega) \subset L^q(\Omega)$ _for any_ q _satisfying_

$$\frac{1}{p} - \frac{1}{n} < \frac{1}{q} \leq \frac{1}{p} ,$$

and for any $u \in H_0^{1,p}(\Omega)$,

$$\|u\|_{0,q} \leq (\omega_n d^n)^{1/q-1/p} \eta^{1/q-1/p+1/n} \left(\frac{1 + 1/q - 1/p}{1 + n/q - n/p}\right)^{1+1/q-1/p} (\|u\|_{0,p} + d\,\|Du\|_{0,p}) \tag{1}$$

where ω_n *is the* $(n - 1)$-*dimensional measure of the unit sphere* S^{n-1} *in* $\mathbb{R}^n$ *and* $\eta = \sup_{x \in \Omega} \eta(x)$, *where* $\eta(x) = \frac{|\Omega \cap B(x,d)|}{|B(x,d)|}$, $B(x, d)$ *denoting the ball centre* x *radius* d *in* $\mathbb{R}^n$ *and* $|\cdot|$ *denoting volume in* $\mathbb{R}^n$.

(ii) *If* $p = n > 1$ *then* $H_0^{1,p}(\Omega) \subset E_\Phi(\Omega)$ *where* Φ *is the Orlicz function*

$$\Phi(t) = |t|^n \exp |t|^{n/n-1}.$$

For $u \in H_0^{1,n}(\Omega)$ *there exists a constant* K, *independent of* u, *such that*

$$\int_\Omega \Phi(\mu |u(x)|)dx \leq \eta \tag{2}$$

with

$$\frac{1}{\mu} = K \|u\|_{1,p} .$$

In particular

$$\|u\|_\Phi \leq K \|u\|_{1,p} . \tag{3}$$

(iii) *If* $p > n$, $H_0^{1,p}(\Omega) \subset L^q(\Omega) \cap L^\infty(\Omega) \cap C(\Omega)$ *for any* $q \geq p$. *For any* $u \in H_0^{1,p}(\Omega)$

$$\|u\|_{0,q} \leq (\omega_n d^n)^{1/q-1/p} \eta^{1/q - 1/p + 1/n} n^{-1/q} \left(\frac{p - 1}{p - n}\right)^{1-1/p} (\|u\|_{0,p} + d\,\|Du\|_{0,p}) \tag{4}$$

$$|u(x)| \leq (\omega_n d^n)^{-1/p} \left(\frac{p - 1}{p - n}\right)^{1-1/p} \eta(x)^{1/n-1/p} (\|u\|_{0,p} + d\,\|Du\|_{0,p}) \tag{5}$$

REMARK. The method in (i) does not allow $1/q = 1/p - 1/n$. However it is well known that $H_0^{1,p}(\Omega)$ is continuously embedded in $L^q(\Omega)$ in this case also.

PROOF. The parts of this theorem are special cases of Theorems 3.3, 3.5 and 5.1 in [7] where similar results for multiplication maps and weighted Sobolev spaces are obtained. We give only a brief outline of the proof, concentrating mainly on the small changes that need to be made to the general results in [7].

The starting point for all the parts is the inequality

$$|u(x)| \leq (\omega_n d)^{-1} \int_{\Omega\cap B(x,d)} (|u(y)| + d|Du(y)|)|x-y|^{1-n}\,dy \tag{6}$$

for any $u \in C_0^1(\Omega)$. Using a generalisation of Holder's inequality it then follows that for some $\beta > 0$ and with $v(y) = |u(y)| + d|Du(y)|$,

$$d\omega_n |u(x)| \leq \left\{ \int_{\Omega\cap B(x,d)} v^p(y)|y-x|^{\beta r(1-n)}dy \right\}^{1/q}$$

$$\left\{ \int_{\Omega\cap B(x,d)} |y-x|^{(\beta r+(1-\beta)\mu)(1-n)}dy \right\}^{1/\mu} \left\{ \int_{\Omega\cap B(x,d)} v_p(y)dy \right\}^{1/\nu} \tag{7}$$

where

$$\frac{1}{\mu} = 1 - \frac{1}{p}, \quad \frac{1}{\nu} = \frac{1}{p} - \frac{1}{q}, \quad \frac{1}{r} = 1 + \frac{1}{q} - \frac{1}{p}.$$

(i) If $p \leq n$ we put $\beta(1-n) = \alpha - n$ where $\alpha > 0$ is required to satisfy $\frac{n-\alpha}{q} + 1 - \frac{n}{p} > 0$ in order for the second integral on the right handside to be finite. These substitutions lead to the inequality

$$\|u\|_{0,q} \leq (\omega_n d^n)^{1/q-1/p}\, \eta^{1/q-1/p+1/n}\, \alpha^{-1/q} \left(\frac{1-1/p}{(n-\alpha)/q+1-n/p}\right)^{1-1/p} \|v\|_{0,p}.$$

If we now minimise the constant on the right hand side (as a function of α) we obtain (i) for $p \leq n$.

(ii) When $p = n > 1$, (1) becomes for $q \geq n$,

$$\|u\|_{0,q}^q \leq (\omega_n d^n)^{1-q/p} \, \eta \left(\frac{q(1 - 1/n) + 1}{n}\right)^{q(1-1/n)+1} \|v\|_{0,p}^q$$

$$\leq \eta \, K^q \{q(1 - 1/n) + 1\}^{q(1-1/n)+1} \|v\|_{0,p}^q$$

where the constant K does not depend on q. Putting $q(1 - 1/n) + 1 = k + n$ and summing over k, $k \geq 0$, we get (see [7] for details)

$$\int_\Omega \Phi(\mu |u(x)|)dx \leq \frac{\eta}{\{K\mu\|v\|_{0,p}\}^{n/n-1}} \sum_{k=0}^{\infty} \frac{1}{k!} \{(K\mu\|v\|_{0,p})^{n/n-1}(k + n)\}^{k+n} .$$

Putting $(K\mu\|v\|_{0,p})^{n/n-1} = A\epsilon$ where $A < 1/e$ and $\epsilon < 1$, so that the series on the right is convergent, we obtain (2) by choosing ϵ to be small enough.

(iii) When $p > n$, we choose $\beta = 0$ in (7) and this leads to (4). Also, (5) is obtained from (6) by a straightforward use of the Hölder inequality. Note that in (iii) any $u \in H_0^{1,p}(\Omega)$ can in fact be identified with a function in the space $C^{1-n/p}(\Omega)$ of Holder continuous functions with exponent $1 - n/p$, as is well known.

As illustrations of the usefulness of Theorem 1 we mention a few simple consequences. For simplicity we restrict ourselves to simple special cases.

COROLLARY 2. _If_ E _is the embedding_ $H_0^{1,p}(\Omega) \subset L^p(\Omega)$, $1 < p < n$, _and_ $d \leq 1$ _then_

$$\|E\| \leq \eta^{1/n}.$$

_If_ $\eta_\infty = \limsup_{\substack{|x|\to\infty \\ x \in \Omega}} \eta(x)$ _then_ E _is a_ k-_set contraction for some_

$$k \leq \eta_\infty^{1/n} .$$

Of particular interest in Corollary 2 are the cases when $\eta < 1$ and $\eta_\infty < 1$. Clearly $\eta < 1$ if Ω is like a cylinder of radius $r < d$ (≤ 1) and also $\eta_\infty < 1$ if Ω is of this form at infinity. In particular, E is a compact map (a 0-set contraction) if $\eta_\infty = 0$. However $\eta_\infty = 0$ is not necessary for E to be compact; for a necessary and sufficient condition see [1].

PROOF. The first part is an immediate consequence of (1). For the second part we put

$$Eu(x) = E\theta_R(x)u(x) + E(1 - \theta_R(x))u(x)$$

where θ_R is a C_0^∞ function satisfying $0 \leq \theta_R(x) \leq 1$ and

$$\theta_R(x) = \begin{cases} 1 & \text{for } x \in B(0,R), \\ 0 & \text{for } x \notin B(0,2R). \end{cases}$$

The map $u \to E\theta_R u$ is compact since the embedding $H_0^{1,p}(\Omega \cap B(0, 2R)) \subset L^p(\Omega \cap B(0, 2R))$ is known to be compact. If we now replace u in (1) by $(1 - \theta_R)u$, in which case we can replace η by

$$\eta_R = \sup_{\substack{x \in \Omega \\ |x| \geq R}} \eta(x)$$

it follows that for any given $\epsilon > 0$ we can choose an R sufficiently large that

$$\|E(1 - \theta_R)\| \leq (\eta + \epsilon)^{1/n}.$$

The result therefore follows.

Similar results follow for the cases $p \geq n$. We mention in particular the following result when $p = n$.

COROLLARY 3. <u>Let</u> Ψ <u>be an Orlicz function which satisfies the conditions</u>:

$$\lim_{t \to 0} \frac{\Phi(\lambda t)}{\Psi(t)} = \infty, \quad \lim_{t \to \infty} \frac{\Phi(\lambda t)}{\Psi(t)} = \infty \quad \underline{\text{for every}}\ \lambda > 0, \ \underline{\text{and suppose that}}\ \eta_\infty = 0.$$

<u>Then the natural embedding</u> $H_0^{1,p}(\Omega \subset E_\Psi(\Omega)$ <u>is compact</u>.

PROOF. The proof again involves writing the embedding E as $E = E\theta_R + E(1 - \theta_R)$. It is then a question of proving that the first map is compact and then showing that $\|E(1 - \theta_R)\| \to 0$ as $R \to \infty$. For the details see Corollary 3.7 in [7].

3. The spaces $H^{1,p}(\Omega)$, $p > 1$.

When Ω is bounded it is necessary that $\partial\Omega$ satisfies some smoothness condition in order that the embedding and compactness theorems of §2 should remain true for the spaces $H^{1,p}(\Omega)$. One such condition is the cone condition which asserts that for any $x \in \Omega$ there exists a cone $k_\Omega(x) \subset \Omega$ with vertex x which is congruent to some fixed cone k_Ω. In this section we shall obtain results similar to those in §2 for $H^{1,p}(\Omega)$ with Ω unbounded under a condition which is similar to the cone condition but in fact weaker than it.

For $x \in \Omega$ let $\Gamma(x,d)$ be the set of points $y \in \Omega \cap B(x,d)$ which are such that the line $\alpha x + (1-\alpha)y$, $0 \leq \alpha \leq 1$, lies in Ω. Writing $y = x + r\xi$, where $r = |y - x|$ and ξ is a unit vector, let the line from x through y in the direction of ξ meet $\partial(\Omega \cap B(x,d))$ at a point distant $\rho(x,\xi)$ from x. If there exist positive numbers d, δ such that $|\Gamma(x,d)| \geq \delta$ for all $x \in \Omega$ then clearly Ω satisfies a cone condition. However we do not need as much as this below.

The results for $H^{1,p}(\Omega)$ follow in much the same way as those for $H_0^{1,p}(\Omega)$ except that in place of (6) we start with the estimate

$$|u(x) - u_\Gamma(x)| \leq \frac{1}{n|\Gamma(x,d)|}\int_{\Gamma(x,d)} |x - y|^{1-n}\rho^n(x,\xi)|Du(y)|dy \tag{8}$$

where

$$u_\Gamma(x) = \frac{1}{|\Gamma(x,d)|}\int_{\Gamma(x,d)} u(y)dy$$

for $u \in C^1(\overline{\Omega})$ (see [7] Lemma 4.1). The basic result is then

THEOREM 4. Let $1/p \geq 1/q > 1/p - 1/n$ and suppose that

$$T_{\alpha,q} = \sup_{y\in\Omega}\left\{\int_{\Gamma(y,d)} |\Gamma(x,d)|^{-q/p}|x - y|^{\alpha-n}dx\right\}^{1/q} < \infty \tag{9}$$

<u>for some</u> α <u>satisfying</u> $\alpha < q(1 - n/p + n/q)$ <u>if</u> $p \leq n$ <u>and</u> $\alpha = n$ <u>if</u> $p > n$. <u>Then</u> $H^{1,p}(\Omega)$ <u>is continuously embedded in</u> $L^q(\Omega)$ <u>and for all</u> u <u>in</u> $H^{1,p}(\Omega)$,

$$\|u - u_\Gamma\|_{0,q} \leq n^{-1/p} \left(\frac{1 - 1/p}{(n-\alpha)/q + 1 - n/p}\right)^{1-1/p} d^{(n-\alpha)/q+1} T_{\alpha,q} \|Du\|_{0,p} \tag{10}$$

From Theorem 4 results analogous to parts (ii) and (iii) of Theorem 1 follow for the special cases $p = n$ and $p > n$. For instance, if $\sup_{q \geq n} T_{\alpha,q} < \infty$ for some $0 < \alpha < n$, $H^{1,n}(\Omega)$ is continuously embedded in the Orlicz space $E_\Phi(\Omega)$ where $\Phi(t) = |t|^n \exp |t|^{n/n-1}$. It is worth digressing at this point to compare the latter result, and that of Theorem 1 (iii), with Trudinger's result [10] for a bounded domain Ω, that $H^{1,n}(\Omega)$ is continuously embedded in $E_{\Phi_0}(\Omega)$ where $\Phi_0(t) = \exp |t|^{n/n-1} - 1$. For large t and $\lambda > 1$

$$\Phi_0(t) \leq \Phi(t) \leq \Phi_0(\lambda t) \tag{11}$$

i.e. $\Phi \sim \Phi_0$ and so for a bounded domain Ω both Φ and Φ_0 give rise to the same Orlicz space $L_\Phi(\Omega)$. However this is not necessarily so for an unbounded Ω as (11) clearly does not hold for all values of t. It can easily be proved that Trudinger's result does not hold for unbounded Ω. For if $\Omega = \mathbb{R}^n \setminus B(0, \delta)$ and $u(x) = |x|^{-\alpha}$ for $1 < \alpha \leq n$, then it is easily seen that $u \in H^{1,n}(\Omega)$ but $u \notin L_{\Phi_0}(\Omega)$. In fact our Orlicz function Φ is best possible in both the exponential term and the power of $|t|$ (see [7] and [10]).

The compactness of the embedding map $H^{1,p}(\Omega) \subset L^q(\Omega)$ is a rare phenomenon for unbounded Ω as was demonstrated by Adams and Fournier in [2]. They proved that for the map to be compact it is necessary that $\lim_{R\to\infty} e^{\lambda R}|\Omega \setminus B(0,R)| = 0$ for all $\lambda > 0$. Using (10) and the same technique as that in the proof of Corollary 2 we can obtain results on the compactness, and more generally the measure of non-compactness of the embeddings. We mention only the following simple case.

THEOREM 5. *Suppose* $\Omega \cap B(0,R)$ *satisfies the cone condition for all* $R > 0$. *Let* $p > n$ *and suppose that* $d \leq n^{1/p}(\frac{p-n}{p-1})^{1-1/p}$ *and*

$\sup_{y \in \Omega} \int_{\Gamma(y,d)} |\Gamma(x,d)|^{-1} dx < \infty$. *Then the natural embedding* $H^{1,p}(\Omega) \subset L^p(\Omega)$

is a k-set contraction for some

$$k \leq k_0 \equiv \limsup_{|y| \to \infty} \int_{\Gamma(y,d)} |\Gamma(x,d)|^{-1} dx.$$

As an example take $n = 2$ and let Ω be the domain in the first quadrant bounded by the x-axis and the curve $y = f(x)$. If $f(x)$ is taken to be (i) $e^{-x/d}$, (ii) $e^{-x^{\alpha}}$, $\alpha > 1$, then it can be shown (see [7]) that k_0 is respectively < 1 and 0 thus giving examples of embedding maps which are respectively k-set contractions for some $k < 1$ and compact.

4. Orlicz-Sobolev spaces.

In [5] embedding theorems are obtained for general Orlicz-Sobolev spaces $H^1L_B(\Omega)$ (they use the notation $W^1L_B(\Omega)$) when Ω is a bounded domain in $\mathbb{R}^n$. They relax the cone condition on Ω, which is usually required for such results, and require only that the Sobolev Embedding Theorem holds for $p = 1$ and $q = n/n-1$, i.e. $H^{1,1}(\Omega)$ is continuously embedded in $L^{n/n-1}(\Omega)$. Such a domain Ω is said to be admissible. In fact a consequence of their result is that if Ω is admissible then the Sobolev Embedding Theorem $H^{1,p}(\Omega) \subset L^q(\Omega)$ holds for all p, q satisfying $1/p - 1/n \leq 1/q \leq 1/p$. A typical result is the following.

Let B be an Orlicz function and define

$$g_B(t) = B^{-1}(t)/t^{1+1/n}, \quad t \geq 0.$$

Assume that $\int_1^{\infty} g_B(t)dt = \infty$ and define the Orlicz function B* by

$$(B^*)^{-1}(|x|) = \int_0^{|x|} g_B(t)dt.$$

Recall that equivalent Orlicz functions give rise to the same Orlicz space and hence, when Ω is bounded, we can suppose that B is such that

$$\int_0^1 g_B(t)dt < \infty.$$

Note that when Ω is unbounded then such a condition would have to be assumed to hold for the reasons mentioned in §1.

THEOREM 6. _Let_ Ω _be a bounded admissible domain in_ $\mathbb{R}^n$.

(i) _If_ $\int_1^\infty g_B(t)dt = \infty$ _then_ $H^1L_B(\Omega)$ _is continuously embedded in_ $L_{B*}(\Omega)$. _If_ C _is an Orlicz function which satisfies the condition_

$$\lim_{x\to\infty} \frac{B^*(\lambda x)}{C(x)} = \infty$$

for every $\lambda > 0$ _then the embedding_ $H^1L_B(\Omega) \subset L_C(\Omega)$ _is compact_.

(ii) _If_ $\int_1^\infty g_B(t)dt < \infty$ _then_ $H^1L_B(\Omega) \subset L_\infty(\Omega) \cap C(\Omega)$.

When $B(t) = |t|^p$, $1 < p < n$ we have

$$B^*(t) = \frac{|t|^{p^*}}{p^*}, \quad \frac{1}{p^*} = \frac{1}{p} - \frac{1}{n},$$

and hence Theorem 6 gives the usual Sobolev Embedding Theorem for $H^{1,p}(\Omega)$ when Ω is bounded. Hence we have the result noted above that if the Sobolev Theorem holds for $p = 1$ it holds in general.

When $B(t) = |t|^n$ then

$$B^*(t) \sim e^{|t|} - |t| - 1$$

so that $H^{1,n}(\Omega) \subset L_{\Phi_1}(\Omega)$ where $\Phi_1(t) = e^{|t|} - |t| - 1$. This is clearly a weaker result than that of Trudinger's discusses in §3.

The above result, together with the other results in [5], and also those in [7] and [10] (the results in [7] being for an arbitrary Ω) have proved to be useful tools in the study of elliptic equations with strong non-linear terms. For details we refer the reader to [4], [7], [8] and [10] where a comprehensive bibliography is given.

REFERENCES.

1. R. A. Adams, Capacity and compact imbeddings, J. Math. Mech. 19 (10) (1970), 923-929.

2. R. A. Adams and J. Fournier, Some imbedding theorems for Sobolev spaces, Can. J. Math. XXIII, (3) (1971), 517-530.

3. M. S. Berger and M. Schechter, Embedding theorems and quasilinear boundary value problems for unbounded domains, Trans. Am. Math. Soc. 172 (1973), 261.

4. T. Donaldson, Nonlinear elliptic boundary-value problems in Orlicz-Sobolev spaces, J. Diff. Equat. 10 (1971), 507-528.

5. T. Donaldson and N. S. Trudinger, Orlicz-Sobolev spaces and imbedding theorems, J. Funct. Anal. 8 (1971), 52-75.

6. D. E. Edmunds and W. D. Evans, Elliptic and degenerate elliptic operators in unbounded domains (to appear).

7. D. E. Edmunds and W. D. Evans, Sobolev spaces and Orlicz spaces in unbounded domains (to appear).

8. J. P. Gosset, Nonlinear elliptic boundary-value problems for equations with rapidly increasing coefficients (to appear).

9. M. Krasnoselskii and Y. B. Rutickii, Convex functions and Orlicz spaces, Noordhoff (Groningen, 1961).

10. N. S. Trudinger, On imbeddings into Orlicz spaces and some applications, J. Math. Mech. 17 (1967) 473-484.

Integral inequalities and spectral theory

W. N. Everitt

Introduction

This paper is in two sections. The first section describes certain quadratic and other inequalities given in Chapter VII of *Inequalities* by Hardy, Littlewood and Polya (Cambridge, 1934) and gives some details of generalisations and recent results in this area. The second section is concerned with certain quadratic inequalities arising from formally symmetric ordinary and partial differential expressions.

Notations

R and C denote the real and complex number fields respectively; R_+ denotes the set of all non-negative real numbers. Open, closed and half-open intervals of R are denoted by (a, b), [a, b] and [a, b) respectively; I denotes an arbitrary interval of R. Lebesgue integration spaces on intervals of R are denoted by $L^k(a, b)$; in all cases $k \geq 1$; the usual norm in $L^k(a, b)$ will be denoted by $\|\cdot\|_k$, with the subscript k omitted when k = 2. A property is 'local' on an interval I of R if it is satisfied on every compact sub-interval of I. AC denotes absolute continuity; $AC_{loc}(I)$ local absolute continuity on I. A prime ' denotes differentiation with respect to an independent variable on R. The symbol '$(f \in D)$' is to be read as 'for all f belonging to D'.

References [1], [10, Chapter VII] etc. refer to the list of references at the end of each section. A reference of the form [24, ...] is to the book *Analytic Inequalities* by Mitrinovic.

Inequalities

In general the inequalities in this report are comparable functions, or functionals, in a suitably extended sense of that given in section 1.6 of *Inequalities* by Hardy, Littlewood and Polya.

Many, but not all, of the inequalities in this report are of the following kind. I is an interval of R; D is a linear manifold of functions with typical element f, say, where $f : I \to R$ or C; F and G are functionals, in general not linear, which map D into R_+ in such a way that

$$F(0) = G(0) = 0, \tag{1}$$

i.e. the null function is mapped onto zero. The inequality then takes the form

$$F(f) \leq KG(f) \qquad (f \in D) \tag{*}$$

where K is independent of f and satisfies $0 \leq K \leq \infty$.

If (*) holds for a finite K, i.e. $K \in R_+$, then it is said to be a valid inequality on D. When this holds it is assumed that K is the best possible, i.e. the smallest, number for which (*) holds; explicitly this means that both the following conditions are satisfied:

$$F(f) = 0 \qquad (f \in D \text{ with } G(f) = 0) \tag{2}$$

<u>and</u>

$$K \stackrel{\text{def}}{=} \sup\{F(f)/G(f) : \text{all } f \in D \text{ with } F(f) > 0\} < \infty. \tag{3}$$

In these circumstances a case of equality for (*) is an element $f \in D$ for which $F(f) = KG(f)$; it is clear from (1) that the null function is always a case of equality; but there may be other elements of D also in this category.

Alternatively if (*) does not hold for any $K \in R_+$ then $K = \infty$ and this is taken to imply that:

<u>either</u> there is an element of $f \in D$, with $f \not\equiv 0$, such that

$$F(f) > 0 \text{ and } G(f) = 0 \tag{4}$$

<u>or</u> (2) above holds but

$$\sup\{F(f)/G(f) : \text{all } f \in D \text{ with } G(f) > 0\} = \infty. \tag{5}$$

In such circumstances (*) is not a valid inequality on D.

Section 1

A number of important integral inequalities involving derivatives in the integrand, are to be found in the chapter devoted to some applications of the calculus of variations in the book Inequalities by Hardy, Littlewood and Polya; see [10, Chapter VIII]. This section of the report discusses a number of these inequalities, in particular those inequalities which in one form or another are concerned with problems in linear analysis.

The following six inequalities are all taken from [10]; many of the results however stem from the earlier paper [9] of Hardy and Littlewood. All these inequalities are of the type (*) as given on page 2 above; all are best possible in the sense there explained.

1. From [10, Theorem 253]: $f \in D$ if $f : [0,\infty) \to R$, $f \in AC_{loc}[0,\infty)$, $f(0) = 0$ and $f' \in L^2(0,\infty)$; then $x^{-1}f \in L^2(0,\infty)$ and

$$(1) \qquad \int_0^\infty x^{-2}f(x)^2\,dx \leq 4\int_0^\infty f'(x)^2\,dx \qquad (f \in D)$$

with equality only when f is null on $[0,\infty)$. (Note: this is a typical example of (*) on page 2 above with

$$F(f) = \int_0^\infty x^{-2}f(x)^2\,dx, \quad G(f) = \int_0^\infty f'(x)^2\,dx \quad (f \in D) \text{ and } K = 4.)$$

2. From [10, Theorem 259]: $f \in D$ if $f : [0,\infty) \to R$, $f' \in AC_{loc}[0,\infty)$, f and $f'' \in L^2(0,\infty)$; then $f' \in L^2(0,\infty)$ and

$$(2) \qquad \left\{\int_0^\infty f'(x)^2\,dx\right\}^2 \leq 4\int_0^\infty f(x)^2\,dx \int_0^\infty f''(x)^2\,dx \qquad (f \in D)$$

with equality if and only if for some $A \in R$ and some $\rho > 0$

$$f(x) = AY(\rho x) \qquad (x \in [0,\infty))$$

where

$$Y(x) = \exp[-\tfrac{1}{2}x]\sin\{\tfrac{1}{2}x\sqrt{3} - \pi/3\} \qquad (x \in [0,\infty)). \tag{3}$$

(Note: this is an example of (*) on page 2 above, with $K = 4$, for which there are non-null cases of equality.)

3. From [10, Theorem 261]: $f \in D$ if $f : (-\infty,\infty) \to R$, $f' \in AC_{loc}(-\infty,\infty)$, f and $f'' \in L^2(-\infty, \infty)$; then $f' \in L^2(-\infty,\infty)$ and

$$\left\{ \int_{-\infty}^{\infty} f'(x)^2 dx \right\}^2 \leqslant \int_{-\infty}^{\infty} f(x)^2 dx \int_{-\infty}^{\infty} f''(x)^2 dx \qquad (f \in D) \tag{4}$$

with equality only when f is null on $(-\infty, \infty)$.

4. From [10, Theorem 168]: $f \in D$ if $f : [0,\infty)$ (or $(-\infty,\infty)$) $\to R$, $f' \in AC_{loc}[0,\infty)$ (or $(-\infty,\infty)$), for some $k \geqslant 1$ both f and $f'' \in L^k(0,\infty)$ (or $(L^k(-\infty,\infty))$; then $f' \in L^k(0,\infty)$ (or $L^k(-\infty,\infty)$) and for all $f \in D$

$$\left\{ \int_0^{\infty} |f'(x)|^k dx \right\}^2 \leqslant K(k) \int_0^{\infty} |f(x)|^k dx \int_0^{\infty} |f''(x)|^k dx \tag{5}$$

(or the integrals over $(-\infty, \infty)$) with $1 \leqslant K(k) < \infty$ for all $k \geqslant 1$.

There seems to be no known simple characterisation of the number $K(k)$, beyond the formal definition (3) on page 2 above, except that $K(2) = 4$, since (5) reduces to inequality (2) when $k = 2$.

5. From [10, Theorem 269]: let $k > 1$ and $l > 1$ be conjugate indices, i.e. $k^{-1} + l^{-1} = 1$, let $f \in D$ if $f : (-\infty,\infty) \to R$, $f' \in AC_{loc}(-\infty,\infty)$, $f \in L^k(-\infty,\infty)$, $f'' \in L^l(-\infty,\infty)$; then $f' \in L^2(-\infty,\infty)$ and

$$\int_{-\infty}^{\infty} |f'(x)|^2 dx \leqslant \left\{ \int_{-\infty}^{\infty} |f(x)|^k dx \right\}^{1/k} \left\{ \int_{-\infty}^{\infty} |f''(x)|^l dx \right\}^{1/l} \qquad (f \in D) \tag{7}$$

with equality only when f is null on $(-\infty, \infty)$.

6. From [10, Theorems 272 and 226]: $f \in D$ if $f : [0,\infty)$ (or $(-\infty,\infty)) \to R$, $f \in AC_{loc}[0,\infty)$ (or $-\infty,\infty$)), xf and f' both $\in L^2(0,\infty)$ (or $L^2(-\infty,\infty)$); then $f \in L^2(0,\infty)$ (or $L^2(-\infty,\infty)$) and

$$\text{(8)} \qquad \left\{ \int_0^\infty f(x)^2 dx \right\}^2 \leq 4\int_0^\infty x^2 f(x)^2 dx \int_0^\infty f'(x)^2 dx \qquad (f \in D)$$

with equality only when for some $A \in R$ and some $\rho > 0$

$$f(x) = A \exp[-\rho x^2] \qquad (x \in [0,\infty))$$

(or similarly on $(-\infty,\infty)$).

Note that following [10, Chapter VIII] the above six inequalities are stated for real-valued functions defined on $[0,\infty)$ or $(-\infty,\infty)$. However the methods of proof given in [10] show that all these inequalities are equally valid when f takes values in the complex field C provided such terms as f^2 are replaced by $|f|^2$, and so on for derivatives.

Work on extensions and generalisations of these inequalities is reported on below. It should be noted that nearly all of these extensions are dependent upon, in one form or another, not only on the calculus of variations but on the spectral theory of differential operators generated by certain ordinary differential expressions. Indeed the original work of Hardy and Littlewood in [9] is remarkable not only for the significance of the results, and for the elegance and classification of the methods of proof, but also because the results exhibit now a link between the calculus of variations and the spectral theory of differential operators. In 1932 it would have been difficult, if not impossible, to see the importance of spectral theory in these results since at that time much was undiscovered which is now known. Nearly all the best possible constants in the L^2 inequalities above are now seen to be determined by properties of the spectrum of certain associated differential operators. Perhaps this is best seen in the many results available in the papers [1], [5], [8], [15], [17], [18] and [19] given in the list of references at the end of this section.

A. Comments on part 1 above

An alternative discussion of the inequality (1) may be found in the paper by Putnam, see [19, section 6]. This type of result stems from a class of inequalities of the form

$$(9) \qquad \int_a^b \{p(x)|f'(x)|^2 + q(x)|f(x)|^2\}dx \geq \mu_0 \int_a^b |f(x)|^2 dx. \quad (f \in D)$$

Basically the coefficients p and q satisfy; $p, q : (a, b) \to R$, with $p \in AC_{loc}(a, b)$, $p > 0$ on (a, b), p' and $q \in L^2_{loc}(a, b)$; $f \in D$ if $f : (a, b) \to C$, $f \in AC_{loc}(a, b)$ and all of $p^{\frac{1}{2}}f'$, $|q|^{\frac{1}{2}}f$, $f \in L^2(a, b)$. The number $\mu_0 \in R$ and is well-determined as a point in the spectrum of a differential operator, generated from the coefficients p and q, in $L^2(a, b)$. (Note: it is possible to recast (9) in a form which then puts it into the class of inequalities given by (*) on page 2; however this disguises the importance of the number μ_0 in the inequality.)

Early work in the regular case of (9), i.e. when the conditions on p and q are satisfied on a compact interval $[a, b]$, is due to Lichtenstein [17]; this was followed in both the regular and singular cases by results of Courant and Hilbert [4, Volume I, Chapter 6]. In the singular case on $[0, \infty)$, with $p(x) = 1$ $(x \in [0, \infty))$, extensive results are given by Putnam in [19]; in particular inequality (1) is discussed in [19, Section 6] and inequality (8) in [19, Section 5].

Both the regular and singular cases of (9), for general coefficients p and q are discussed by Bradley and Everitt in [1]. Two examples from [1, Section 7] are:

(i) let $a = 1$, $b = \infty$ and let $\tau \in (0, \infty)$; let $p(x) = x^{\tau}$ $(x \in [1, \infty))$ and $q(x) = 0$ $(x \in [1, \infty))$; then

$$\int_1^{\infty} x^{\tau} |f'(x)|^2 dx \geq \mu_0(\tau) \int_1^{\infty} |f(x)|^2 dx \qquad (f \in D(\tau))$$

where $\mu_0(\tau) = 0$ $(\tau \in (0, 2))$, $\mu_0(2) = \frac{1}{4}$, $\mu_0(\tau) > \frac{1}{4}$ and is strictly increasing on $(2, \infty)$; here $D(\tau)$ is a linear manifold of $L^2(1, \infty)$ for which the left-hand integral is finite. There is equality for $\tau \in (0, 2]$ only when f is null on $[1, \infty)$, and for $\tau \in (2, \infty)$ only when f is a multiple of an eigenvector of a well-determined differential operator in $L^2(1, \infty)$.

(ii) let $a = 0$, $b = \infty$ and $p(x) = 1$ $(x \in [0, \infty))$; let $q \in L(0, \infty)$; then

$$\int_0^{\infty} \{ |f'(x)|^2 + q(x) |f(x)|^2 \} dx \geq - \left\{ \int_0^{\infty} |q(x)| dx \right\}^2 \int_0^{\infty} |f(x)|^2 dx \quad (f \in D)$$

where D is a linear manifold of $L^2(0, \infty)$ for which both integrals on the left-hand side are defined and finite. Nothing is known about cases of equality, nor if this inequality is best possible when taken over all $q \in L(0, \infty)$.

B. Comments on parts 2, 3, 4 above

As mentioned in [10, Section 7.8] the inequality (2) was originally suggested by the Laudau inequality discussed in [14, Section 2.22].

The possibility of a generalisation of inequalities (2), (4) and (5) to higher order derivatives seems to have been first noted by Halperin, see [8]. Using the norm notation $\|\cdot\|_k$ in $L^k(0, \infty)$ (or $L^k(-\infty, \infty)$) this takes the form given by Hille in [11, Pages 20-32, Sections 1 and 4]

$$(10) \qquad \|f^{(r)}\|_k \leq C_{n,r} \|f\|_k^{(n-r)/n} \|f^{(n)}\|_k^{r/n}$$

for integers r and n satisfying $0 < r < n$, and for suitable functions $f \in L^k$ with $f^{(n)} \in L^k$. Inequalities of type (10) also appear as special cases of the general results of Ljubič in [18]; see in particular [18, Theorem 3] where an upper estimate is given for $C_{n,r}$. (Note: the results of Ljubič, which are dependent on the spectral theory of operators in Banack and Hilbert spaces, seem to have been largely overlooked by other workers in this field.)

An important generalisation of inequality (5) is due to Kallman and Rota; see [14, Page 197] where it is shown, for a wide class of Banach function spaces, that

$$\|f'\|^2 \leq 4\|f\|\|f''\| \tag{11}$$

for vectors f in a suitably chosen linear manifold. The proof of this inequality requires certain properties of the infinitesimal generator of continuous semi-groups of contraction operators. This result shows that in (5) the positive number $K(k)$ satisfies $K(k) \leq 4^k$ $(k \in (1,\infty))$. The Kallman-Rota inequality has been extensively discussed by several authors including Hille in [11, Pages 20-32] and [12], Kurepa in [16], Trebels and Westphall in [20, Pages 115-119], Gindler and Goldstein in [23]. All these results include the inequality (5) in the L^k spaces as a special case. Recent results on an upper bound for $K(k)$ are to be found in the paper [13] of Holbrook. The work of Gindler and Goldstein contains much interesting information about bounds for $K(k)$; in particular they conjecture, see [23, Section 5], that $K(k) = 2^{4-4/k}$ for $2 \leq k \leq \infty$, and $K(k) = 2^{4/k}$ for $1 \leq k \leq 2$.

Trebels and Westphall in [20] extended the Hille inequality (10) to the case when the numbers r and n may take arbitrary real values in the range $0 < r < n$. As is made clear by the authors in [20] the results do not give precise values for $C_{n,r}$ nor cases of equality, but this is due to the nature of the problem rather than the analysis itself.

With respect to this last statement the result of Kato is exceptional. Kato in [15] considers the Kallman-Rota inequality in the special case when the Banach space is a Hilbert space. In these circumstances Kato showed that the constant 4 in (11) can be replaced by 2 and also gives necessary and sufficient conditions for equality. In particular the Kato analysis gives a new proof of the Hardy-Littlewood inequality (2), together with all the cases of equality given in (3).

In the special case of (10) when $k = 2$ and the interval is $(0, \infty)$, the value of $C_{n,r}$ may be determined, at least theoretically, from the general analysis (which in fact holds for an arbitrary Hilbert space) of Ljubič [18, Theorem 6]; also the cases of equality are characterised in [18, Theorem 8]. As pointed out by Ljubič the method used in this part of [18] depends on the idea of quadratic functionals first employed by Hardy and Littlewood in [9].

When $r = 1$, $n = 2$ the Ljubič analysis gives another proof of the Hardy-Littlewood inequality (2), together with all the cases of equality. The analysis becomes harder to apply when n increases and in the case $n = 4$ an alternative approach, again based on the idea of quadratic functionals in [9], is given by Bradley and Everitt in [2]; here $C_{4,2}$ is characterised as a zero of an algebraic function which gives the bounds $2.96 < C_{4,2} < 2.99$. The analysis in [2] also yields all the cases of equality; these are similar to (3) for the inequality (2). With similar analysis, reported on briefly in [2], Bradley and Everitt have shown that $2.25 < C_{4,1} = C_{4,3} < 2.28$. For the case $n = 3$ Ljubič has shown, see [18, Page 75], that $C_{3,1} = C_{3,2} = 3^{\frac{1}{2}}[2(2^{\frac{1}{2}} - 1)]^{-\frac{1}{3}}$; this has an approximate value of 1.84. In this way it may be said that the inequality (10) has been fully analysed in the space $L^2(0, \infty)$ for the cases $0 < r < n \leqslant 4$.

Inequality (10) in the case when the interval is $(-\infty, \infty)$ and $k = 2$ provides an extension of inequality (4) to higher derivatives. Here again the analysis of Ljubič in [18] applies and shows that $C_{n,r} = 1$ for all integers r and n satisfying $0 < r < n$; also that there are no cases of equality except the null function on $(-\infty, \infty)$.

An extension of inequaltiy (2), in a different direction to the Ljubič-Kallman-Rota-Hille type inequality, is given by Everitt in [5]. Given $a \in R$ and two coefficients $p, q : [a, \infty) \to R$ such that $p \in AC_{loc}[a, \infty)$, p is positive and q bounded below on $[a,\infty)$, the inequality considered in [5] is

$$(12)\quad \left\{\int_a^\infty \{p(x)\,|f'(x)|^2 + q(x)\,|f(x)|^2\,dx\right\}^2 \leq K(p,\,q)\int_a^\infty |f(x)|^2\,dx \int_a^\infty |M[f](x)|^2\,dx \qquad (f \in D)$$

where

$$M[f](x) = -\,(p(x)f'(x))' + q(x)f(x) \qquad (x \in [a,\infty))$$

and D is the maximal linear manifold of all $f : [a,\infty) \to C$ for which $M[f]$ is defined on $[a,\infty)$ and $f,\ M[f] \in L^2(a,\infty)$. It is known that the conditions on p and q imply that $p^{\frac{1}{2}}f'$ and $|q|^{\frac{1}{2}}f$ are both in $L^2(a,\infty)$ for all $f \in D$. (Note: the differential expression $M[\cdot]$ is formally symmetric on $[a,\infty)$; this is essential to the analysis of inequality (12).)

Inequality (12) is an example of the general inequality (*) on page 2 above. $K(p,\,q)$ is independent of f but depends on the coefficients p and q. All the remarks on pages 2 and 3 about valid inequalities and cases of equality hold for (12).

It should be noted that the form of the left-hand side of (12), which is a Dirichlet type integral, excludes the possibilty in general of applying the methods of [14], [15] and [18], which all depend on the left-hand side of the inequality being of the form $\|Tf\|^2$ where T is a linear operator. The only case of (12) which falls under these methods is essentially $p(x) = 1$ and $q(x) = 0 \quad (x \in [a,\infty))$ when (12) reduces to (2).

The analysis of the inequality (12) depends on the spectral properties of the differential equation $M[y] = \lambda y$ on $[a,\infty)$ in the space $L^2(a,\infty)$; in particular the analysis depends upon the existence of solutions of this equation in $L^2(a,\infty)$ and the m-coefficient of Weyl; details are given in [5]. Two general results are given here:

(i) in all cases of the coefficient p and q, see [5, Section 13],

$$K(p, q) \geq 4$$

(ii) if $q \geq 0$ but is not null on $[a, \infty)$ then $K(p, q) = \infty$ and there is no valid inequality; [5, Section 15].

The application of the Liouville transformation, for second-order differential expressions, to the inequality (12) is considered by Everitt in [22].

Two examples of (12) are given here; see also [5, Section 17]:

(i) Let $a = 0$, $p(x) = 1$ and $q(x) = \mu$ $(x \in [0, \infty))$ where $\mu \in R$; in this case write $K(p, q) = K(\mu)$; then the inequality (12) takes the form

$$(13) \quad \left\{ \int_0^\infty \{ |f'(x)|^2 + \mu |f(x)|^2 \} dx \right\}^2 \leq K(\mu) \int_0^\infty |f(x)|^2 dx \int_0^\infty |f''(x) - \mu f(x)|^2 dx$$

for all $f \in L^2(0, \infty)$ with $f'' \in L^2(0, \infty)$. Full details of this example are given in the paper by Brodlie and Everitt [3] where it is shown that

$$K(\mu) = 4 \ (\mu \in (-\infty, 0]) \text{ and } K(\mu) = \infty \quad (\mu \in (0, \infty)).$$

When $\mu = 0$ the inequality (13) reduces to (2) and the analysis in [5] gives yet another proof of the Hardy-Littlewood inequality (2), together with all the cases of equality. When $\mu \in (-\infty, 0)$ it is shown in [3] that there is equality in (13) only when f is null on $[0, \infty)$. When $\mu \in (0, \infty)$ there is no valid inequality; in fact (4) on page 3 holds in this case.

(ii) Let $a > 0$ and $\tau \in R$; let $p(x) = x^\tau$ and $q(x) = 0$ $(x \in [a, \infty))$; in this example (12) becomes, writing $K(p, q) = K(\tau)$,

$$(14) \quad \left\{ \int_a^\infty x^\tau |f'(x)|^2 dx \right\}^2 \leq K(\tau) \int_a^\infty |f(x)|^2 dx \int_a^\infty |(x^\tau f'(x))'|^2 dx$$

for all $f \in L^2(a, \infty)$ with $(x^\tau f')' \in L^2(a, \infty)$. The analysis of [5, Section 17] shows that

$$K(\tau) < \infty \ (\tau \in (-\infty, 1)) \text{ and } K(\tau) = \infty \quad (\tau \in [1, \infty)).$$

Recently a complete analysis of the inequality (14) has been made by Everitt and Jones and details will appear in [21]. This has shown that K(•) in (14) may be characterised by

$$\begin{aligned} K(\tau) &= 4 && (\tau \in (-\infty, 0]) \\ &= [\cos\{(3-\tau)^{-1}\pi\}]^{-2} && (\tau \in [0, 1)) \\ &= \infty && (\tau \in [1, \infty)); \end{aligned}$$

these results give the best possible values of K(•). For $\tau \in (-\infty, 0) \cup (0, 1)$ the only case of equality is the null function. For $\tau = 0$ the inequality (14) reduces to the Hardy-Littlewood inequality (2) with the cases of equality given in (3). There is no valid inequality when $\tau \in [1, \infty)$; when $\tau \in [1, 3/2]$ it is interesting to note that the inequality is not valid, not because of (4) of the Introduction but because (5) holds.

Finally in this part mention is made of a 'finite' form of the Hardy-Littlewood inequality (2). From (2) it follows that, for all $\epsilon > 0$,

$$(15) \qquad \int_0^\infty f'(x)^2\,dx \leq \epsilon^{-1}\int_0^\infty f(x)^2\,dx + \epsilon\int_0^\infty f''(x)^2\,dx;$$

conversely if (15) is known to hold for all $\epsilon > 0$ then (2) follows, although without any knowledge of the cases of equality. A finite form of (15) is given in [24, Section 2.23, (10)], i.e. for all $X > 0$ and all $\epsilon > 0$

$$(16) \qquad \int_0^X f'(x)^2\,dx \leq (\epsilon^{-1} + 12X^{-2})\int_0^X f(x)^2\,dx + \epsilon\int_0^X f''(x)^2\,dx;$$

this inequality is best possible. For all $f \in D$, the linear manifold for the inequality (2), the inequality (15) follows from (16) on letting X tend to ∞; inequality (2) then follows, also with the cases of equality.

C. Comments on part 5 above

A discussion of the inequality (7) but on the half-line $[0, \infty)$ instead of $(-\infty, \infty)$ is given by Everitt and Giertz in [7], i.e. if $k > 1$, $l > 1$ with $k^{-1} + l^{-1} = 1$ then

$$(17) \qquad \int_0^\infty |f'(x)|^2 dx \leq K(k, l)\left\{\int_0^\infty |f(x)|^k dx\right\}^{1/k}\left\{\int_0^\infty |f''(x)|^l\right\}^{1/l}$$

for all $f \in L^k(0, \infty)$ with $f'' \in L^l(0, \infty)$. The analysis in [7] shows that $1 \leq K(k, l) < \infty$; numerical upper bounds for $K(k, l)$ are obtained. There is no known simple characterisation of the value of $K(k, l)$ except for the self-conjugate case when $k = l = 2$ and then (17) reduces to the Hardy-Littlewood inequality (2).

D. Comments on part 6 above

There are inequalities of the form (8) for higher derivatives; only one such result is given here due to Everitt [6]. Let $f \in D$ if $f : (0, \infty) \to C$, $f' \in AC_{loc}(0, \infty)$, both $x^{-1}f$ and $xf'' \in L^2(0, \infty)$; then $f' \in L^2(0, \infty)$ and

$$(18) \qquad \left\{\int_0^\infty |f'(x)|^2 dx\right\}^2 \leq \int_0^\infty x^{-2} |f(x)|^2 dx \int_0^\infty x^2 |f''(x)|^2 \, dx \qquad (f \in D)$$

with equality only when f is null.

The inequality (8) is called Weyl's inequality and is given in [24, Section 2.19] where it is derived from a 'finite' form by a limit process. Although this establishes (8) some additional analysis is required to determine the cases of equality.

References

1. Bradley, J. S. and Everitt, W. N. : Inequalities associated with regular and singular problems in the calculus of variations. Trans. Amer. Math. Soc. 182, 303-321 (1973).

2. Bradley, J. S. and Everitt, W. N. : On the inequality $\|f''\|^2 \leqslant K\|f\|\|f^{(4)}\|$. To appear in Quart. J. Math. Oxford Ser. (2).

3. Brodlie, K. W. and Everitt, W. N. : On an inequality of Hardy and Littlewood. To appear in Proc. Royal Soc. Edinburgh (A).

4. Courant, R. and Hilbert D. : Methoden der Mathematischen Physik; Erster Band. Springer, Berlin 1931.

5. Everitt, W. N. : On an extension to an integro-differential inequality of Hardy, Littlewood and Polya. Proc. Royal Soc. Edinburgh (A) 69 295-333 (1972).

6. Everitt, W. N. : Unpublished result of 1973.

7. Everitt, W. N. and Giertz, M. : On the integro-differential inequality $\|f'\|_2^2 \leqslant K\|f\|_p\|f''\|_q$. J. Math. Anal. and Appl. 45, 639-653 (1974).

8. Halperin, I. : Closures and adjoints of linear differential operators. Ann. Math. 38, 880-919 (1937).

9. Hardy, G. H. and Littlewood, J. E. : Some integral inequalities connected with the calculus of variations. Quart. J. Math. Oxford Ser. (2) 3 241-252 (1932).

10. Hardy, G. H., Littlewood, J. E. and Polya, G. : Inequalities. Cambridge, 1934.

11. Hille, E. : Generalizations of Landau's inequality to linear operators. Linear Operators and Approximation (Edited by P L Butzer et al). Birkhäuser Verlag, Basel and Stuttgart, 1972.

12. Hille, E. : On the Landau-Kallman-Rota inequality. J. of Approx. Theory 6, 117-122 (1972).

13. Holbrook, J. A. R. : A Kallman-Rota-Kato inequality for nearly euclidean spaces. To be published.

14. Kallman, R. R. and Rota, G.-C. : On the inequality $\|f'\|^2 \leqslant 4\|f\|\|f''\|$. Inequalities - II (edited by O. Shisha). Academic Press, New York, 1970.

15. Kato, T. : On an inequality of Hardy, Littlewood and Polya. Advances in Math. 7, 217-218 (1971).

16. Kurepa, S. : Remarks on the Landau inequality. Aequationes Math. 4 240-241 (1970).

17. Lichtenstein, L : Zur Variationsrechnung. Kgl. Ges. Wiss. Nach. Math.-Phys. Kl. 2, 161-192 (1919).

18. Ljubič, Ju. I. (or Lyubich, Yu.) : On inequalities between the powers of a linear operator. Translations Amer. Math. Soc. Ser. (2) 40, 39-84 (1964); translated from Izv. Akad. Nauk. SSSR Ser. Mat. 24, 825-864 (1960).

19. Putnam, C. R. : An application of spectral theory to a singular calculus of variations problem. Amer. J. Math. 70, 780-803 (1949).

20. Trebels, W and Westphal, V. I. : A note on the Landau-Kallman-Rota-Hille inequality. Linear Operators and Approxamination (Edited by P L Butzer et al). Birkhäuser Verlag, Basel and Stuttgart, 1972.

21. Everitt, W. N. and Jones, D. S. : On an integral inequality. To be published.

22. Everitt, W. N. : Integral inequalities and the Liouville transformation. To appear in the Proceedings of the Conference on the Theory of Ordinary and Partial Differential Equations, University of Dundee, March 1972; to be published in the series Lecture Notes in Mathematics, Springer-Verlag, Berlin.

23. Gindler, H. A. and Goldstein, J. A. : Dissipative operator versions of some classical inequalities. To be published.

24. Mitrinovic, D. S. : Analytic Inequalities. Springer-Verlag, Berlin, 1970.

Section 2

This section is concerned with inequalities involving integrals of functions and their derivatives, arising from ordinary and partial differential expressions of the second order. These inequalities are due to Everitt and Giertz; many of them are taken from the survey article [2]; some are of the type (*) given in the Introduction.

Let $M[\cdot]$ denote the ordinary, formally symmetric differential expression defined by

$$M[f] = -(pf')' + qf \quad \text{on } I \subseteq R \tag{1}$$

where I is an interval of the real line R, and p and q are real-valued coefficients defined on I.

Let $P[\cdot]$ denote the partial, formally symmetric differential expression defined by

$$P[f] = -\Delta f + qf \quad \text{on } G \subseteq R^n \tag{2}$$

where G is an open set of euclidean n-space R^n, Δ is the Laplacian in R^n, and q is a real-valued coefficient defined on G.

In this section $\|\cdot\|$ will denote the usual norm in $L^2(I)$ or $L^2(G)$.

A. The ordinary case

Basic conditions on the coefficients p and q; let $p, q : I \to R$, both p and $q \in AC_{loc}(I)$; let $p(x) > 0$ $(x \in I)$ and $p' \in L^2_{loc}(I)$; let $q(x) \geq 0$ $(x \in I)$.

Let $f \in D$, a linear manifold of $L^2(I)$, if $f : I \to C$, $f' \in AC_{loc}(I)$, both f and $M[f] \in L^2(I)$.

(a) The case $I = (-\infty, \infty)$. Suppose that for some $\varepsilon \in (0, 1)$ the coefficients p and q satisfy

$$(3) \qquad \{p(x)\}^{1/2} |q'(x)| \leq (1 - \epsilon) \{q(x)\}^{3/2} \qquad (x \in (-\infty, \infty))$$

then (i) $(pf')'$, $\{pq\}^{1/2}f'$, $qf \in L^2(-\infty, \infty)$ $(f \in D)$

and (ii) the following inequality holds

$$(4) \qquad \|(pf')'\|^2 + (1 + \epsilon) \|\{pq\}^{1/2}f'\|^2 + \epsilon \|qf\|^2 \leq \|M[f]\|^2 \qquad (f \in D);$$

for the proof see [1, Theorem 1]. The condition $q \geq 0$ on $(-\infty, \infty)$ implies that $M[\cdot]$ is limit-point at both $\pm\infty$ and this result is critical to the proof of (4). It is now known if (4) is best possible so that cases of equality have not been considered.

Suppose now that $p(x) = 1$ $(x \in (-\infty, \infty))$ and that q satisfies (3), i.e.

$$(5) \qquad |q'(x)| \leq c\{q(x)\}^{3/2} \qquad (x \in (-\infty, \infty))$$

for some $c \in (0, 2)$; then

(i) f'', $q^{1/2}f'$, $qf \in L^2(-\infty, \infty)$ $(f \in D)$

and (ii) the following inequalities hold

$$(6) \qquad (1 - c^2/4) \|qf\| \leq \|M[f]\| \qquad (f \in D)$$

$$(7) \qquad (2 - c) \|q^{1/2}f'\| \leq \|M[f]\| \qquad (f \in D)$$

$$(8) \qquad \min\{1, 4c^{-2} - 1\} \|f''\| \leq \|M[f]\| \qquad (f \in D)$$

where the numbers appearing on the left-hand sides are all best possible, but no cases of equality are known other than the null function.

If in (5) the number $c \in (0, \sqrt{2})$ and if $\delta \in [\frac{1}{2}c, c^{-1}]$ then

$$(9) \quad (1 - \delta c) \|qf\|^2 + (2 - c\delta^{-1}) \|q^{1/2}f'\|^2 + \|f''\|^2 \leq \|M[f]\|^2 \qquad (f \in D).$$

It is not known if this inequality is best possible.

The inequalities (6) to (9) are all taken from the detailed account in [3].

(b) The case $I = [a, \infty)$. Suppose that for some $\epsilon \in (0, 1)$ the coefficients p and q satisfy the condition (3) but now on the half-line $[a, \infty)$; additionally suppose that the coefficient p is such that the following inequality holds

$$(10) \quad \left\{ \int_a^\infty p(x) |f'(x)|^2 dx \right\}^2 \leqslant K \int_a^\infty |f(x)|^2 dx \int_a^\infty |(p(x)f'(x))'|^2 dx \quad (f \in D)$$

for some $K \in (0, \infty)$ (note that (10) is inequality (12) of Section 1 with $q(x) = 0$ ($x \in [a, \infty)$)); then for any $\delta \in (0, 1)$ the following inequality is valid

$$(11) \quad \delta \|(pf')'\|^2 + (1 + \epsilon) \|\{pq\}^{1/2} f'\|^2 + \epsilon \|qf\|^2$$
$$\leqslant \|M[f]\|^2 + (1 + \sqrt{K})^2 q(a)^2 (1 - \delta)^{-1} \|f\|^2 \quad (f \in D).$$

The proof of (11) is given in [1, Theorem 3]. It should be noted that the basic conditions and condition (3) on p and q imply that $q(a) > 0$, so that the second term on the right-hand side of (11) is essential to the inequality. It is not known if (11) is best possible.

B. The partial case

Inequalities similar to (4) may be obtained for the partial differential expression $P[\cdot]$ given in (2), provided certain conditions are satisfied by the coefficient q.

The basic condition is that $q \in L^2_{loc}(G)$. Additionally q has to satisfy a condition similar to (5); in the case when q has classical partial derivatives $\partial_k q (k = 1, 2, \ldots, n)$ of the first order on G, this takes the form

$$(12) \quad |\partial_k q| \leqslant c_k q^{3/2} \quad \text{on } G \quad (k = 1, 2, \ldots, n)$$

where c_k $(k = 1, 2, \ldots, n)$ are non-negative real numbers. It is then possible to prove inequalities of the form

$$(13) \quad a_0 \|qf\|^2 + \sum_{k=1}^{n} a_k \|q^{1/2} \partial_k f\|^2 + a_{n+1} \|\Delta f\|^2 \leq \|P[f]\|^2 \qquad (f \in D)$$

where a_k $(k = 1, 2, \ldots, n, n + 1)$ are non-negative real numbers not all zero, and D is a linear manifold of functions $f : G \to C$ such that $P[f]$ is defined and both f and $P[f] \in L^2(G)$.

Full details of these inequalities are given in [4].

References

1. Everitt, W. N. and Giertz, M. : Some inequalities associated with certain ordinary differential operators. Math. Zeit. 126, 308-326 (1972).

2. Everitt, W. N. and Giertz, M. : On limit-point and separation criteria for linear differential expressions. Proceedings of Equadiff III (Czechoslovak conference on differential equations and their applications). J. E. Purkyne University, Brno, Czechoslovakia, 1972.

3. Everitt, W. N. and Giertz, M. : Inequalities and separation for certain ordinary differential operators. P.London Math. Soc. (3) 28, 352-372 (1974).

4. Everitt, W. N. and Giertz, M. : Indqualities and separation for certain partial differential expressions. To be published.

ON THE DEFICIENCY INDICES OF POWERS OF FORMALLY SYMMETRIC DIFFERENTIAL EXPRESSIONS

W.N. EVERITT and M. GIERTZ

This lecture is concerned with part of the deficiency index problem of ordinary differential expressions. We refer to [2] and [9] for details of standard results quoted in the lecture and for a general account of the theory of deficiency indices.

We shall consider powers, and in particular the squares, of formally symmetric second order differential expressions $M[\cdot]$, where

$$M[f] = -(pf')' + qf \tag{1}$$

acting on complex-valued functions f defined on the half line $[0,\infty)$. As usual, the ' here denotes differentiation with respect to the independent variable $x \in [0,\infty)$. Derivatives of order greater than two will be denoted by $f^{(3)}$, $f^{(4)}$, and so on. Since we deal only with one fixed interval $[0,\infty)$ we shall use abbreviated notations such as $L^2 = L^2(0,\infty)$, $AC_{loc} = AC_{loc}[0,\infty)$ etc. when referring to standard subspaces of functions $f: [0,\infty) \to C$.

Throughout the lecture we assume that the coefficients p and q defining $M[\cdot]$ satisfy the following basic conditions:

(i) p *and* q *are defined and real-valued on* $[0,\infty)$,

(ii) $p(x) > 0 \qquad (x \in [0,\infty))$, (2)

(iii) $q' \in AC_{loc}$ *and* $p'' \in AC_{loc}$.

The last condition (iii) ensures that the formal square $M^2[\cdot]$ of $M[\cdot]$, defined by

$$M^2[f] = M[M[f]] = [p^2f'']'' - [(2pq-pp'')f']' + (q^2-pq'' - p'q')f, \tag{3}$$

exists as a differential expression; that is, (3) defines a function on $[0,\infty)$ whenever f is a function for which $f^{(3)}$ exists and is in AC_{loc}.

Higher powers of $M[\cdot]$, defined recursively by

$$M^n[\cdot] = M[M^{n-1}[\cdot]],$$

may also exist as differential expressions provided p and q satisfy additional regularity requirements. The first part of the lecture contains a survey of what is known at present about the deficiency indices of such powers. In this part we make the tacit assumption that p and q are of sufficient regularity in order to consider certain powers of $M[\cdot]$. More precisely, whenever we speak of a power $M^n[\cdot]$ we require both $q^{(2n-3)}$ and $p^{(2n-2)}$ to be in AC_{loc}. Analogously, when we speak of $M^n[f]$ we always implicitly assume that $f^{(2n-1)}$ is in AC_{loc} so that $M^n[f]$ is a function on $[0,\infty)$.

In the second part of the lecture we shall present two recent results concerning the deficiency indices of $M^2[\cdot]$.

The basic conditions on p and q imply that both $M[\cdot]$ and $M^2[\cdot]$ are formally symmetric, and this holds true as well for any higher power of $M[\cdot]$ that exists as a differential expression. Thus the deficiency indices of the minimal closed symmetric operator T_1 generated by $M[\cdot]$ in L^2, that is the closure of the restriction to C_0^∞, are either (1,1) or (2,2). The first of these two possibilities corresponds to the limit-point case of Hermann Weyl [11], and the second to the limit-circle case.

In general, the minimal closed operator T_n generated by $M^n[\cdot]$ in L^2 has deficiency indices (r,r), where r is an integer satisfying $n \leq r \leq 2n$, see [10, § 17.4]. When the deficiency indices are (r,r) we shall say that $M^n[\cdot]$ *is in the limit-*r *case at infinity*, or simply refer to $M^n[\cdot]$ as limit-r. We shall also say that $M^n[\cdot]$ is in the *minimal condition* when $r=n$, and in the *maximal condition* when $r=2n$.

It is well known that when λ is a complex non-real number then the number r of linearly independent L^2 - solutions of the equation

$$M^n[f] = \lambda f \tag{4}$$

does not depend on λ. This number r is, in view of (5) below by definition, precisely the r in the limit-r classification of $M^n[\cdot]$. Thus this classification depends on p and q only. When λ is real and $M^n[\cdot]$ is limit-r, the number of linearly independent solutions of (4) which are in L^2 is at most r, see [3, Ch XIII, § 6.9], but there are examples where it is less than r.

The maximal operator generated in L^2 by $M^n[\cdot]$ is T^*_n . Its domain, which is given by

$$D(T^*_n) = \{f;\ f \in L^2 \text{ and } M^n\,[f] \in L^2\}\ , \tag{5}$$

contains the domains of all linear operators which may be generated in L^2 by $M^n[\cdot]$,see [9 § 17].

The earliest results on the problem of the limit-r classification of powers of $M[\cdot]$ were published in 1969 by Chaudhuri and Everitt [1]. They considered $M^2[\cdot]$ and proved that

(a) $M^2[\cdot]$ is limit-4 if and only if $M[\cdot]$ is limit-circle (that is, limit-2),

(b) $M^2[\cdot]$ is limit-2 if and only if $D(T^*_2\ ) \subset D(T^*_1)$,

(c) $M^2[\cdot]$ is limit-2 when p = 1 and q satisfies certain specified conditions,

(d) There exist coefficients p and q for which $M^2[\cdot]$ is limit-3.

These results have since been extended to higher powers of $M[\cdot]$, so that we at present have the following information.

(a): The proof of (a) given in [1] extends directly, see [12], to

Theorem 1. Assume that there exists an n *for which* $M^n[\cdot]$ *is limit*-2n.

Then, whenever it exists, $M^m[\cdot]$ *is limit*-2m.

In particular, all existing powers of M[·] are in the maximal condition when M[·] is limit-circle. Since this result completely describes the limit-r classification of powers of M[·] in the limit-circle case we assume, for the rest of this lecture, that

$$M[\cdot] \text{ is limit-point at } \infty.$$

(b): In view of (5), the result in (b) is equivalent to the statement that $M^2[\cdot]$ is in the minimal condition precisely if

$$f \in L^2,\ M^2[f] \in L^2 \Rightarrow M[f] \in L^2.$$

The following extension to higher powers was given by Everitt and Giertz in 1972, see [4].

Theorem 2. Assume that $M^n[\cdot]$ *exists. Then* $M^n[\cdot]$ *is limit*-n *if and only if*

$$f \in L^2,\ M^n[f] \in L^2 \Rightarrow M^m[f] \in L^2 \qquad (m = 1,2,\ldots,n-1). \tag{6}$$

Following the terminology in [4], we shall say that $M^n[\cdot]$ is *partially separated* in L^2 when (6) is satisfied. Theorem 2 is known to hold true also for higher order differential expressions. In fact*, when the n:th power of an arbitrary formally symmetric differential expression L[·] exists, then $L^n[\cdot]$ is partially separated if and only if every (real or complex) n:th order polynomial in L[·] is in the minimal condition. For this result, see Kauffman [7] and Zettl [12].

(c): Since the publication of [1], a number of conditions have been given which ensure that certain powers of M[·] are in the minimal condition or, equivalently, partially separated. We have:

Theorem 3. Assume that $M^n[\cdot]$ *exists. Then* $M^n[\cdot]$ *is limit*-n *if either*

$$p = 1 \text{ and } \int^\infty |q+k| < \infty \text{ for some number } k, \tag{7}$$

or, if for some positive numbers k, K

$$q + k > 0 \text{ and } \log(p(x)) \leq K \int_0^x \{(q+k)/p\}^{1/2} \tag{8}$$

on a set of infinite measure, where $K < [n(n-1)]^{-1/2}$

---

* The 'only if' half of this statement is not correct; for the correct version see theorem 2 of the addendum by A. Zettl to this paper.

It is rather striking that the conditions in Theorem 3 impose restrictions on p and q only, and not on the derivatives of these coefficients, whereas the conclusion, concerning all existing powers of $M[\cdot]$, involves differential expressions defined in terms of such derivatives.

The condition (7) is given by Everitt and Giertz in [4], and (8), which is based on an earlier result in [4], is due to Read, see [10]. The result corresponding to (7) has been generalised by Zettl [12] to formally symmetric differential expressions of arbitrary order. The result corresponding to (8) includes the case

$$p = 1 \textit{ and } q \textit{ is bounded below,} \tag{9}$$

which improves previous conditions given by Everitt and Giertz [4] and Kauffman [7], requiring some restrictions on the derivatives of q in addition to (9).

When p and q satisfy either (7) or (8) it is clear that if a given power $M^n[\cdot]$ is in the minimal condition, then so are all smaller powers of $M[\cdot]$ as well. This is true in general, see [7]:

Theorem 4. When $M^n[\cdot]$ *is limit*-n , *then* $M^m[\cdot]$ *is limit*-m *for* $m = 1,2, \ldots, n-1$.

(d): A set of examples, constructed along the same lines as the example in [1], which put $M^n[\cdot]$ in the limit-(2n-1) case for all integers $n \geq 1$ is given in [5]. In particular, this set contains all expressions $M[\cdot]$ where the coefficients p and q are defined in terms of the parameter $\varepsilon > 0$ by

$$p(x) = (x+1)^{2+\varepsilon}, \quad q(x) = -\frac{1}{4}(1 + 2\varepsilon - \varepsilon^2)(x+1)^{\varepsilon} \qquad (x \in [0,\infty)) \tag{10}$$

Let us illustrate the above results in the following diagram indicating the possible numbers r of linearly independent L^2-solutions of $M^n[f] = \lambda f \quad (\operatorname{Im} \lambda \neq 0)$,

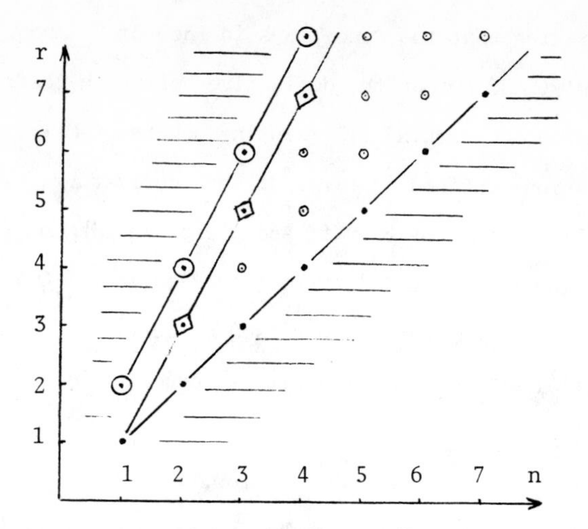

and summarize what we know in sections which correspond to (a), (b), (c) and (d) above:

(a) If we are on the upper line —⊙——⊙——⊙—, that is in the maximal condition, for some $n \geq 1$, then we are on this line for all m such that $M^m[\cdot]$ exists as a differential expression.

(b) We are on the lower line —·——·——·—, that is in the minimal condition, for a certain n if and only if $M^n[\cdot]$ is partially separated.

(c) If we are on the lower line for a certain n then we are on it for $m = 1,2,\ldots, n$; we are on this line when $p = 1$ and q satisfies either (7) or (8) above.

(d) There are examples where we start with $M[\cdot]$ in the limit-point case, and then leave the lower line. In these examples $M^n[\cdot]$ is limit-$(2n-1)$ for all positive integers n, that is, we stay on the line —◇——◇—. In all these cases $x^{-2}\, p(x) \to \infty$ as $x \to \infty$, in fact $\int^{\infty} p^{-1/2} < \infty$.

As for the remaining points (n,r), that is $3 \leq n < r < 2n-1$, in the unshaded area, it is not known if there exists an expression $M[\cdot]$ such that $M^n[\cdot]$ is limit-r.

Attempts to find coefficients p and q where p(x) increases less rapidly than x^2, and in particular when p = 1, prompted the authors of [1] to put the following

Question: Assuming that p=1, *does there exist a coefficient* q *which puts* $M^2[\cdot]$ *in the limit-3 case* - or is $M^2[\cdot]$ always limit-2 when $M[\cdot]$ is limit-point?

Since then, considerable effort has been spent in trying to construct such a coefficient q, or to prove the converse result when p = 1 (or $\int^{\infty} p^{-1/2} = \infty$).

The remaining part of the lecture deals with some recent results on the classification of $M^2[\cdot]$ when $M[\cdot]$ is limit-point. We begin with the following theorem due to Everitt and Giertz [5].

Theorem 5. *Let* p *and* q *satisfy the basic conditions in* (2), *and assume that there exist positive numbers* k,K *and* x_0 *and a number* $\alpha \in (-\infty,2]$ *such that*

$$0 < p(x) < K\, x^{\alpha} \qquad q(x) > -k\, x^{2-\alpha} \qquad (x \in (x_0,\infty)) \tag{11}$$

Then $M[\cdot]$ *is limit-point and* $M^2[\cdot]$ *is limit-2.*

We note the following two special cases:

$\alpha = 0$: This choice gives the well known condition

$$0 < p(x) \leq K \text{ and } q(x) > -kx^2 \tag{12}$$

for $M[\cdot]$ to be limit-point. Thus this classical condition also implies that $M^2[\cdot]$ is limit-2. This generalises an earlier result given by Kumar [8]. It also indicates that an affirmative answer to the first formulation of the above question would require a coefficient q of a rather artificial nature. For every k > 0 it must dip below $-kx^2$ for arbitrarily large x, in order to break (12), and yet keep $M[\cdot]$ in the limit-point case.

$\alpha = 2$: At this critical value we obtain

$$0 < p(x) \leq Kx^2 \text{ and } q(x) \geq -k\,. \tag{13}$$

It is clear that any coefficients p, q satisfying (13) satisfy also the Read condition (8), but only just so. Both (8) and (13) fail when $p(x) = x^{2+\varepsilon}$ $(\varepsilon>0)$ and q is bounded below.

The examples in (10) show that there is not room for much improvment in (13). Even if theese examples do break both conditions in (13), they do so with very little margin on taking ε to be positive but small.

Finally, let us return to the question quoted above from [1]. As it turns out the answer is yes; when p=1 there exists a coefficient q for which $M[\cdot]$ is limit-point and $M^2[\cdot]$ is limit-3. We obtain such a coefficient by constructing an infinitely differentiable function ψ in L^2 such that ψ''/ψ is also in C^∞ and such that there exists a function φ with the properties

(i) $\psi\varphi' - \psi'\varphi = 1$ on $[0,\infty)$

(ii) $\varphi \notin L^2$

(iii) $\Phi \in L^2$ where $\Phi(x) = \psi(x)\int_0^x \varphi^2 - \varphi(x)\int_x^\infty \psi\varphi$ $\quad (x \in [0,\infty))$

(iv) $\Psi \in L^2$ where $\Psi(x) = \psi(x)\int_0^x \varphi\psi + \varphi(x)\int_x^\infty \psi^2$ $\quad (x \in [0,\infty))$.

We then define q and $M[\cdot]$ by

$$q(x) = (\psi''/\psi)(x) \qquad M[f] = -f'' + qf,$$

so that ψ and φ are two linearly independent solutions of $M[f] = 0$, with ψ in L^2 and φ not in L^2. This puts $M[\cdot]$ in the limit-point case. Since $M[\Psi] = \psi$ and $M[\Phi] = \varphi$, it follows easily that ψ, φ, Ψ and Φ are linearly independent solutions of $M^2[f] = 0$. Three of these are in L^2, so $M^2[\cdot]$ must be limit-3 or limit-4 according to the result quoted from [3] earlier. One is not in L^2 so $M^2[\cdot]$ can not be limit-4. Thus $M^2[\cdot]$ is in the limit-3 case at ∞.

For any $\varepsilon>0$, it is possible to construct such a coefficient q which satisfies

$$q(x) \geq -x^{2+\varepsilon} \qquad (x \in [0,\infty)),$$

see [6].

This shows that the condition on q in (12) is best possible when p=1.

REFERENCES

1. Chaudhuri, Jyoti and Everitt, W.N.: On the square of a formally self-adjoint differential expression. J. London Math. Soc. (2) 1 (1969) 661-673

2. Coddington, E.A. and Levinson, N.: Theory of ordinary differential equations. McGraw-Hill, New York and London, 1955.

3. Dunford, N. and Schwartz, J.T.: Linear operators; Part II: Spectral theory. Interscience, New York, 1963.

4. Everitt, W. N. and Giertz, M.: On some properties of the powers of a formally self-adjoint differential expression. Proc. London Math. Soc. (3) 24 (1972) 149-170.

5. Everitt, W. N. and Giertz, M.: On the integrable-square classification of ordinary symmetric differential expressions. (to appear in J. Lond. Math. Soc.).

6. Everitt, W. N. and Giertz, M.: Examples concerning the integrable-square classification of ordinary symmetric differential expressions (to appear).

7. Kauffman, R. M.: Polynomials and the limit point condition. (To appear in Trans. Amer. Math. Soc.)

8. Kumar, Krishna V.: A criterion for a formally symmetric fourth-order differential expression to be in the limit-2 case at ∞. J. London Math. Soc., (2) 8 (1974).

9. Naimark, M. A.: Linear differential operators; Part II. Ungar, New York, 1968.

10. Read, T.T.: On the limit point condition for polynomials in a second order differential expression. Chalmers University of Technology and the University of Göteborg, Department of Mathematics No. 1974-13.

11. Weyl, H.: Über gewöhnliche Differentialgleichungen mit Singularitäten und die zugehörigen Entwicklungen willkürlicher Funktionen. Math. Annalen 68 (1910) 220-269.

12. Zettl, A.: The limit point and limit circle cases for polynomials in a differential operator. (To appear in Proc. Royal Soc. Edinburgh.)

to

On the deficiency indices of powers of formally symmetric differential expressions

Dr T. T. Read, visiting the University of Dundee for the academic year 1974-75, has made the following observation on the manuscript:

"A family of differential expressions given in 10 has the property that for any positive integer N there is an expression $M[\cdot]$ such that $M^n[\cdot]$ is limit-n for all $n \leq N$ and $M^n[\cdot]$ is limit-$(2n - N)$ for all $n > N$. Thus in the above diagram these examples show that we are on the lower line for $n \leq N$ and on a line parallel to the upper line for $n \geq N$. In particular given (n, r) with $n < r \leq 2n - 1$, we obtain an expression $M[\cdot]$ such that $M^n[\cdot]$ is limit-r by taking $N = 2n - r$.

The family consists of the expressions $M[\cdot]$ with $p(x) = q(x) = e^{\alpha x}$, α a positive constant. Since q is positive and bounded away from 0, it follows that 0 is not in the essential spectrum of $M^n[\cdot]$ for any n. In this case the deficiency indices of the minimal closed operator T_n generated by $M^n[\cdot]$ in L^2 are (r, r) where r is the number of linearly independent L^2-solutions of $M^n[f] = 0$. For the above expression the solutions of $M[f] = 0$

are of the form e^{sx} where $s(s + \alpha) - 1 = 0$. For any positive integer n the solutions of $M^n[f] = 0$ which are not solutions of $M^{n-1}[f] = 0$ are again of the form e^{sx} where $(s + n\alpha)(s + (n - 1)\alpha - 1 = 0$. Thus if

$$N(N + 1)^{-1/2} < \alpha \leq N(N - 1)^{-1/2},$$

then $M^n[f] = 0$ has n linearly independent L^2-solutions when $n \leq N$ and $2n - N$ linearly independent L^2- solutions when $n > N$ and the deficiency indices of the powers of $M[\cdot]$ are as asserted above.

T. T. Read

16 September 1974."

Addenda

to

On the deficiency indices of powers of

formally symmetric differential expressions

Dr R M Kauffman, visiting the University of Dundee for the academic year 1974-75, has made the following observation on the manuscript:

"There is a rule (details of which will be published elsewhere) which relates the deficiency index of higher powers of a formally symmetric differential expression, with real and sufficiently differentiable coefficients, to that of lower powers.

Let L be such a differential expression, necessarily of even order 2n (say). Let d(L) denote the deficiency index of the minimal closed symmetric operator in $L^2(0,\infty)$ generated by L on $[0,\infty)$. Then the rule states (note $d(L^0) = 0$),

(i) for any positive integer $k \geq 1$

$$d(L^k) - d(L^{k-1}) \leq 2n$$

and (ii) if the integer $j \geq k$ then

$$d(L^j) - d(L^{j-1}) \geq d(L^k) - d(L^{k-1}).$$

The second part (ii) of the rule may be restated as follows: "In the graph of deficiency indices $d(L^k)$ against powers k (see the diagram above) the slope of the graph is non-decreasing as the power k increases."

This rule can be quite helpful in calculations. Suppose L is of the second-order, i.e. n = 1, and suppose that for some integer $j \geq 2$ it is known $d(L^j) = k$ (say) $\neq j$, then it is possible to use the above rule to calculate the deficiency index of any other power of L. For example if $d(L^4) = 5$, then $d(L) = 1$, $d(L^2) = 2$, $d(L^3) = 3$, $d(L^5) = 7$, $d(L^6) = 9$ and so on. If L is of higher order than the second, the rule does not give such complete information but nevertheless it narrows considerably the possibilities for the deficiency indices of the powers L^k.

Examples constructed by T T Read (to be published elsewhere) show that any possibility not prohibited by the above rule can actually occur. In these examples of Read the differential expression L can be of any even order."

R M Kauffman

October 1974.

Professor Anton Zettl, visiting the University of Dundee, under the auspices of the Science Research Council of the United Kingdom, for the academic year 1974-75, has made the following observation on the manuscript.

"Let M be a symmetric differential expression, i.e. $M = M^+$ where

$$My = a_m(t)y^{(m)} + a_{m-1}y^{(m-1)} + \dots + a_1(t)\, y' + a_0(t)y$$

and

$$M^+y = (-1)^m(\overline{a}_m(t)y)^{(m)} + (-1)^{m-1}(\overline{a}_{m-1}(t)y)^{(m-1)} + \dots - (\overline{a}_1(t)y)' + \overline{q}_0(t)y.$$

We assume that the coefficients a_j are complex-valued functions defined on $0,\infty)$ and sufficiently differentiable so that powers of $M : M, M^2, M^3, \dots, M^k$ can be formed. In this situation when m is even, say $m = 2r$, a_m must be real and we assume further that $a_m(t) \neq 0$ for $t \geq 0$. For $m = 2r + 1$, $a_m = i\, b_m$ where b_m is real and we assume $(-1)^r\, b_m > 0$, $r > 0$.

Let $N_+(L)$ and $N_-(L)$ denote the deficiency indices of the symmetric expression L on $[0,\infty)$ associated with the upper and lower half-planes respectively. Theorem 1 below is contained in 'Deficiency indices of polynomials in symmetric differential expressions" which will appear as part of the Proceedings of the 1974 Dundee Conference on Ordinary and Partial Differential Equations, to be published by Springer-Verlag in the Lecture Notes in Mathematics series. The second result is contained in 'Deficiency indices of polynomials in symmetric differential expressions II" which is currently being prepared for publication.

Theorem 1. (a) *Suppose* k *is even, say* $k = 2r$; *then*

$$N_+(M^k),\ N_-(M^k) \geq r\left[N_+(M) + N_-(M)\right].$$

(b) *Suppose* k *is odd, say* $k = 1$; *then*

$$N_+(M^k) \geq (r + 1)N_+(M) + rN_-(M) \quad \textit{and}$$

$$N_-(M^k) \geq rN_+(M) + (r + 1)N_-(M).$$

Note in particular if $N_+(M) = N_-(M)$ (this is always the case when all the coefficients of M are real) then $N_+(M^k),\ N_-(M^k) \geq kN_\pm(M)$.

Corollary 1. *If* $N_+(M^n)$ *or* $N_-(M^n)$ *is minimal for some positive integer* n, *then both* $N_+(M)$ *and* $N_-(M)$ *are minimal, i.e.* M *is in the limit-point case.*

Corollary 2. *If* M *is limit-circle, i.e. one (and hence both) of* $N_+(M)$ *and* $N_-(M)$ *are maximal, then* M^k *is limit-circle.*

It is known - see [12] *above - that the converse of corollary 2 also holds.*

If the deficiency indices of M (or of M^k) are nether maximal nor minimal, then theorem 1 does not determine precisely those of M^k (or M) but rather restricts the possibilities, i.e. theorem 1 shows that certain values - allowed by the general classification results - cannot occur.

In general strict inequality can occur in theorem 1 - see the example of Chaudhuri/Everitt in [1]. The next result gives a necessary and sufficient condition for equality to hold in theorem 1. This condition is known as partial separation: We say that M^k is partially separated in $L^2(0,\infty)$ if $f \in L^2(0,\infty)$, $f^{(km-1)}$ locally absolutely continuous, and $M^k f$ in $L^2(0,\infty)$ together imply that $M^r f \in L^2(0,\infty)$ for all $r = 1, 2, \ldots, k - 1$.

<u>Theorem 2. A necessary and sufficient condition that</u>

1. $N_+(M^k) = r\left[N_+(M) + N_-(M)\right] = N_-(M^k)$ <u>if</u> $k = 2r$ <u>and</u>

2. $N_+(M^k) = (r+1)N_+(M) + rN_-(M)$ <u>and</u>

   $N_-(M^k) = rN_+(M) + (r+1)N_-(M)$ <u>if</u> $k = 2r+1$

<u>is that</u> M^k <u>is partially separated in</u> $L^2(0,\infty)$. <u>In particular if</u> $N_+(M) = N_-(M) = q$ <u>then</u> $N_+(M^k) = kq = N_-(M^k)$ <u>if and only if</u> M^k <u>is partially separated in</u> $L^2(0,\infty)$,

Anton Zettl

October 1974

On the Spectral Theory of Schrödinger and Dirac Operators with Strongly Singular Potentials

H. Kalf , U.-W. Schmincke

J. Walter , R. Wüst

Introduction

In the present paper we should like to give a survey of the work on the spectral theory of Schrödinger and Dirac operators that has been done first at Professor G. Hellwig's institute at the Technical University of Berlin and since 1966 at his Institute of Mathematics in Aachen. For lack of space we shall not enter into any results on the one-dimensional case. Moreover, we shall confine ourselves to the very first spectral problem [1] (see [50, p.302 ff.]):

(Q.1) Does the minimal operator to be associated with the Schrödinger or Dirac expression have a unique self-adjoint extension (in which case the minimal operator is called essentially self-adjoint [76, p.51]), or what is equivalent to this, is the spectrum of its closure a subset of the real line?

(Q.2) If its smallest closed extension is not self-adjoint, does there exist a physically distinguished extension which is self-adjoint?

---

[1] More detailed information about the nature of the spectrum of Schrödinger operators is to be found in [23,58,62,86,87,88] (regularity of eigenfunctions and eigenpackets), [63,64] (decay properties of eigenfunctions or C^∞- vectors), and [25, 31,60] (absence of eigenvalues)

Concerning the motivation and importance of these questions we must refer to the literature on the foundations of quantum mechanics (e.g.,[26,42,81] ; see also [85]).

It is the uniqueness, not the existence of some self-adjoint extension (which is guaranted by a theorem of J. v. Neumann [76, pp.339,361]), that is doubtful. Quite often physicists forget (see, e.g., [1;41, p.1665 ff.]) that their speaking of the spectrum of the Hamiltonian remains highly ambiguous unless at least (Q.2) is answered in the affirmative. While examples where the answer even to (Q.2) is negative [2)] can certainly be regarded as pathological, it should be noted that in the cases of direct physical interest the answer to (Q.1) is always negative for the ground state of the radial part of the Schrödinger expression arising after separation of the variables (cf.[2] in this context).

The Hamiltonians occuring in nonrelativistic quantum mechanics are always bounded from below. (The Hamiltonian commonly used for the Stark effect is unphysical for the very reason that it is unbounded from below as was first observed by Oppenheimer [44, p.74] . See also [37, p.404; 38, p.256; 46; 56, p.10].) We therefore start with a study of such operators.

[2)] For Schrödinger operators $-\Delta + q$ this occurs when the local negative singularities of q destroy the semiboundedness of the operator or when the local singularities of q are so delicately distributed that $D(-\Delta) \cap D(q)$ is no longer dense in L^2 [71, p.27 f.] . For Dirac operators see footnote 4.

A fundamental role is played in this § 1 by the $\frac{1}{r^2}$ - potential [3], which is sometimes too hastily dismissed as uninteresting in the physical literature [37, p.198; 41, p.1667] (For example, the Hamiltonian for a spin zero particle in a Coulomb field gives rise to a Schrödinger operator involving a $\frac{1}{r^2}$ - potential [5].) Since all Schrödinger operators of direct physical interest (for which (Q.1) was positively answered by Kato [34]) do not yet display the peculiarity of the Dirac operator with the physically interesting $-\frac{\mu}{r}$ -potential, namely that the answer to (Q.1), (Q.2) depends critically on the value of the constant μ [4] (such potentials are called "strongly singular" in [28,30,32,65,66] , "transitional potentials" in [9]) we think it advisable to study this phenomenon in the simpler case of a Schrödinger operator first. In fact, it turns out that a certain modification of the usual perturbation method invented in this context (in [65]) can be carried over to Dirac operators (see § 4).

---

[3] As it represents one of the few examples for which Schrödinger's equation can be solved in terms of well-known higher transcendental functions, it was already discussed in the early times of quantum mechanics [74,p.24 ff.; 45;69]. A rigorous treatment of the inverse square potential was given by Meetz [40].

[4] It seems to have passed unnoticed in the physical literature (with the possible exceptions of [16;10, p.1091; 51]) that for hydrogen-like atoms with atomic number Z the answer to (Q.1) is affirmative if and only if $Z \leqslant 118$, whereas it is conspicuous from the eigensolutions of the separated Dirac equation that the range $Z \geqslant 138$ is critical (the answer to (Q.2) is negative then).

The relativistic Hamiltonians for particles with half-integral spin are never semibounded. Thus Friedrichs's method of constructing a physically distinguished self-adjoint extension can no longer be applied to such operators. However, an analogue of an explicit characterization of the Friedrichs extension of certain Schrödinger operators (§ 2) turns out to be well-adapted to define a physically distinguished self-adjoint extension for a large class of Dirac operators including hydrogen - like atoms with atomic number $Z \leqslant 137$ (§ 5).

The idea of constructing a distinguished self-adjoint extension by means of cut-off potentials is frequently encountered in physics ([38,39,40]; [47,48,51] work in the range $Z \geqslant 138$ where no such distinguished self-adjoint extension exists). In § 6 it is shown that this idea can be made precise for Dirac operators with a spectral gap, in particular for hydrogen - like atoms with $Z \leqslant 137$.

The multi-dimensional extension of an inequality of Hardy's (Lemma 1) will be of vital importance throughout this paper. An elegant proof Shortley [69] gave of it seems to have escaped notice in the literature hitherto. It is therefore reproduced here. The proofs of Theorems 2 and 4 are partly new. Theorem 3 is new in the restricted sense that is has been distilled from the work of Friedrichs [12,13] and [28] .

1. Essential Self-Adjointness of Schrödinger Operators Bounded from Below

In order to keep this presentation as transparent as possible we shall only consider Schrödinger operators without magnetic fields. The incorporation of vector potentials does not present any serious additional difficulties once those arising from singularities of the electrostatic potential have been overcome.

Let $q \in L^2_{loc}(\mathbb{R}^n)$ $(n \geqslant 2)$ be real-valued. Then the operator defined by

$$T_o u = Du := -\Delta u + qu \ , \quad D(T_o) = C_o^\infty(\mathbb{R}^n)$$

is a symmetric (in particular, densely defined) operator in the Hilbert space $H = L^2(\mathbb{R}^n)$ (norm $\|\cdot\|$ and salar product $(\cdot,\cdot)$ as usual). If T_o is an operator arising from problems in nonrelativistic quantum mechanics it will be bounded from below. Does this property suffice already to ensure its essential self-adjointness? The following example shows that the answer is no.

Example 1a. Let $q(x) = \frac{\beta}{|x|^2}$, $n \geqslant 5$. Then T_o is essentially self-adjoint if and only if $\beta \geqslant \beta_o := 1 - (\frac{n-2}{2})^2$. T_o is bounded from below if and only if $\beta \geqslant -(\frac{n-2}{2})^2$.

The first assertion follows from separation of the variables, the second from Hardy's inequality (Lemma 1 and Remark 1) below.

Povzner [49, p.31] and Wienholtz [83] (cf. also [14, p.58]) proved that the semiboundedness of T_o does imply its essential self-adjointness if $q \in C^o(\mathbb{R}^n)$. [5)]

---

5) Rellich's corresponding question at the International Congress of Mathematicians in Amsterdam [55] had therefore been positively answered by Povzner already.

This contains as a special case the first essential self-adjointness criterion for T_o, which is due to Carleman [4] and Friedrichs [12] (Satz 2, p.691, together with footnote 6, p.690), viz. $q \in C^o(\mathbb{R}^n)$ and bounded from below. Later, Stetkær-Hansen [75] proved that $q \in C^o(\mathbb{R}^n)$ can be replaced by the much milder local condition $q \in Q_{\alpha,loc}(\mathbb{R}^n)$. This class of functions had been introduced by Stummel [77] when extending Carleman's characterization of $D(T_o^*)$ to singular q. Weakening of the condition $q \in Q_{\alpha,loc}(\mathbb{R}^n)$ is impeded by Example 1a (Simon [72] was the first to use Example 1a in this context).

For $n \leqslant 4$ the potential given in Example 1a is no longer in $L^2_{loc}(\mathbb{R}^n)$. In order to deal with such examples for all dimensions $n \geqslant 2$ the idea to replace $\mathbb{R}^n$ temporarily by a general domain Ω suggests itself. In view of the Dirac operator to be considered in §§ 4-6 the case $\Omega = \mathbb{R}^n_+ := \mathbb{R}^n \setminus \{0\}$ with the operator T defined by

$$(1.1) \qquad Tu = Du \qquad , \qquad D(T) = C_o^\infty(\mathbb{R}^n_+)$$

for real-valued $q \in L^2_{loc}(\mathbb{R}^n_+)$ will be of primary interest.

<u>Example 1b.</u> Let $q(x) = \frac{\beta}{|x|^2}$, $n \geqslant 2$. Then the assertions in Example 1a also hold good for T. For $n \geqslant 5$ we have $\overline{T} = \overline{T}_o$ [72], i.e., $C_o^\infty(\mathbb{R}^n_+)$ is a core [35, p.166] for $\overline{T}_o$.

The semiboundedness of the operators in Examples 1a,b is a consequence of the following inequality of Hardy's (see [32] where some historical remarks are also given), which will be used in various places of this paper.

<u>Lemma 1.</u> (1.2) $$\int_{\mathbb{R}^n} \left|\frac{\partial}{\partial r}u(x)\right|^2 dx \geqslant \left(\frac{n-2}{2}\right)^2 \int_{\mathbb{R}^n} \frac{|u(x)|^2}{|x|^2}\,dx \quad (u \in C_o^\infty(\mathbb{R}^n_+))$$

<u>Proof</u> [69]. For $f \in C^\infty((0,\infty))$ let $f(r)$ or f denote the operator of multiplication by $f(|x|)$ with $C_o^\infty(\mathbb{R}^n_+)$ as domain of definition, and p_r the radial momentum operator defined by

$$p_r u = \frac{1}{i} r^{-\frac{n-1}{2}} \frac{\partial}{\partial r} (r^{\frac{n-1}{2}} u), \; D(p_r) = C_o^\infty(\mathbb{R}^n_+) .$$

p_r is clearly symmetric and satisfies the commutation relation

$$(p_r f - f p_r)u = - i f' u \qquad (u \in C_o^\infty(\mathbb{R}^n_+)) .$$

Putting $A_s := p_r - i s r^{-1} \quad (s \in \mathbb{R})$
we therefore have

$$(p_r^2 u,u) + s(s+1)(r^{-2}u,u) = (A_s^* A_s u,u) .$$

Hence $$\|p_r u\|^2 - \frac{1}{4}\|r^{-1}u\|^2 = \|A_{-\frac{1}{2}} u\|^2 \geqslant 0 .$$

Because of $$\|p_r u\|^2 = \left\|\frac{\partial u}{\partial r}\right\|^2 - \frac{(n-1)(n-3)}{4}\|r^{-1}u\|^2$$

this is the assertion. ∎

<u>Remark 1.</u> Approximating $u \in C_o^\infty(\mathbb{R}^n)$ by $C_o^\infty(\mathbb{R}^n_+)$- functions Lemma 1 is easily seen to hold for $u \in C_o^\infty(\mathbb{R}^n)$. A second limit process shows its validity for $u \in H^1(\mathbb{R}^n)$, the usual Sobolev space of L^2- functions whose first generalized derivatives belong to $L^2(\mathbb{R}^n)$ (cf. [35, p.345]). The constant $(\frac{n-2}{2})^2$ is the best possible. Moreover, there is equality in (1.2) if and only if $u = 0$.

In order to give a second application of Lemma 1 let us turn for a moment to the operator $S := (-\Delta + q)\restriction C_o^\infty(\Omega)$ where

$$\text{(i)} \quad q \in Q_{\alpha,loc}(\Omega) \quad .$$

Let ϱ be a non-negative function with $\varrho(x) \longrightarrow \infty$ as $x \longrightarrow \partial\Omega$ (if Ω is not bounded ∞ is regarded as a point of $\partial\Omega$) and $(\nabla\varrho)^2 \leq \varphi^2(\varrho)$ for some $\varphi > 0$. Then Stetkær-Hansen [75] showed that S is essentially self-adjoint when bounded from below provided $\int^{\infty} \frac{dt}{\varphi(t)} = \infty$. On the other hand, it became apparent from the work of [27,78,79] that something stronger than semiboundedness had to be assumed to ensure essential self-adjointness if $\int^{\infty} \frac{dt}{\varphi(t)} < \infty$. The following theorem, which was proved in [80], combines these two cases. In reproducing its proof we shall be partly more explicit than [80].

Theorem 1 [80] Let $\varrho, \sigma \geq 0$ be smooth functions [6)] with the properties

$$\text{(ii)} \quad \lim_{x \to \partial\Omega} [\varrho(x) + \sigma(x)] = \infty; \quad \text{(iii)} \quad (\nabla\varrho)^2 \leq \varphi^2(\varrho), \ (\nabla\sigma)^2 \leq \psi^2(\sigma)$$

where $0 < \varphi, \psi \in C^0[0,\infty)$ are suitable functions satisfying $\int^{\infty} \frac{dt}{\varphi(t)} = \infty$, $\int^{\infty} \frac{dt}{\psi(t)} < \infty$. Suppose besides (i) that there is a $\delta > 0$ such that

$$(Su,u) \geq (1+\delta) \left(\left(\int_{\sigma(\cdot)}^{\infty} \frac{dt}{\psi(t)} \right)^{-2} u,u \right) \qquad (u \in C_0^{\infty}(\Omega)).$$

Then S is essentially self-adjoint.

[6)] It is actually sufficient that ϱ and σ satisfy a uniform Lipschitz condition on every compact subdomain of Ω so that they are differentiable a.e. In applications it sometimes convenient to consider non-smooth σ (see Theorem 2).

Proof. To show that $\overline{R(S)} = L^2(\Omega)$ [23, p.177 f.] let $h \in L^2(\Omega)$ (without loss of generality real-valued) satisfy $(h,Su) = 0$ $(u \in C_o^\infty(\Omega))$. Thus $h \in D(S^*)$. Because of (i), $D(S^*)$ can be characterized by [24]

$$(1.3) \quad D(S^*) = \left\{ u \mid u \in H^2_{loc}(\Omega) \cap L^2(\Omega) \quad , \quad Du \in L^2(\Omega) \right\} \text{ (real valued)}$$

$H^2_{loc}(\Omega)$ denotes the set of all L^2_{loc}- functions whose generalized derivatives up to the second order belong to $L^2_{loc}(\Omega)$. Let $K \subset\subset \Omega$. Then there exists a sequence of real-valued functions $h_n \in C_o^\infty(K)$ $(n \in \mathbb{N})$ with [24]

$$(1.4) \quad \lim_{n\to\infty} \| h_n - h \|_{L^2(K)} = \lim_{n\to\infty} \| \Delta h_n - \Delta h \|_{L^2(K)} = \lim_{n\to\infty} \| q\, h_n - q\, h \|_{L^2(K)} = 0$$

The identity [7] [83, p.60] (see also [36, p.138])

$$(1.5) \quad \eta u(-\Delta + q)\eta u = \eta^2 u(-\Delta + q)u - \tfrac{1}{2} \nabla (u^2 \nabla \eta^2) + u^2 (\nabla \eta)^2 \qquad (\eta\, , \; u \in C^2(\Omega))$$

with $u = h_n$ and $\eta \in C_o^\infty(\Omega)$ yields, thanks to (1.4) and $(-\Delta + q)h = 0$, after integration

$$\lim_{n\to\infty} (S\eta u_n, \eta u_n) = \| h \nabla \eta \|^2 \geqslant (1 + \delta) \left\| \left(\int_{\sigma(\cdot)}^{\infty} \frac{dt}{\psi(t)} \right)^{-1} h\eta \right\|^2$$

Now we choose $\eta(x) = f(\rho(x))\, g(\sigma(x))$ $(x \in \Omega)$ where f,g are smooth functions defined on $[0,\infty)$ with compact support.

---

[7] This method of multiplicative variation, which was introduced into the calculus of variations by Jacobi, is employed in the 2nd ed. of Courant-Hilbert [6, p.398] to show that the ground state of $\overline{S}$ is characterized by the absence of nodes.

Because of (ii) η has compact support in Ω. Assumption (iii) and a simple inequality lead us to

$$(1 + \frac{1}{\varepsilon}) \| hf' g \varphi \|^2 + (1 + \varepsilon) \| hfg' \psi \|^2 \geqslant$$
$$\geqslant (1 + \delta) \left\| \left(\int_{G(\cdot)}^{\infty} \frac{dt}{\psi(t)} \right)^{-1} hfg \right\|^2 \qquad (\varepsilon > 0) .$$

This gives the desired result $h = 0$ if f and g can be chosen to satisfy

$$(1.6) \qquad f = 1 \quad , \quad g' \psi = 1 \quad .$$

It is true that the choice (1.6) is not immediately compatible with f,g to have compact support. It is, however, not difficult to construct functions that have compact support and enjoy the properties (1.6) after a limiting process (see [80]). ∎

Putting $\Omega = \mathbb{R}^n$, $\rho(x) = |x|$, $G(x) = 0$ $(x \in \mathbb{R}^n)$ in Theorem 1 we obtain the above mentioned result of Stetkær-Hansen. Furthermore, Theorem 1 enables us to prove the following criterion.

<u>Theorem 2</u>. Suppose, $q = q_1 + q_2$ where $q_1 \in Q_{\alpha,loc}(\mathbb{R}^n_+)$, $q_2 \in L^\infty(\mathbb{R}^n)$,

$$q_1(x) \geqslant \frac{\beta_0}{|x|^2} \qquad (x \in \mathbb{R}^n_+) .$$

Then T is essentially self-adjoint. The constant β_0 (from Examples 1a,b) is the best possible.

<u>Proof.</u> 1st step. Choose

$$\rho(x) = |x| \quad (x \in \mathbb{R}^n_+), \quad G(x) = \begin{cases} -\log|x| & \text{if } 0 < |x| < 1 \\ 0 & \text{if } \quad |x| \geqslant 1 . \end{cases}$$

Then
$$\int_{G(x)}^{\infty} \frac{dt}{\psi(t)} = e^{-G(x)} \quad (x \in \mathbb{R}^n_+) .$$

Without loss of generality we may assume $q_2 = 0$ [23, p.182] .

Now suppose $q(x) \geqslant \frac{\beta}{|x|^2}$ where $\beta > \beta_o$ is a suitable number.
As a consequence of Hardy's inequality (1.2)

$$((T + (1+\beta - \beta_o)I)\, u,u) \geqslant \left(\beta + \left(\tfrac{n-2}{2}\right)^2\right) \|r^{-1}u\|^2 + (1+\beta-\beta_o)\, \|u\|^2$$

$$\geqslant (1+\beta-\beta_o)\left(\int_{|x|<1} \frac{|u(x)|^2}{|x|^2}\, dx + \int_{|x|\geqslant 1} |u(x)|^2 dx\right)$$

$$= (1+\beta-\beta_o)\,(e^{2\sigma}u,u) \qquad (u \in C_o^\infty(\mathbb{R}^n_+))$$

holds, so that Theorem 1 implies the essential self-adjointness of $T + (1+\beta-\beta_o)I$. Hence T is essentially self-adjoint [23, p.181]

2nd step. Let f be as in the proof of Lemma 1. The anti-commutation relation

$$(Tf^2 + f^2T)u = 2(fTf - f'^{\,2})u \qquad (u \in C_o^\infty(\mathbb{R}^n_+))$$

which is the abstract version of (1.5), is immediately verified. Thus

$$\|(T + f^2)u\|^2 - \|f^2u\|^2 = \|Tu\|^2 + ((Tf^2 + f^2T)u,u)$$

$$\geqslant 2((fTf - f'^{\,2})u,u).$$

For $\beta \geqslant \beta_o$ Hardy's inequality yields

$$(Tfu,\ fu) \geqslant \|r^{-1}\, fu\|^2 .$$

Hence $\qquad \|(T + f^2)u\|^2 - \|f^2u\|^2 \geqslant 2((f^2r^{-2} - f'^{\,2})u,u).$

Put $f(r) = ar^{-1}$ $(a > 0)$, so that

$$\|f^2u\|^2 \leqslant \|(T + f^2)u\|^2 . \tag{1.7}$$

It was shown in step 1 that $T + f^2$ is essentially self-adjoint. In view of (1.7)

$$(T + f^2) + (-f^2) = T$$

is therefore essentially self-adjoint according to the Kato-Rellich perturbation theorem [35, p.289] (for an extension see [89]). That the constant β_o is sharp follows from Example 1b.

The first step in the proof of Theorem 2 was, among other things, given in [32]. The idea of applying a perturbation theorem to the sum of T and a suitably chosen symmetric "intercalary" operator f^2 is due to [65]. (An unnecessary growth restriction on q to be found there was eliminated by Grütter [17].) This modification of the usual perturbation argument where q is regarded as a "small" perturbation of $-\Delta$ is required by the following two facts. i) $H_o := -\Delta \restriction C_o^\infty(\mathbb{R}^n_+)$ is essentially self-adjoint if and only if $n \geqslant 4$. ii) $\beta_o r^{-2}$ is H_o - bounded if and only if $n \geqslant 5$ (the relative bound being 1; see equ. (6) with s = 0 in [65]). - In [30] Theorem 2 was proved by showing the symmetry of T^* extending an argument of Carleman [3, p.176 ff.; 4] and Friedrichs [13] (cf. Remark 2 at the end of § 2).

The recent work of Simon [72] and Kato [36] made it apparent that as far as the positive local singularities of q are concerned, the minimum requirement to define T_o densely in $L^2(\mathbb{R}^n)$, viz. $q \in L^2_{loc}(\mathbb{R}^n)$, suffices already to establish its essential self-adjointness. One therefore suspects that the condition $q_1 \in Q_{\alpha,loc}(\mathbb{R}^n_+)$ in Theorem 2 may be weakened to $q_1 \in L^2_{loc}(\mathbb{R}^n_+)$. This conjecture was proved by Simon [73] (for a variant of his proof see [33]). Theorem 2 also lends itself to various other generalizations, for which we refer to a paper by Simader [70] appearing before long. In particular, q may be allowed to fall-off as rapidly as $-|x|^2$ at infinity (cf. also [30]), which leads to theorems of the type considered in § 3. Above all, Simader gives interesting ab ovo proofs of his results not depending, as we for simplicity do here, on previous regularity results such as those of [24].

2. An Explicit Characterization of the Friedrichs Extension

Every semibounded operator A acting in some Hilbert space H possesses the following natural extension besides its closure. Let

$$\mathfrak{A} := \{ u \mid u \in H, \text{ there exists } \{u_j\}_{j \in \mathbb{N}} \subset D(A) \text{ such that } s\text{-}\lim_{j\to\infty} u_j = u \text{ and } \lim_{j,k\to\infty} (A(u_j - u_k), u_j - u_k) = 0 \}.$$

Then $A_F := A^* \restriction \mathfrak{A}$ is a self-adjoint extension of A (the Friedrichs extension) with the same lower bound as A (Freudenthal [11]; for Friedrichs's original construction see [12, Satz 9; 23, Section 12.3]). It should be noted that there may be self-adjoint extensions different from A_F with the same lower bound as A. For two properties each of which is characteristic of A_F we refer the reader to [35, pp.326,331].

Taking for A the differential operator T of (1.1), the domain of definition of its Friedrichs extension can be characterized quite explicitly for a large class of potentials. It turns out that $D(T_F)$ consists exactly of those elements of $D(T^*)$ that have finite energy integrals. The distinguished role T_F plays among the self-adjoint extensions of T is therefore physically obvious. To formulate this as a theorem we define for $q \in L^2_{loc}(\mathbb{R}^n_+)$ an operator $\hat{T}$ by

$$\hat{T}\,u = Du\,,$$

$$D(\hat{T}) = \left\{ u \,\middle|\, u \in H^2_{loc}(\mathbb{R}^n_+) \cap L^2(\mathbb{R}^n),\ \int_{\mathbb{R}^n} \frac{|u(x)|^2}{|x|^2}\,dx < \infty,\quad Du \in L^2(\mathbb{R}^n) \right\}$$

and introduce the notations

$$\varphi_u(t) := \left(\int_{|\xi|=1} |u(t\xi)|^2 d\sigma_n \right)^{1/2} \qquad (t \in (0,\infty)), \tag{2.1}$$

$$q_\pm := \frac{1}{2}\,(|q| \pm q)\,.$$

(2.1) makes sense for every "distinguished" representative of the class $u \in H^2_{loc}(\mathbb{R}^n_+)$ (see [28, p.236 f.] and the literature cited there).

<u>Theorem 3</u>. Suppose, $q = q_1 + q_2$ where $q_1 \in Q_{\alpha,loc}(\mathbb{R}^n_+)$, $q_2 \in L^\infty(\mathbb{R}^n)$ and

$$q_1(x) \geqslant \frac{\beta}{|x|^2} \qquad (x \in \mathbb{R}^n_+)$$

for some $\beta \in \mathbb{R}$. Then the following assertions hold,

a) $u \in D(\hat{T}) \Longrightarrow |q|^{1/2}u,\ \nabla u \in L^2(\mathbb{R}^n)$;

b) $\hat{T}$ is symmetric;

c) $\hat{T}$ is bounded from below if $\beta \geqslant -\left(\frac{n-2}{2}\right)^2$;

d) $\hat{T} = T_F$ if $\beta > -\left(\frac{n-2}{2}\right)^2$.

<u>Proof.</u> Ad a): (For the line of reasoning cf. [30].) Let $0 < \rho < R < \infty$, $u \in D(\hat{T})$ real-valued (without loss of generality). By means of Gauss's theorem and the identity

$$\begin{aligned} &\int_{\rho<|x|<R} |x|^{-s} \left[\nabla(|x|^{s/2} u(x))\right]^2 dx \\ &= \int_{\rho<|x|<R} (\nabla u(x))^2 dx + \frac{s}{2}\left[\frac{s}{2} - (n-2)\right] \int_{\rho<|x|<R} \frac{u^2(x)}{|x|^2}\,dx \\ &\qquad + \frac{s}{2}\, t^{n-2} \varphi_u^2(t)\Big|_\rho^R \qquad (s \in \mathbb{R}) \end{aligned} \tag{2.2}$$

we obtain

$$(2.3)\quad \int_{\varrho<|x|<R} (-\Delta u(x) + q(x)u(x))u(x)dx =$$

$$= -\frac{1}{2}\Big[(s-n)t^{n-2}\varphi_u^2(t) + t^{-1}(t^n\varphi_u^2(t))'\Big]_\varrho^R$$

$$+ \int_{\varrho<|x|<R} |x|^{-s}\big[\nabla(|x|^{s/2}u(x))\big]^2 dx$$

$$+ \int_{\varrho<|x|<R} \Big\{q(x) - \frac{s}{2}\Big[\frac{s}{2} - (n-2)\Big]|x|^{-2}\Big\}u^2(x)\,dx .$$

Because of $u \in L^2(\mathbb{R}^n)$ there exists a null sequence on which $(t^n\varphi_u^2(t))' \geqslant 0$ holds. Putting $s = n$ in (2.3) and observing $q_-^{1/2}\,u \in L^2(\mathbb{R}^n)$ this enables us to conclude

$$\int_{|x|<R} |x|^{-n}\big[\nabla(|x|^{n/2}u(x))\big]^2 dx < \infty$$

and

$$(2.4)\quad \int_{|x|<R} q_+(x)u^2(x)dx < \infty ,$$

for the left-hand side of (2.3) remains finite as $\varrho \to 0+$.

Going back to (2.2) with $s = n$ and noting

$$\liminf_{t\to 0+} t^{n-2}\varphi_u^2(t) = 0$$

(because of $r^{-1}u \in L^2(\mathbb{R}^n)$) we see

$$(2.5)\quad \int_{|x|<R} (\nabla u(x))^2 dx < \infty .$$

Since $u \in L^2(\mathbb{R}^n)$ also implies the existence of a sequence tending to infinity on which

$$(2.6)\quad (\varphi_u^2(t))' \leqslant 0$$

holds, we infer from (2.3) putting $s = 0$ that (2.4) and (2.5) are also valid with $R = \infty$.

Ad b): Let $0 < \varrho < R < \infty$, $u \in D(\hat{T})$. Then

$$\int\limits_{\varrho<|x|<R} (-\Delta u(x) + q(x)u(x))\,\overline{u(x)}\,dx = \chi_u(\varrho) + \chi_u(R) + \int\limits_{\varrho<|x|<R} (|\nabla u(x)|^2 + q(x)|u(x)|^2 dx$$

where

$$(2.7) \qquad \chi_u(t) := -\int\limits_{|x|=t} \overline{u}\,\frac{\partial u}{\partial n}\,dS \qquad (t \in (0,\infty)).$$

As a consequence of a)

$$\chi_0 := \lim_{\varrho \to 0+} \chi_u(\varrho) \quad \text{and} \quad \chi_\infty := \lim_{R\to\infty} \chi_u(R)$$

exist. Owing to Schwarz's inequality and $\nabla u \in L^2(\mathbb{R}^n)$ we have

$$\int\limits_0^\infty \frac{|\chi_u(t)|}{t}\,dt < \infty .$$

Hence $\chi_0 = \chi_\infty = 0$, so that $(\hat{T}u,u)$ is real.

Ad c): This assertion is trivial now because of b), Lemma 1, and Remark 1.

Ad d): From b) we observe that it is sufficient to prove $T_F \subset \hat{T}$. In view of (1.3) we therefore only need to show

$$u \in D(T_F) \quad \Rightarrow \quad r^{-1}u \in L^2(\mathbb{R}^n) .$$

Let $u \in D(T_F)$. Then there exists a sequence $\{u_j\}_{j\in\mathbb{N}} \subset C_0^\infty(\mathbb{R}^n_+)$ such that

$$s - \lim_{j\to\infty} u_j = u$$

and $$\lim_{j,k\to\infty} (T(u_j-u_k), u_j-u_k) = 0 .$$

Now, $$(T(u_j-u_k),u_j-u_k) \geq [\beta + (\tfrac{n-2}{2})^2]\ \|r^{-1}(u_j-u_k)\|^2$$

on account of Hardy's inequality (1.2). $\{r^{-1}u_j\}_{j\in\mathbb{N}}$ is thus a Cauchy sequence. It is clear that $r^{-1}u$ is its limit. ■

In developing the theory of sesquilinear forms in Hilbert space, Friedrichs [12] assuming $q \in C^1(\mathbb{R}^3)$ characterized $D(T_F)$ by those elements $u \in D(T^*)$ that satisfy ∇u, $q_+^{1/2} u \in L^2(\mathbb{R}^3)$. In [28] , dealing with singular q and more general operators, it was shown that the theory of sesquilinear forms can be avoided and that the condition $q_+^{1/2} u \in L^2(\mathbb{R}^n)$, which has to be assumed from the start when forms are considered, is a consequence of the other conditions to be imposed on u [8]. In the presentation above we have assumed $r^{-1}u \in L^2(\mathbb{R}^n)$ instead of $\nabla u \in L^2(\mathbb{R}^n)$ as in [28] in order to have a closer analogy with the corresponding result for Dirac operators in § 5.

Remark 2. The condition in item d) of Theorem 3 requires $q(x)$ to tend to $+\infty$ as $|x| \to 0+$ if $n = 2$. In [28, p.248] a result can be found where q may go to $-\infty$ even for $n = 2$.
A simple argument for which we refer the reader to [30] shows that in the case $q \geqslant \beta_0 r^{-2}$, $r^{-1}u \in L^2(\mathbb{R}^n)$ $(u \in D(T^*))$ need not be assumed but follows from (2.2) and (2.3). Upon that, $|q|^{1/2}u$, $\nabla u \in L^2(\mathbb{R}^n)$ together with the symmetry of T^* can be concluded. This provides another proof of Theorem 2.

---

[8] However, when working with forms [35, p.349; 71, p.43 f.] one can allow local singularities even stronger than those admissible to define T (cf. footnote 2; the remark in [28, p.235], l.-12, is therefore not appropriate).

3. Schrödinger Operators not Necessarily Semibounded

The proof of Theorem 2 mentioned in Remark 2 relies on the familiar fact [35, p.269] that a symmetric operator is essentially self-adjoint if and only if its adjoint is symmetric. To characterize the adjoint of the differential operator T_o (or T, but we shall confine ourselves to $\mathbb{R}^n$ as the basic domain in this section) we need the notion of generalized derivative (cf. 1.3)). It is natural to ask whether it does not suffice to show the symmetry of the largest classical differential operator that can be associated with $Du := -\Delta u + qu$ within $L^2(\mathbb{R}^n)$ in order to secure the essential self-adjointness of T_o, that is to show the symmetry of

$$T_1 u = Du \ , \ D(T_1) = \left\{ u \mid u \in C^2(\mathbb{R}^n) \cap L^2(\mathbb{R}^n),\ Du \in L^2(\mathbb{R}^n) \right\} .$$

We continue, of course, to assume

(i) $$q \in L^2_{loc}(\mathbb{R}^n)$$

to render T_1 a densely defined operator. It was shown by Wienholtz [84;23, p.189] that

(3.1) $$T_1 \text{ is symmetric} \iff T_o \text{ is essentially self-adjoint}$$

provided $q \in C^o(\mathbb{R}^n)$. The '$\Longrightarrow$' part ('$\Longleftarrow$' is trivial) is proved by establishing $\overline{R(T_o \pm iI)} = L^2(\mathbb{R}^n)$ for hölder continuous potentials by means of a regularity result for weak solutions of elliptic partial differential equations (a so-called Weyl's Lemma). A perturbation argument allows transition to $q \in C^o(\mathbb{R}^n)$.

Wienholtz [84] proved the symmetry of T_1 for potentials that may fall-off as rapidly as $-|x|^2$ at infinity thus carrying Levinson's well-known limit-point criterion over to the multi-dimensional case. His result was subsequently generalized in a series of papers by B. Hellwig [20,21,22] and Rohde [57,59] (the papers of J. Walter [79] have already been mentioned in § 1 in a different context). Rohde's paper [59] also contains an extended version of the equivalence (3.1).

We do not aim at generality of results here. Instead we shall give a new and short proof of the following significant special case of a theorem of B. Hellwig's.

Theorem 4 [22]. Assume, in addition to (i), that there exist numbers $t_o, M_o, M_1 > 0$ and a function $M \in C^1[t_o, \infty)$ with $M \geqslant M_o$ such that the following conditions are satisfied,

(ii) $\quad q(x) \geqslant -M(|x|) \qquad (x \in \mathbb{R}^n,\ |x| \geqslant t_o)$;

(iii) $\quad \left|\left(\frac{1}{\sqrt{M(t)}}\right)'\right| \leqslant M_1 \qquad (t \geqslant t_o)$;

(iv) $\quad \int_{t_o}^{\infty} \frac{dt}{\sqrt{M(t)}} = \infty$.

Then T_1 is symmetric. Moreover, T_o is essentially self-adjoint if $q \in C^o(\mathbb{R}^n)$ holds besides (ii) - (iv).

The proof of Theorem 4 is greatly facilitated by the following lemma, which is due to N. Nilsson [43] (for generalizations see [23, p.88; 22]), and which will be proved in the same manner as in [30]. Nilsson established the essential self-adjointness of T_o for singular q with a $-|x|^2$ fall-off at infinity by showing the symmetry of T_o^* with the help of cut-off functions (which are avoided in [30]). His and Wienholtz's paper [84] appeared almost simultaneously.

Lemma 2. Under the assumptions (i) - (iii) of Theorem 4

$$\int_{|x|>t_0} \frac{|\nabla u(x)|^2}{M(|x|)}\,dx < \infty$$

holds for all $u \in D(T_1)$.

Proof. We may assume $u \in D(T_1)$ to be real-valued. Let $R > t_o$; then, using

$$2\,\frac{|u||\nabla u|}{\sqrt{M}} \leqslant \frac{1}{2M_1}\,\frac{|\nabla u|^2}{M} + 2M_1\,|u|^2 ,$$

we obtain

$$(3.2)\quad \int_{t_o<|x|<R} (-\Delta u + qu)\,\frac{u}{M}\,dx = -\frac{1}{2}\left[\frac{t^{n-1}}{M(t)}\,(\varphi_u^2(t))'\right]_{t_o}^{R}$$

$$+ \int_{t_o<|x|<R} \frac{(\nabla u)^2 + qu^2}{M}\,dx + 2\int_{t_o<|x|<R} \frac{1}{\sqrt{M}}\,u\,\frac{\partial u}{\partial n}\left(\frac{1}{\sqrt{M}}\right)'\,dx$$

$$\geqslant -\frac{1}{2}\left[\frac{t^{n-1}}{M(t)}\,(\varphi_u^2(t))'\right]_{t_o}^{R} + \frac{1}{2}\int_{t_o<|x|<R} \frac{(\nabla u)^2}{M}\,dx$$

$$- \delta \int_{t_o<|x|<R} u^2 dx$$

where $\delta := 2\,M_1^2 + 1$. Now the assertion follows when we consider (3.2) on a sequence tending to infinity on which (2.6) holds taking into account that the left-hand side of (3.2) remains finite as $R \to \infty$.

Proof of Theorem 4. Let $u \in D(T_1)$. With the notation

$$\psi_u(t) := \int_{|x|<t} (-\Delta u + qu)u\,dx \qquad (t>0)$$

and (2.7) we have

$$\operatorname{Re}\psi_u(R) + i\operatorname{Im}\psi_u(R) = \chi_u(R) + \int_{|x|<R} (|\nabla u|^2 + q\,u^2)dx \quad (R>0).$$

Lemma 2 implies
$$\int_{t_0}^{\infty} \frac{|\chi_u(t)|}{\sqrt{M(t)}}\, dt < \infty ,$$
so that
$$\lim_{t\to\infty} \inf |\chi_u(t)| = 0$$
because of (iv). Hence there exists a sequence $\{R_n\}_{n\in\mathbb{N}}$ tending to infinity with
$$\lim_{n\to\infty} \operatorname{Im} \psi_u(R_n) = 0$$

Since $\operatorname{Im} \psi_u(\infty)$ clearly exists, $(T_1 u,u) = \operatorname{Re}\ \psi_u(\infty)$ (this kind of argument is also to be found in [79]) is real. The second assertion of Theorem 4 follows from (3.1). ∎

Remark 3. It should be noted that we did not prove that

$$(3.3) \qquad \lim_{R\to\infty} \int_{|x|<R} (|\nabla u|^2 + q|u|^2)dx$$

exists. As a matter of fact, (3.3) will generally fail to exist under the assumptions of Theorem 4 even in the one-dimensional case as was shown recently by Everitt, Giertz, and McLeod [8].

We conclude this section with the remark that the equivalence (3.1) does not hold for strongly singular potentials. This follows from Example 1a and (in comparing with [35, p.299] the reader should bear in mind that Kato deals loc. cit. with $n = 3$ only)

Example 1c. Let $n \geqslant 5$ and $q(x) = \frac{\beta}{|x|^2}$ where β is an arbitrary real number. Then T_1 is symmetric (choose M equal to a constant in Theorem 4).

4. Essential Self-Adjointness of Dirac Operators

In the Hilbert space

$$H = [L^2(\mathbb{R}^3)]^4 := \left\{ u \;\middle|\; u = \begin{pmatrix} u_1 \\ u_2 \\ u_3 \\ u_4 \end{pmatrix} : \mathbb{R}^3 \to \mathbb{C}^4 \;,\; \int_{\mathbb{R}^3} |u(x)|^2 dx < \infty \right\}$$

with the scalar product

$$(u,v) = \int_{\mathbb{R}^3} [u(x)]^t \, \overline{v(x)} \, dx \qquad (u^t \text{ be the transpose of } u)$$

the formal part of a Dirac operator describing the behaviour of a spin $1/2$ particle with nonzero rest mass under the influence of an electrostatic potential q is given by

$$\tau := \vec{\alpha} \cdot \vec{p} + \beta + q$$

where $\quad \vec{p} := \frac{1}{i} \nabla \qquad\qquad \vec{\alpha} := (\alpha_1, \alpha_2, \alpha_3).$

$\alpha_1, \alpha_2, \alpha_3, \alpha_4 := \beta$ are Hermitian 4 x 4 matrices satisfying the commutation relations

$$\alpha_j \alpha_k + \alpha_k \alpha_j = 2 \delta_{jk} I \qquad (j,k \in \{1,2,3,4\})$$

(I denotes the 4 x 4 unit matrix which will be identified with the unit operator in H; similarly the matrices α_j are interpreted as operators $\alpha_j : H \to H$), and $q \in L^2_{loc}(\mathbb{R}^3_+)$ is a real-valued function (we shall also employ the symbol "q" for any multiplication operator to be associated with q in H).

There are two important examples where the answer to (Q.1) (p.1) is known to be affirmative for the minimal operator

$$(4.1) \qquad T := \tau \restriction D_o \;, \quad D_o := [C_o^\infty(\mathbb{R}^3_+)]^4 \;.$$

Example 2. $q = 0$. The minimal operator for a free particle $T_o := (\vec{\alpha}\cdot\vec{p} + \beta)\restriction D_o$ is essentially self-adjoint (Rellich [54, Kapitel III] ; Kato [35, p.306] shows the essential self-adjointness of the larger operator $(\vec{\alpha}\cdot\vec{p} + \beta)\restriction [\, C_o^{\infty}(\mathbb{R}^3)\,]^4$).

Example 3. $q(x) = -\dfrac{\mu}{|x|}$. Then T is essentially self-adjoint if and only if $|\mu| \leqslant \frac{1}{2}\sqrt{3}$ (Rellich [53, p.722 f. ; 54, Kapitel III]). Rellich's proof depends on separation of the variables and the (at that time unproved) possibility of carrying the Weyl-Stone theory for Sturm-Liouville operators over to the systems generated by the radial part of τ (cf. [61,82]). If $\mu = Z\alpha$ ($\alpha = 1/137.038...$ is the fine structure constant) T is the Dirac operator of a hydrogen-like atom (one electron in a Coulomb field of a nucleus with atomic number Z) and thus essentially self-adjoint if and only if $Z \leqslant 118$.

In this section we should like to give a criterion for the essential self-adjointness of (4.1) covering Example 3 (in the range $Z \leqslant 118$), but also potentials q which are not necessarily spherically symmetric. For simplicity we therefore concentrate on the case where q is continuous throughout $\mathbb{R}^3$ with the possible exception of the origin.

The first result obtained without using separation is due to Kato [34, p.205 f.; 35, p.307] . He showed that T is essentially self-adjoint if $q \in C^o(\mathbb{R}^3_+)$ is real-valued and satisfies

$$(4.2) \qquad |q(x)| \leqslant \frac{\mu}{|x|} \qquad (x \in \mathbb{R}^3_+)$$

for some $\mu < \frac{1}{2}$ (this contains hydrogen-like atoms with $Z \leqslant 68$).

In fact, (4.2) implies

$$\|qu\| \leq 2\mu\|\nabla u\| \leq 2\mu((\vec{p}^{\,2}u,u) + \|u\|^2)^{1/2}$$
$$= 2\mu\|T_o u\| \qquad (u \in D_o,\ \mu > 0)$$

on account of Hardy's inequality (1.2). $T = T_o + q$ is therefore essentially self-adjoint for $\mu<\frac{1}{2}$ according to the Kato-Rellich perturbation theorem [35, p.289]. (By means of his concept of a pseudo-Friedrichs extension [35, p.341 ff.] Kato was able to improve the number $\frac{1}{2}$ to $\frac{2}{\pi}$ [corresponding to $Z \leq 87$], but he could not show that the pseudo-Friedrichs extension was the only self-adjoint extension of T.) Using a "mixed" method (perturbation in reducing subspaces) Rejto [52] proved that T with the potential of Example 3 is essentially self-adjoint if $\mu < \frac{3}{4}$ ($Z \leq 102$). All results mentioned so far [9] are strictly contained in

<u>Theorem 5</u> [66] . Assume,

$$\text{(I)}\quad \begin{cases} q \text{ can be expressed as } q = q_1 + q_2 \text{ with real-valued func-} \\ \text{tions } q_1 \in C^o(\mathbb{R}^3_+),\quad q_2 \in L^\infty(\mathbb{R}^3) \text{ satisfying} \\ |q_1(x)|^2 \leq \dfrac{a}{|x|^2}\, f'(|x|) \qquad (x \in \mathbb{R}^3_+)\,, \\ \text{where } a \in (0, \frac{1}{4}) \text{ is a suitable number and } f \in C^\infty((0,\infty)) \text{ a} \\ \text{monotone function with } t < f(t) \leq 3t \quad (t > 0). \end{cases}$$

Then T is essentially self-adjoint.

<u>Corollary</u>. If $q \in C^o(\mathbb{R}^3_+)$ is real-valued and satisfies (4.2) for some $\mu \in (0, \frac{1}{2}\sqrt{3})$, then T is essentially self-adjoint.

---

[9] as well as that of Gustafson and Rejto [18] who assume (4.2) with q spherically symmetric and $\mu < \frac{1}{2}\sqrt{3}$

Proof of the Corollary. Let $\mu \in (0, \frac{1}{2}\sqrt{3})$ and choose numbers $a' \in (0, \frac{1}{4})$, $c \in (1,3]$ such that $a'c = \mu^2$. Then (I) holds with $a = a'$, $f(t) = ct \quad (t > 0)$.

Theorem 5, which is to some extent sharp as Example 3 shows, is proved by means of a perturbation argument. However, q can no longer be viewed as a "small" perturbation of T_o (this method was exhausted by Kato since Hardy's inequality is sharp). Instead T is written as $T = (T_o + U) + (q - U)$ where U is chosen in such a manner that $q - U$ is a "small" perturbation of $T_o + U$. In contrast to the nonrelativistic case dealt with on p.10f, U can no longer be chosen as a symmetric operator, so that the Kato-Rellich perturbation theorem is not applicable to $q - U$. The ensuing difficulties are overcome by the following

<u>Theorem 6</u> [66]. Let H be a Hilbert space and A,B,C,S_1,S_2 symmetric operators in H with D(A) as common domain of definition. Moreover, assume that S_1,S_2 are bounded and $T_o := A + S_1$ is essentially self-adjoint. Let $z \in \mathbb{C}$ with $0 < |z| < 1$, $k > 1$. Suppose, writing

$$F := T_o + zC \quad , \quad G := B + S_2 + zC$$

that the following inequalities hold for $u \in D(A)$,

$$(4.3) \qquad \|T_o u\|^2 - \|u\|^2 \geqslant \|Au\|^2 \geqslant \|Cu\|^2 \ ;$$

$$(4.4) \qquad \|Fu\|^2 \geqslant k\,\|Gu\|^2 \ .$$

Then $T := T_o - B$ is essentially self-adjoint.

Theorem 6 is proved with the help of results on the stability of closedness and the stability of bounded invertibility under "small" perturbations [35, pp.190,196]. Theorem 5 can be proved by means of Theorem 6 as follows. Using the notations of (I) we put

$$k := \frac{2}{1+4a} \ , \quad s := \frac{8a}{1+4a} \ , \quad g(t) := \frac{1}{4t}\left(\frac{f(t)}{t} - 1\right) \qquad (t>0)$$

and make the following identifications (all operators having D_o as common domain of definition; for $h \in C^\infty((0,\infty))$ $h(r)$ or h denotes the operator of multiplication by $h(|x|)$ defined on D_o):

$$A = \vec{\alpha}\cdot\vec{p} \ , \quad B = -q \ , \quad C = g(r)r^{-1}\vec{\alpha}\cdot\vec{x} \quad {}^{10)}$$

$$S_1 = \beta \ , \quad S_2 = \frac{1}{k}(q^2+s^2g^2)^{-1}\, sg(sg + iqr^{-1}\vec{\alpha}\cdot\vec{x})\beta \ ,$$

$$z = is \ , \quad k = k \ .$$

The boundedness of S_2 can be verified by a simple but somewhat lengthy calculation. (4.3) is a direct consequence of Hardy's inequality. (4.4) is less easier seen to be valid. An important tool in deriving (4.4) is besides Lemma 1 the inequality

$$(4.5) \qquad (H(r)\,L_\sigma u,u) \geqslant -(H(r)\vec{L}^2u,u) \qquad (u \in D_o)$$

which holds for any $0 < H \in C^\infty((0,\infty))$. Here $\vec{L} := \vec{x}\times\vec{p}$ is the orbital angular momentum operator ("$\times$" denotes the vector product in this context and)

$$L_\sigma := \vec{\sigma}\cdot\vec{L} \qquad \text{with} \qquad \vec{\sigma} := \frac{1}{2i}\vec{\alpha}\times\vec{\alpha}$$

the spin-orbit coupling operator.

---

10) $\vec{x} := (x_1,x_2,x_3)$, x_j being the position operator (with domain D_o)

5. A Distinguished Self-Adjoint Extension of Dirac Operators

Let us consider Example 3 again. The radial part of τ is in the limit-point case at zero if $|\mu| \leq \frac{1}{2}\sqrt{3}$ and in the limit circle case if $|\mu| > \frac{1}{2}\sqrt{3}$; there is limit-point case at ∞ for $\mu \in \mathbb{R}$ [82, 29]. However, it can be seen from the eigensolutions of the radial Dirac equation that we have a nonoscillatory limit-circle case in the transition interval $\frac{1}{2}\sqrt{3} < |\mu| < 1$ (corresponding to $119 \leq Z \leq 137$). In the analogous case of the Schrödinger operator $(-\Delta + \beta r^{-2}) \restriction C_0^\infty(\mathbb{R}_+^3)$ the transition interval is $-\frac{1}{4} < \beta < \frac{3}{4}$ ([40]; see our Theorems 2 and 3). Hence a suitable boundary condition eliminating the "nonprincipal solution" (cf. [19, p.355]) defines a distinguished self-adjoint realisation of τ (which is usually tacitly used in physics). It is only in the range $|\mu| > 1$ ($Z \geqslant 138$) that we have an oscillatory limit-circle case at zero, so that the answer to (Q.2) (p.1) becomes negative (cf. [5; 39, p.140; 47,48,51] in this context).

In the nonrelativistic case it is the Friedrichs extension that represents the desired distinguished self-adjoint extension of the minimal operator (see § 2). Dirac operators, however, are never bounded from below, so that Friedrichs's procedure is not applicable to such operators. For a class of potentials containing the Coulomb potential $-\frac{\mu}{r}$ in the range $|\mu| < 1$ a distinguished self-adjoint extension of Dirac operators was first given in [67]. Generalizing the method of intercalating a suitable operator into $T = T_0 + q$ described in § 4 the desired extension of T was obtained by means of an "intercalary multiplicator" G as the closure of $\overline{G^{-1}}\ \overline{GT}$ (G is a multiplication operator generated by some power of $|x|$). More precisely, the following theorem holds.

Theorem 7 [67]. Suppose,

$$(\mathrm{II})\begin{cases} q \text{ can be written as } q = q_1 + q_2 \text{ where } q_1 \in C^0(\mathbb{R}^3_+), \\ q_2 \in L^\infty(\mathbb{R}^3) \text{ are real-valued functions satisfying} \\ \frac{1}{|x|^2}\left(h(|x|)+\frac{s}{2}\right)^2 \leqslant k\left(|q_1(x)|^2 + \frac{1}{|x|^2}h^2(|x|)\right) \leqslant \\ \qquad \leqslant \frac{1}{|x|^2}\left(h(|x|)+\frac{s+1}{2}\right)^2 + \frac{1}{|x|}h'(|x|) \qquad (x \in \mathbb{R}^3_+) \\ \text{with appropriate numbers } s \in [0,1),\ k>1,\ c>1 \text{ and a} \\ \text{suitable function } h \in C^\infty((0,\infty)) \text{ with } 0 < h(t) \leqslant \frac{1-s}{2c} \quad (t>0). \end{cases}$$

Then there exists a bounded symmetric operator S such that

$$T_G := \overline{r^{-s/2}}\ \overline{r^{s/2}(T-S)} + \overline{S}$$

is an essentially self-adjoint extension of T.

Remark 4. If (II) holds with s=0 then T is essentially self-adjoint. This is a slight generalization of Theorem 5. In fact, the inequality occuring in (II) with s=0 is equivalent to

$$(5.1)\quad k|x|^2|q_1(x)|^2 \leqslant \frac{1}{4}\frac{\partial}{\partial|x|}\left[|x|(4h(|x|)+1)\right] - (k-1)h^2(|x|) \qquad (x \in \mathbb{R}^3_+).$$

If we have k=c and $[t(4k\ h(t)+1)]' \geqslant 0$ $(t>0)$ a sufficient (but not a necessary) condition for (5.1) is (note $0 < h \leqslant \frac{1}{2k}$)

$$(5.2)\quad k|x|^2|q_1(x)|^2 \leqslant \frac{1}{4}\ \frac{k+1}{2k^2}\ \frac{\partial}{\partial|x|}\left[|x|(4k\ h(|x|)+1)\right] \qquad (x \in \mathbb{R}^3_+)$$

Now let (I) hold. Then there exists a number $k>1$ such that $\frac{1}{4}\ \frac{k+1}{2k^3} = a$. Putting $c := k$, $s := 0$, $h(t) := \frac{1}{4k}\left(\frac{f(t)}{t} - 1\right)$ $(t>0)$ we have $(\mathrm{I}) \Rightarrow (5.2) \Rightarrow (\mathrm{II})$.

Remark 5. If q is in particular a Coulomb potential, $q(x) = -\frac{\mu}{|x|}$, (II) is satisfied provided

$$(5.3)\qquad 1-4\mu^2 \leq (1-s)^2 < 4(1-\mu^2) \qquad (s\in[0,1) \text{ suitably})$$

holds. Indeed, the right inequality in (5.3) implies

$$\mu^2 < 1 - \frac{(1-s)^2}{4} = \frac{1-s^2}{2} + \frac{(s+1)^2}{4} \quad .$$

Thus there exist numbers $k > 1$, $c > 1$ with the property

$$(5.4)\qquad \mu^2 \leq \frac{1-s^2}{2kc} + \frac{(s+1)^2}{4k} - \frac{k-1}{k}\,\frac{(1-s)^2}{4c^2} \quad .$$

(5.4) is equivalent to

$$k\left[\mu^2 + \left(\frac{1-s}{2c}\right)^2\right] \leq \left(\frac{1-s}{2c} + \frac{s+1}{2}\right)^2 .$$

On the other hand we have

$$\begin{aligned} k\left[\mu^2 + \left(\frac{1-s}{2c}\right)^2\right] - \left(\frac{1-s}{2c} + \frac{s}{2}\right)^2 &\geq k\mu^2 - \left[\frac{s(1-s)}{2c} + \frac{s^2}{4}\right] \\ &\geq \mu^2 - \left[\frac{s(1-s)}{2} + \frac{s^2}{4}\right] \\ &= \mu^2 + \frac{(1-s)^2}{4} - \frac{1}{4} \geq 0, \end{aligned}$$

using the left inequality in (5.3). The inequality in (II) is therefore valid with $h(t) := \frac{1-s}{2c}$ $(t > 0)$.

If $|\mu| < \frac{1}{2}\sqrt{3}$, (5.3) holds good with s=0 which means that T is essentially self-adjoint . In the transition interval $\frac{1}{2}\sqrt{3} < |\mu| < 1$ one has to choose $s > 0$ (T is no longer essentially self-adjoint as we know), whereas (5.3) cannot be satisfied for $s \in \mathbb{R}$ any more if $|\mu| > 1$.

Like the Friedrichs extension of the analogous nonrelativistic case, the essentially self-adjoint extension of T given in Theorem 7 can explicitly be characterized by restricting T^* in terms of a boundary condition making its distinction physically conspicuous (finiteness of the potential energy; cf. § 2).

We point out that under very general conditions on q, in particular those of Theorem 7, T^* can be characterized as follows [7],

$$D(T^*) = \left\{ u \mid u \in H,\ u_j \in H^1_{loc}(\mathbb{R}^3_+),\ j \in \{1,2,3,4\}\ ,\ \ \tau u \in H \right\},$$
$$T^* u = \tau u\ .$$

H^1_{loc} is the usual Sobolev space of L^2_{loc} - functions possessing local derivatives of first order in the distributional sense.

Theorem 8 [67]. Suppose, q satisfies hypothesis (II). Let

$$D_\sigma := D(T^*) \cap D(\overline{r^{-\sigma}}),\quad T_\sigma := T^* \restriction D_\sigma \qquad (\sigma \in \mathbb{R})\ .$$

Then

$$\overline{T_G} = \overline{T^*_\sigma} \qquad (\sigma \in [\ \tfrac{1}{2},\ 1-\tfrac{s}{2}\ )\ .$$

Especially $T^*_{1/2}$, the restriction of T^* to the subspace

$$D_{1/2} = \left\{ u \mid u \in H,\ u_j \in H^1_{loc}(\mathbb{R}^3_+),\ j \in \{1,2,3,4\}\ ,\ \int_{\mathbb{R}^3} \frac{|u(x)|^2}{|x|}\, dx < \infty,\ \ \tau u \in H \right\},$$

is an essentially self-adjoint extension of T.

The proof of Theorem 7 is like that of Theorem 5 achieved by means of an abstract perturbation theorem. Those parts of Theorem 7 that can be proved by purely abstract Hilbert space methods are collected in the following theorem for the proof of which we must refer the reader to [67].

Theorem 9 [67] . Let A_o,B,C,G be operators in a Hilbert space H.

A_o be essentially self-adjoint and B, G symmetric. Assume that the following hypotheses are fulfilled:

(i) $D_o := D(A_o) = D(B) = D(C) \subset D(G) =: D_1$,

(ii) $R(A_o) \cup R(B) \cup R(C) \subset D_1$;

(iii) $R(G) = D_1$, $R(G \restriction D_o) = D_o$;

(iv) there are numbers $c_o > 0$, $c_1 > 0$, $c_2 > 1$, $c_3 > 1$, $c_4 > 1$ such that the following inequalities hold for $u \in D_o$,

$\| A_o u \| \geqslant c_o \| u \|$, $\| A_o G^2 u \| \geqslant c_1 \| u \|$,

$\| A_o G u \| \geqslant c_2 \| (A_o G - G A_o) u \|$,

$\| G A_o u \| \geqslant c_3 \| G C u \|$, $\| G(A_o + C) u \| \geqslant c_4 \| G(B+C) u \|$.

Then G is invertible, $G(A_o - B)$ is closable, and

$$A_G := \overline{G^{-1}}\ \overline{G(A_o - B)}$$

is an essentially self-adjoint extension of $A_o - B$.

The validity of Theorem 7 is now established by specializing the operators occurring in Theorem 9 to (D_o is the subspace of (4.1))

$$A_o = (\vec{\alpha} \cdot \vec{p} + \beta) \restriction D_o \ , \quad B = -q + S_o \qquad (D(B) = D_o)$$

$$C = r^{-1} h(r)\, r^{-1}\, \vec{\alpha} \cdot \vec{x} \quad (D(C) = D_o), \ G = r^{s/2} \quad (D(G) = [C_o^1(\mathbb{R}^3_+)]^4)$$

where

$$S_o = \frac{1}{k} (q_1^2 + r^{-2} h^2)^{-1} (\tfrac{1}{2} r^{-1} s + r^{-1} h)(r^{-1} h + i q_1\, r^{-1} \vec{\alpha} \cdot \vec{x}) \beta, \ D(S_o) = D_o$$

is a bounded symmetric operator. Granted that the assumptions of Theorem 9 are satisfied, $\overline{G^{-1}}\ \overline{G(A_o - B)}$ is an essentially self-adjoint extension of $A_o - B = T - (S_o + q_2)$ whence the assertion of Theorem 7 follows with $S := S_o + q_2$. In verifying the hypotheses of Theorem 9 it is in particular the last two inequalities of (v) that require a great deal of technicalities. These relations can be shown to hold by means of a generalization of Hardy's inequality and inequality (4.5), together with various commutation relations.

Theorem 8 can be proved by showing $D(T_G) \subset D(\overline{r^{-G}})$ first. Because of $T_G \subset T^*$ this implies $\overline{T_G} \subset \overline{T_G^*}$. The inclusion $\overline{T_G^*} \subset \overline{T_G}$ follows from the symmetry of T_G^* which in turn is a simple consequence of $r^{-G} u \in H \quad (u \in D(T_G^*))$.

6. Construction of a Distinguished Self-Adjoint Extension of Dirac Operators by Means of Cut-Off Potentials

In this section we shall present a second method to gain distinguished self-adjoint realizations of Dirac expressions as extensions of minimal operators T_o that are not necessarily essentially self-adjoint. This method is inspired by the intention frequently encountered in physics to motivate by means of a cut-off procedure the choice of the "principal solution" [19, pp.355,402] among the eigensolutions of the Schrödinger or Dirac equation [38,39]. A characteristic feature of the Dirac operators considered here is the gap in their spectrum[11].

---

[11] Not to be confused with the gap (-1,1) in their essential spectrum. Taking Example 3, every self-adjoint extension of T has the gap (-1,1) in its essential spectrum irrespective of $\mu \in \mathbb{R}$ [82]. $\overline{T}$ has a spectral gap, however, if and only if $|\mu| < 1$.

This <u>spectral gap</u>, though not a proper substitute for the lacking semiboundedness of these operators, allows us to prove some propositions which are well known in the case of semibounded operators, for instance

$$A_1 \geqslant A_2 > 0 \implies A_1^{-1} \leqslant A_2^{-1}$$

A_1, A_2 being self-adjoint operators in a Hilbert space H [35, p.330].

Let a Dirac expression with a nonoscillatory potential q becoming singular at the origin be given (e.g., Example 3 with $|\mu|<1$). With the help of a family $\{q_t\}_{t \geqslant t_o}$ of bounded functions q_t arising from q by means of a suitable cut-off we define a monotone family of self-adjoint Dirac operators. We show that their resolvents (at the point $\lambda = 0$) are also monotone, that they are strongly convergent, and that their limit operator is invertible. The inverse of this limit operator is then a self-adjoint extension of the operator T defined in (4.1). The convergence theorem used in this context is

<u>Theorem 10</u> [90]. Let $\{T_t\}_{t \in J}$ ($J := [t_o, \infty)$ where $t_o \in \mathbb{R}$ is an appropriate number) be a family of self-adjoint operators. Assume,

(i) $D(T_t) = \text{const.} =: D \qquad (t \in J)$;

(ii) there is a subset $D_1 \subset D$, dense in H, such that $\{T_t \restriction D_1\}$ is strongly convergent;

(iii) $T_t - T_s \geqslant 0$ for all $t,s \in J$ with $t \geqslant s$ (or for all $t,s \in J$ with $s \geqslant t$);

(iv) $\overline{T_t - T_s}$ is bounded (and everywhere defined) in H $(t, s \in J)$ and

$$\lim_{t \to s} \|\overline{T_t - T_s}\| = 0 \qquad (s \in J);$$

(v) there exist numbers $\lambda \in \mathbb{R}$, $\alpha > 0$ with

$$\|(T_t - \lambda I)\, u\| \geqslant \alpha \|u\| \qquad (t \in J,\ u \in D).$$

Then the following propositions hold,

a) $\lambda \in \bigcap_{t \in J} \varrho(T_t)$;

b) $s\text{-}\lim_{t\to\infty} (T_t - \lambda I)^{-1} =: R(\lambda)$ exists, and $R(\lambda)$ is invertible [12];

c) $T_g := R(\lambda)^{-1} + \lambda I$ is a self-adjoint extension of $s\text{-}\lim_{t\to\infty} T_t \upharpoonright D_1$;

d) $D(T_g) = \left\{ u \,|\, u \in H, \text{ there exists } \{u_t\}_{t \in J} \subset D \text{ such that } s\text{-}\lim_{t\to\infty} u_t = u \text{ and } \{T_t u_t\} \text{ is convergent} \right\}$ [12], and for all $u \in D(T_g)$ we have $T_g u = s\text{-}\lim_{t\to\infty} T_t u_t$ (u_t $(t \in J)$ being an appropriate element of D).

---

12) To every family of operators $\{A_t\}_{t \in J}$ we can attach the set $G := \left\{ \langle u,v \rangle \;|\; \langle u,v \rangle \in H \times H, \text{ there exists } \{u_t\}_{t \in J} \text{ with } u_t \in D(A_t)\ (t \in J) \text{ and } \lim_{t\to\infty} (\|u_t - u\| + \|A_t u_t - v\|) = 0 \right\}$. If G is the graph of an operator A (for which in the case of symmetric operators A_t condition (ii) of Theorem 10 is sufficient) then $\{A_t\}_{t \in J}$ is called strongly graph convergent [15]. For the interdependence of strong graph and strong resolvent convergence (generalized strong convergence in the terminology of Kato [35; VIII, § 1.1]) see [90] .

Now take the example

$$q(x) = -\frac{\mu}{|x|} \qquad (x \in \mathbb{R}^3_+ \; ; \quad \mu \in (0,1) \text{ suitably})$$

again. Let $\{q_t\}_{t \geq t_o}$ be a family of bounded functions continuous in t (locally uniformly with respect to x) with

$$q_t(x) = q(x) \qquad \text{for} \quad |x| \geq \frac{c}{t} \qquad (t \geq t_o)$$

and
$$q_t(x) \leq q_s(x) \qquad (x \in \mathbb{R}^3, \; t \geq s \geq t_o)$$

where t_o, c are suitably chosen. Taking for T_t the closure of $(\vec{\alpha}\cdot\vec{p} + \beta + q_t) \upharpoonright D_o$, conditions (i) - (iv) are obviously satisfied with $D_1 := D_o$.

Thus there remains the hard analytic problem of showing the validity of (v). It is solved with the help of the following

<u>Theorem 11</u> [68]. Suppose,

(III) $q(\cdot) = Q(|\cdot|)$ can be written as $Q = Q_1 + Q_2$ with real-valued functions Q_1, $Q_2 \in C^1((0,\infty))$ satisfying

$$r^2 \, Q_1^2(r) \leq a \, f'(r) \quad , \quad |Q_2(r)| \leq \text{const.} \qquad (r > 0)$$

where $a \in (0, \frac{1}{4})$ is a suitable number and $f \in C^\infty((0,\infty))$ a monotone function with $r < f(r) \leq 3r \quad (r > 0)$;

(IV)
$$\mu := \sup_{r>0} |r\,Q(r)| < 1 \,,$$
$$\nu := \sup_{r>0} |r\,(rQ(r))'| < 2(1-\mu)^2 \,.$$

Assertion: $1 - \left[\mu + \frac{\nu}{2(1-\mu)}\right]^2 > 0$, and the spectrum of $\overline{T}$ is a subset of

$$\left\{ \lambda \mid \lambda \in \mathbb{R} \,, \; \lambda^2 \geq 1 - \left[\mu + \frac{\nu}{2(1-\mu)}\right]^2 \right\}.$$

Remark 6. Theorem 11 is to some extent sharp as Example 3 with $|\mu| < \frac{1}{2}\sqrt{3}$ shows. Theorem 11 implies $\sigma(\overline{T}) \subset \mathbb{R} \setminus (-\sqrt{1-\mu^2}, \sqrt{1-\mu^2})$. On the other hand, it is known by separation that $\sqrt{1-\mu^2}$ for $\mu > 0$ [electron] ($-\sqrt{1-\mu^2}$ for $\mu < 0$ [positron]) is the lowest (highest) eigenvalue of $\overline{T}$ and that the essential spectrum of $\overline{T}$ equals $\mathbb{R} \setminus (-1,1)$ (cf. [39, p.142; 82]).

While condition (III) implies the essential self-adjointness of T (see Theorem 5), condition (IV) is sufficiently general to cover for example the Dirac operators of hydrogen-atoms with $Z \leqslant 137$, i.e. operators which are not necessarily essentially self-adjoint. Theorem 11 enables us to prove a result which together with Theorem 10 establishes a distinguished self-adjoint extension of such operators.

Theorem 12 [68]. Let q satisfy (IV) but not necessarily (III). Moreover, assume sign $Q(r) \neq 0$ ($0 < r < r_o$, $r_o > 0$ suitably), and let ε be number with $0 < \varepsilon < 2(1-\mu)^2 - \nu$. Then there exists a family of real-valued functions $q_t \in C^1(\mathbb{R}^3_+)$ with

$$q_t(x) = q(x) \quad \text{for } |x| \geqslant \frac{c}{t}, \qquad |q_t(x)| \leqslant t \qquad (x \in \mathbb{R}^3_+)$$

for all $t \geqslant t_o$ (t_o, c positive numbers) such that $\{T_t\}_{t \geqslant t_o}$ with

$$T_t := \overline{(\vec{\alpha}\cdot\vec{p} + \beta + q_t) \upharpoonright D_o}$$

satisfies the conditions of Theorem 10 with $\lambda = 0$, $D_1 := D_o$. The limit operator T_g defined in Theorem 10 is a self-adjoint extension of T with the property

$$\sigma(T_g) \subset \left\{ \lambda \mid \lambda \in \mathbb{R},\ \lambda^2 \geqslant 1 - \left[\mu + \frac{\nu+\varepsilon}{2(1-\mu)}\right]^2 \right\}.$$

To prove Theorem 12, a family of mollified cut-off potentials

$$q_t(x) := Q_t(|x|) \qquad (x \in \mathbb{R}^3_+ \, , \, t \geq t_o)$$

is constructed for which

$$|Q_t(r)| \leq t \qquad (r > 0) \, ,$$

$$\mu_t := \sup_{r>0} |rQ_t(r)| \leq \mu \, ,$$

$$\nu_t := \sup_{r>0} |r(rQ_t(r))'| \leq \nu + \varepsilon$$

holds ($t \geq t_o$). Since the Q_t are bounded, Theorem 11 yields

$$\| T_t \, u \| \geq \alpha \| u \| \qquad (u \in D(T_t))$$

where

$$\alpha := \left\{ 1 - \left[\mu + \frac{\nu + \varepsilon}{2(1-\mu)} \right]^2 \right\}^{1/2} > 0 \, ,$$

which is assumption (v) of Theorem 10.

In accordance with what has been said at the beginning of § 5, this construction of a distinguished self-adjoint extension of T fails for Coulomb potentials with $|\mu| > 1$ because of the absence of a spectral gap of such operators.

References

1. Barut, A.O.: Some unusual applications of Lie algebra representations in quantum theory. SIAM J. Appl. Math. 25, 247-259 (1973)

2. Behncke, H.: Some remarks on singular attractive potentials. Nuovo Cimento 55, 780-785 (1968)

3. Carleman, T.: Sur les équations intégrales singulières à noyau réel et symétrique . Uppsala universitets årsskrift 1923.

4. Carleman, T.: Sur la théorie mathématique de Schrödinger. Arkiv för mat., astr. och fysik 24 B, N:o 11(1934) (= Edition complète des articles de Torsten Carleman: Malmö 1960)

5. Case, K.M.: Singular potentials. Phys. Rev. 80, 797-806 (1950)

6. Courant, R., und Hilbert, D.: Methoden der Mathematischen Physik I. 2. Auflage. Berlin: Springer 1931

7. Evans, W.D.: On the unique self-adjoint extension of the Dirac operator and the existence of the Green matrix. Proc. London Math. Soc. (3) 20, 537-557 (1970)

8. Everitt, W.N., Giertz, M., and McLeod, J.B.: On the strong and weak limit-point classification of second-order differential expressions. The University of Wisconsin Technical Summary Report # 1338, September 1973

9. Frank, W.M., Land, D.J., and Spector, R.M.: Singular potentials. Rev. Mod. Physics 43, 36-98 (1971)

10. Frank, Ph., und v.Mises, R.: Die Differential- und Integralgleichungen der Mechanik und Physik II. 2. Auflage. Braunschweig: Vieweg 1934

11. Freudenthal, H.: Über die Friedrichssche Fortsetzung halbbeschränkter Hermitescher Operatoren. Nederl. Akad. Wetensch. Proc. 39, 832-833 (1936)

12. Friedrichs, K.: Spektraltheorie halbbeschränkter Operatoren und Anwendung auf die Spektralzerlegung von Differentialoperatoren I,II. Math. Ann. 109, 465-487, 685-713 (1933/34) (Berichtigung, Math. Ann. 110, 777-779 (1934/35))

13. Friedrichs, K.: Über die ausgezeichnete Randbedingung in der Spektraltheorie der halbbeschränkten gewöhnlichen Differentialoperatoren zweiter Ordnung. Math. Ann. 112, 1-23 (1935/36)

14. Glazman, I.M.: Direct methods of qualitative spectral analysis of singular differential operators. Jerusalem: Israel program for scientific translations 1965

15. Glimm, J., and Jaffe, A.: Singular perturbations of selfadjoint operators. Comm. Pure Appl. Math. 22, 401-414 (1969)

16. Gordon, W.: Die Energieniveaus des Wasserstoffatoms nach der Diracschen Theorie des Elektrons. Z. Physik 48, 11-14 (1928)

17. Grütter, A.: Wesentliche Selbstadjungiertheit eines Schrödinger-Operators. Math. Z. 135, 289-291 (1974)

18. Gustafson, K.E., and Rejto, P.A.: Some essentially self-adjoint Dirac operators with spherically symmetric potentials. Israel J. Math. 14, 63-75 (1973)

19. Hartman, Ph.: Ordinary Differential Equations. New York: John Wiley 1964

20. Hellwig, B.: Ein Kriterium für die Selbstadjungiertheit elliptischer Differentialoperatoren im R_n. Math. Z. 86, 255-262 (1964)

21. Hellwig, B.: Ein Kriterium für die Selbstadjungiertheit singulärer elliptischer Differentialoperatoren im Gebiet G. Math. Z. 89, 333-344 (1965)

22. Hellwig, B.: A criterion for self-adjointness of singular elliptic differential operators. J. Math. Anal. Appl. 26, 279-291 (1969)

23. Hellwig, G.: Differential operators of mathematical physics. Reading: Addison-Wesley 1967

24. Ikebe, T., and Kato, T.: Uniqueness of the self-adjoint extensions of singular elliptic differential operators. Arch. Rational Mech. Anal. 9, 77-92 (1962)

25. Jansen, K.-H.: Neue Kriterien für das Fehlen von L^2- Lösungen für $-\Delta v = f(x,v)$ im R_n unter besonderer Berücksichtigung des linearen Falles. Jber.Deutsch.Math.Verein. 72, 190-204 (1971).

26. Jauch, J.M.: Foundations of quantum mechanics. Reading: Addison-Wesley 1968

27. Jörgens, K.: Wesentliche Selbstadjungiertheit singulärer elliptischer Differentialoperatoren zweiter Ordnung in $C_o^\infty(G)$. Math. Scand. 15, 5-17 (1964)

28. Kalf, H.: On the characterization of the Friedrichs extension of ordinary or elliptic differential operators with a strongly singular potential. J. Functional Anal. 10, 230-250 (1972)

29. Kalf, H.: A limit-point criterion for separated Dirac operators and a little known result on Riccati's equation. Math. Z. 129, 75-82 (1972)

30. Kalf, H.: Self-adjointness for strongly singular potentials with a $-|x|^2$ fall-off at infinity. Math. Z. 133, 249-255 (1973)

31. Kalf, H.: The quantum mechanical virial theorem and the absence of positive energy bound states of Schrödinger operators. Submitted to J. Analyse Math.

32. Kalf, H., and Walter, J.: Strongly singular potentials and essential self-adjointness of singular elliptic operators in $C_o^\infty(\mathbb{R}^n \setminus \{0\})$. J. Functional Anal. 10, 114-130 (1972)

33. Kalf, H., and Walter, J.: Note on a paper of Simon on essentially self-adjoint Schrödinger operators with singular potentials. Arch. Rational Mech. Anal. 52, 258-260 (1973)

34. Kato, T.: Fundamental properties of Hamiltonian operators of Schrödinger type. Trans. Amer. Math. Soc. 70, 195-211 (1951)

35. Kato, T.: Perturbation theory for linear operators. Berlin-Heidelberg-New York: Springer 1966

36. Kato, T.: Schrödinger operators with singular potentials. Israel J. Math. 13, 135-148 (1972)

37. Kemble, E.C.: The fundamental principles of quantum mechanics. New York: Dover 1958

38. Landau, L.D., and Lifshitz, E.M.: Quantum Mechanics. Non-relativistic theory. London: Pergamon 1959

39. Landau, L.D., und Lifschitz, E.M.: Relativistische Quantentheorie. Berlin: Akademie-Verlag 1971

40. Meetz, K.: Singular potentials in nonrelativistic quantum mechanics. Nuovo Cimento 34, 690-708 (1964)

41. Morse, Ph.M., and Feshbach, H.: Methods of theoretical physics II. New York: Mc Graw-Hill 1953

42. v. Neumann, J.: Mathematische Grundlagen der Quantenmechanik. Berlin: Springer 1932

43. Nilsson, N.: Essential self-adjointness and the spectral resolution of Hamiltonian operators. Kungl. Fysiogr. Sällsk. i Lund Förh. Bd. 29, Nr. 1 (1959)

44. Oppenheimer, J.R.: Three notes on the quantum theory of aperiodic effects. Phys. Rev. 31, 66-81 (1928)

45. Oseen, C.W.: Über die Fundamentalintegrale einiger wellenmechanischen Differentialgleichungen. Akiv för mat., astr. och fysik. 22 A, N:o 2 (1930)

46. Oseen, C.W.: Deux remarques sur la méthode des perturbation dans la mécanique ondulatoire Arkiv för math. astr. och fysik 25 A , N:o 2 (1934)

47. Popov, V.S.: " Collapse to the center " at $Z > 137$ and critical nuclear charge. Soviet J. Nuclear Phys. 12, 235-243 (1971) (english translation of Yad. Fiz. 12, 429-447 (1970))

48. Popov, V.S.: On the properties of the discrete spectrum for Z close to 137. Soviet Phys. JETP 33, 665-673 (1971) (english translation of Ž. Eksp. Teor. Fiz. 60, 1228-1244 (1971))

49. Povzner, A.Ya.: The expansion of arbitrary functions in terms of eigenfunctions of the operator $-\Delta u + cu$. Amer. Math. Soc. Transl. (2) 60, 1-49 (1967) (english translation of Mat. Sb. 32, 109-156 (1953))

50. Reed, M., and Simon, B.: Methods of modern mathematical physics I: Functional analysis. New York: Academic Press 1972

51. Rein, D.: Über den Grundzustand überschwerer Atome. Z. Physik 221, 423-430 (1969)

52. Rejto, P.A.: Some essentially self-adjoint one-electron Dirac operators. Israel J. Math. 9, 144-171 (1971)

53. Rellich, F.: Die zulässigen Randbedingungen bei den singulären Eigenwertproblemen der mathematischen Physik. Math. Z. 49, 702-723 (1943/44)

54. Rellich, F.: Eigenwerttheorie partieller Differentialgleichungen II. Vervielfältigtes Vorlesungsmanuskript: Göttingen 1953

55. Rellich, F.: Halbbeschränkte Differentialoperatoren höherer Ordnung. Proc. Internat. Congress Math. Amsterdam 1954. Vol. 3, 243-250

56. Rellich, F.: Perturbation theory of eigenvalue problems. New York: Gordon and Breach 1969

57. Rohde, H.-W.: Über die Symmetrie elliptischer Differentialoperatoren. Math. Z. 86, 21-33 (1964)

58. Rohde, H.-W.: Regularitätsaussagen mit Anwendungen auf die Spektraltheorie elliptischer Differentialoperatoren. Math.Z. 91, 30-49 (1966)

59. Rohde, H.-W.: Kriterien zur Selbstadjungiertheit elliptischer Differentialoperatoren I,II. Arch. Rational Mech. Anal. 34, 188-201, 202-217 (1969)

60. Rohde, H.-W.: Ein Kriterium für das Fehlen von Eigenwerten elliptischer Differentialoperatoren. Math. Z. 112, 375-388 (1969)

61. Rohde, H.-W.: Die Weyl-Stonesche Theorie für Systeme gewöhnlicher Differentialoperatoren. RWTH Aachen 1969 (unveröffentlicht)

62. Rohde, H.-W., und Wienholtz, E.: Ein Regularitätssatz zur schwachen 2. Randbedingung mit Anwendungen auf elliptische Differentialoperatoren. Math. Z. 98, 9-26 (1967)

63. Schmincke, U.-W.: Über das Verhalten der Eigenfunktionen eines singulären elliptischen Differentialoperators. Math. Z. 111, 267-288 (1969)

64. Schmincke, U.-W.: Über die Potenzräume eines Schrödinger-Operators. Math. Z. 114, 349-360 (1970)

65. Schmincke, U.-W.: Essential selfadjointness of a Schrödinger operator with strongly singular potential. Math. Z. 124, 47-50 (1972)

66. Schmincke, U.-W.: Essential selfadjointness of Dirac operators with a strongly singular potential. Math. Z. 126, 71-81 (1972)

67. Schmincke, U.-W.: Distinguished selfadjoint extensions of Dirac operators. Math. Z. 129, 335-349 (1972)

68. Schmincke, U.-W.: A spectral gap theorem for Dirac operators with central field. Math. Z. 131, 351-356 (1973)

69. Shortley, G.H.: The inverse-cube central force field in quantum mechanics. Phys. Rev. 38, 120-127 (1931)

70. Simader, C.G.: Bemerkungen über Schrödinger-Operatoren mit stark singulären Potentialen. Math. Z. (erscheint demnächst)

71. Simon, B.: Quantum mechanics for Hamiltonians defined as quadratic forms. Princeton University Press 1971

72. Simon, B.: Essential self-adjointness of Schrödinger operators with positive potentials. Math. Ann. 201, 211-220 (1973)

73. Simon, B.: Essential self-adjointness of Schrödinger operators with singular potentials. Arch. Rational Mech. Anal. 52, 44-48 (1973)

74. Sommerfeld, A.: Atombau und Spektrallinien. Wellenmechanischer Ergänzungsband zur 4. Aufl. Braunschweig: Vieweg 1929

75. Stetkær - Hansen, H.: A generalization of a theorem of Wienholtz concerning essential self-adjointness of singular elliptic operators. Math. Scand. 19, 108-112 (1966)

76. Stone, M.H.: Linear transformations in Hilbert space and their applications to analysis. New York: Amer. Math. Soc. Colloq. Publ. 1932

77. Stummel, F.: Singuläre elliptische Differentialoperatoren in Hilbertschen Räumen. Math. Ann. 132, 150-176 (1956)

78. Triebel, H.: Erzeugung nuklearer lokalkonvexer Räume durch singuläre Differentialoperatoren zweiter Ordnung. Math. Ann. 174, 163-176 (1967)

79. Walter, J.: Symmetrie elliptischer DifferentialoperatorenI,II. Math. Z. 98, 401-406 (1967); 106, 149-152 (1968)

80. Walter, J.: Note on a paper by Stetkær - Hansen concerning essential self-adjointness of Schrödinger operators. Math. Scand. 25, 94-96 (1969)

81. Watanabe, S.: Knowing and guessing. New York: John Wiley 1969

82. Weidmann, J.: Oszillationsmethoden für Systeme gewöhnlicher Differentialgleichungen. Math. Z. 119, 349-373 (1971)

83. Wienholtz, E.: Halbbeschränkte partielle Differentialoperatoren zweiter Ordnung vom elliptischen Typus. Math. Ann. 135, 50-80 (1958)

84. Wienholtz, E.: Bemerkungen über elliptische Differentialoperatoren. Arch. Math. 10, 126-133 (1959)

85. Wightman, A.S.: The problem of existence of solutions in quantum field theory. Proc. 5th Annual Eastern Theor. Phys. Conference 1966. New York: Benjamin 1967

86. Witte, J.: Über die Regularität der Spektralschar eines singulären elliptischen Differentialoperators. Math. Z. 107, 116-126 (1968)

87. Witte, J.: Über das Verhalten der Spektralschar eines elliptischen Differentialoperators in der Umgebung der Singularität des Potentials $q(x) = |x|^{-\alpha}$. Math. Z. 115, 140-152 (1970)

88. Witte, J.: Über Regularitätseigenschaften der Potenzräume eines Schrödinger-Operators mit singulärem Potential. Math. Z. 128, 199-205 (1972)

89. Wüst, R.: Generalizations of Rellich's theorem on perturbation of (essentially) self-adjoint operators. Math. Z. 119, 276-280 (1971)

90. Wüst, R.: A convergence theorem for selfadjoint operators applicable to Dirac operators with cutoff potentials. Math. Z. 131, 339-349 (1973)

Scattering theory for differential operators, III; exterior problems

S. T. Kuroda

I. Introduction

In the present note we intend to study some spectral properties of exterior problems for selfadjoint elliptic operators by applying an abstract stationary method in the scattering theory developed in [10]. For second order operators, exterior problems have been investigated extensively. Here, we quote only Ikebe[7], Shenk and Thoe[12] for Schrödinger operators, and Birman[2], Mochizuki[11] for more general operators. We note that the almost best possible condition $O(|x|^{-\delta})$, $\delta > 1$, for the decay of perturbing coefficients was used in [11].

For higher order operators, however, the study of exterior problems seems not to have been so complete. In this note we shall show that some spectral properties, such as the principle of limiting absorption, the discreteness of the singular spectrum (modulo a finite number of accumulation points), and the existence and the completeness of wave operators, can be established for higher order exterior problems rather straightforwardly by applying the abstract method of [10].

The two Hilbert space theory of Belopol'skiǐ and Birman[1] can also be applied to exterior problems of higher order operators or systems (cf. Deǐč[3]). It seems, however, that the discreteness of the singular spectrum was not proved and that the scope of applicability of two approaches is different.

The present note is a continuation of our previous work [10], the first and the second part of which will be referred to as (I) and (II), respectively. The method given in (I) is a one Hilbert space method. The key tool in applying it to exterior problems is provided by an idea given by Birman in [2] (cf.also Kato[8]). Suggested by Birman's work we argue as follows. We make the direct sum of a given exterior problem with an interior problem and compare it with the direct sum of exterior and interior Dirichlet problems. The theory of M. G. Kreĭn[9 concerning extensions of positive Hermitian forms then gives a useful formula for the square root of the difference of resolvents of these two direct sums (Lemma 2.1). Combining it with a simple estimate whic we prove in this note (Lemma 2.2), we see that a function in the range of that square root belongs to a space smaller than L^2. This makes possible the application of perturbation method of (I) to the resolvents. In a similar way the direct sum of two Dirichlet problems can be compared in its turn with a problem in the entire space.

Main results are stated in Theorems 2.2 - 2.4. Problems in the entire space as were treated in (II) are included in our results as a special case; but stronger conditions on the regularity of coefficient are required than in (II). Some problems in which the exterior and the interior domains are not separated (e.g. a problem with a "jump condition" across the boundary) are also included in our results.

2. Notations and theorems

2.1. We shall consider the following two differential operators and associated forms in R^n.

$$L_0u = \sum_{|\alpha|,|\beta|\le m} D^\alpha a^{(1)}_{\alpha\beta} D^\beta u,$$

$$Lu = \sum_{|\alpha|,|\beta|\le m} D^\alpha (a^{(1)}_{\alpha\beta} + a_{\alpha\beta}(x)) D^\beta u,$$

$$h_0[u,v] = \sum_{|\alpha|,|\beta|\le m} (a^{(1)}_{\alpha\beta} D^\alpha u, D^\beta v),$$

$$h[u,v] = \sum_{|\alpha|,|\beta|\le m} ((a^{(1)}_{\alpha\beta} + a_{\alpha\beta}(x)) D^\alpha u, D^\beta v),$$

and a domain $\Omega\subset R^n$ such that: i) $K = R^n\backslash\Omega$ is compact; and ii) the boundary $\partial\Omega$ of Ω consists of a finite number of mutually disjoint closed sufficiently smooth surfaces. We regard that L_0, L, h_0, and h are formal expressions which can be applied to any u, v, whenever the expression makes sense. Operators with specified domain of definition will be introduced later by designating (rather abstractly) the behaviour of functions near $\partial\Omega$.

We first list up some notations to be used frequently:

$\mathcal{R} = R^n \backslash \partial\Omega = \Omega \cup \Omega'$, where Ω' is the interior of $K = R^n \backslash \Omega$;

the Schwartz space $\mathfrak{D}(G) = C_0^\infty(G)$ is written as $\mathcal{D}(G)$ for convenience;

$H^m(G)$ is the Sobolev space of order m consisting of all functions in $L^2(G)$ having L^2-derivatives up to order m (inclusive);

$\mathring{H}^m(G)$ is the completion of $\mathcal{D}(G)$ in $H^m(G)$; in particular, $H^m(\mathcal{R}) = H^m(\Omega) \oplus H^m(\Omega')$, $\mathring{H}^m(\mathcal{R}) = \mathring{H}^m(\Omega) \oplus \mathring{H}^m(\Omega')$;

$|u|_{m,G}$ denotes the norm of $H^m(G)$; in particular $|u|_{0,G}$ is the norm of $L^2(G)$; we put $|u|_m = |u|_{m,R^n}$; in particular $|u|_0$ is the norm of $L^2(\mathbf{R}^n)$;

$D_j = -i\partial/\partial x_j$ and $D^\alpha = D_1^{\alpha_1} \cdots D_n^{\alpha_n}$, where $\alpha = (\alpha_1, \cdots, \alpha_n)$ is a multi-index; $|\alpha| = \alpha_1 + \cdots + \alpha_n$.

When G is either R^n, Ω , or $\mathcal{R}$, the following weighted spaces are also used, where s is a real number:

$$L_s^2(G) = \{u \mid (1+|x|^2)^{s/2} u(x) \in L^2(G)\} ,$$

$$H_s^m(G) = \{u \mid D^\alpha u \in L_s^2(G), \ |\alpha| \le m\}$$

with the respective norm defined and denoted as

$$|u|_{L_s^2(G)} = |u|_{(s),G} = |(1+|x|^2)^{s/2} u(x)|_{0,G},$$

$$|u|_{H_s^m(G)} = |u|_{m,(s),G} = \Big(\sum_{|\alpha| \le m} |D^\alpha u|_{(s),G}^2 \Big)^{1/2} .$$

We put $|u|_{(s)} = |u|_{(s),R^n}$ and $|u|_{m,(s)} = |u|_{m,(s),R^n}$. The spaces $L^2(R^n)$ etc. are sometimes abbriviated to L^2 etc.

The domain of operators and forms are denoted by the letter D as $D(T)$ for operators and $D(h)$ for forms. The restriction of an operator T to a set D is denoted by $T_{|D}$. Finally, $B(X,Y)$ is the spaces of all bounded linear operators T from X to Y with $D(T) = X$ and $B_\infty(X,Y) \subset B(X,Y)$ is the set of all compact operators in $B(X,Y)$. We write $B(X) = B(X,X)$ and $B_\infty(X) = B_\infty(X,X)$.

2.2. We assume that the coefficients $a_{\alpha\beta}^{(1)}$ and $a_{\alpha\beta}(x)$ of L satisfy the following conditions (C.1) to (C.4).

(C.1) $a_{\alpha\beta} \in C^m(\mathcal{R}) \cap L^\infty(\mathcal{R})$ and $a_{\alpha\beta}$ with $|\alpha|=|\beta|=m$ are uniformly continuous in R^n; furthermore, $D^\gamma a_{\alpha\beta}(x)$, $|\alpha|$, $|\beta|$, $|\gamma| \le m$, are bounded in $\mathcal{R}\setminus U$, where U is a neighbourhood of $\partial\Omega$.

(C.2) $a_{\alpha\beta}^{(1)} = \overline{a_{\beta\alpha}^{(1)}}$, $a_{\alpha\beta}(x) = \overline{a_{\beta\alpha}(x)}$.

(C.3) There exists $c_1 \ge 0$ such that for $p = 0$ and 1 one has

$$\sum_{|\alpha|=|\beta|=m} (a_{\alpha\beta}^{(1)} + p a_{\alpha\beta}(x))\xi^{\alpha+\beta} \ge c_1 |\xi|^{2m}, \quad \xi \in R^n, \quad x \in \mathcal{R}.$$

(C.4) There exist $\delta > 1$ and $c_2 \ge 0$ such that

$$|a_{\alpha\beta}(x)| \le c_2(1+|x|^2)^{-\delta/2}, \quad |\alpha|, |\beta| \le m, \quad x \in \mathcal{R},$$

$$|D^\gamma a_{\alpha\beta}(x)| \le c_2(1+|x|^2)^{-\delta/4}, \quad |\alpha|, |\beta| \le m, \quad 0<\gamma\le\alpha, \quad x \in \mathcal{R}.$$

By virtue of Gårding's inequality L is bounded below on $\mathcal{D}(R^n)$ and on $\mathcal{D}(\mathcal{R})$. Let H_1, H_2, and H_3 be the Friedrichs extensions of $L_{0|\mathcal{D}(R^n)}$, $L_{|\mathcal{D}(R^n)}$, and $L_{|\mathcal{D}(\mathcal{R})}$, respectively, and let h_j be the Hermitian form associated with H_j. Gårding's inequality shows that h_1, h_2, and h_3 are the restrictions of h_0, h, and h, respectively, to $H^m(R^n)$, $H^m(R^n)$, and $\mathring{H}^m(\mathcal{R})$.

We next consider a selfadjoint operator H_4 satisfying the following conditions (C.5) and (C.6).

(C.5) H_4 is a selfadjoint extension of $L_{|\mathcal{D}(\mathcal{R})}$ and is bounded below.

Let h_4 be the Hermitian closed form associated with H_4 and let

$H_4-\lambda \geq c > 0$. Let D_4 be the Hilbert space $D(h_4)$ with norm $(h_4-\lambda)[u]$. Let $\rho \in \mathcal{D}(R^n)$ be equal to 1 in a neighbourhood of $\partial\Omega$.

(C.6) The mapping $u \to \rho u$ from D_4 to $L^2(R^n)$ is compact.

Condition (C.6) holds for any λ and ρ, if it holds for one λ and one ρ. This can be seen by the following proposition.

<u>Proposition 2.1.</u> $D(h_3) \subset D(h_4) \subset H^m(\mathcal{R}\backslash U)$, where U is an arbitrary neighbourhood of $\partial\Omega$.

$D(H_3) \subset D(H_4)$ is obvious. The rest of the proposition will be proved at the end of § 3.

Main results of the present note are concerned with spectral properties of H_4 and are given in the following theorems. Theorem 2.2 has a form analogous to Theorems 1.5-1.8 of (II) and is given in a somewhat abridged form. We put $P_1(\xi) = \sum_{|\alpha|,|\beta|\leq m} a^{(1)}_{\alpha\beta}\xi^{\alpha+\beta}$.

<u>Theorem 2.2.</u> Let conditions (C.1) - (C.6) be satisfied. Let e_1 be the set of all critical values (cf. Definition 1.4 of (II)) of $P_1(\xi)$ and let $\lambda_{min} = \inf_{\xi\in R^n} P_1(\xi)$. Then, the following assertions hold. i) The set $\{\lambda_n\}$ of all eigenvalues of H_4 in $I_0 \equiv (\lambda_{min},\infty)\backslash e_1$ has no points of accumulation in I_0. Each λ_n is of finite multiplicity. ii) For non-real ζ the resolvent $R_4(\zeta) = (H_4-\zeta)^{-1}$ determines a bounded operator $\tilde{R}_4(\zeta)$ from $L^2_{\delta/2}(R^n)$ to $L^2_{-\delta/2}(R^n)$ and the operator valued function $\tilde{R}_4(\zeta)$ of ζ can be extended by continuity up to the upper and the lower banks of $I_1 \equiv I_0\backslash\{\lambda_n\}$. The resulting operator valued function is locally Hölder continuous with respect to the operator norm ($L^2_{\delta/2} \to L^2_{-\delta/2}$). iii) In particular, let $v \in L^2_{\delta/2}(R^n)$ and let $u_{\lambda\pm i\varepsilon}$, $\lambda \in I$, $\varepsilon > 0$, be a unique solution of $(H_4-(\lambda\pm i\varepsilon))u_{\lambda\pm i\varepsilon} = v$ in $L^2(R^n)$. Then, the limit $u_{\lambda\pm i0} = \lim_{\varepsilon\downarrow 0} u_{\lambda\pm i\varepsilon}$ exists in $L^2_{-\delta/2}(R^n)$. $u_{\lambda\pm i0}$ satisfies for any $\phi \in \mathcal{D}(\mathcal{R})$

(2.1) $$\sum_{|\alpha|,|\beta|\le m} (D^{\alpha}(a^{(1)}_{\alpha\beta}+a_{\alpha\beta}(x))D^{\beta}\phi,u_{\lambda\pm i0}) - \lambda(\phi,u_{\lambda\pm i0}) = (v,u_{\lambda\pm i0}),$$

where (,) denotes the duality between $L^2_{\delta/2}$ and $L^2_{-\delta/2}$. iv) The part of $H_4 = \int \lambda dE_4(\lambda)$ in $E_4(I_1)L^2(R^n)$ is absolutely continuous and is unitarily equivalent to H_1 . v) The wave operators

$$W_{\pm}(H_4,H_1) = \underset{t\to\pm\infty}{\text{s-lim}}\ e^{itH_4}e^{-itH_1}$$

exist and give unitary equivalence mentioned in iv). The principle of invariance for wave operators (cf. Theorem 3.13 of (I)) holds as well.

We next apply Theorem 2.2 to exterior problems. We first assume (C.5') $H_{4,e}$ is a selfadjoint extension of $L|_{D(\Omega)}$ and is bounded below.

Let $h_{4,e}$ be the closed Hermitian form in $L^2(\Omega)$ associated with $H_{4,e}$. Let $\mathbf{D}_{4,e}$ be the Hilbert space $\mathbf{D}(h_{4,e})$ with norm $(h_{4,e}-\lambda)[u]$, $h_{4,e}-\lambda \ge c > 0$. Letting ρ be as in (C.6), we assume (C.6') the mapping $u \to \rho u$ from $\mathbf{D}_{4,e}$ to $L^2(\Omega)$ is compact.

<u>Theorem 2.3.</u> Let conditions (C.1) - (C.4), (C.5'), and (C.6') be satisfied. Then, assertions i) - iv) of Theorem 2.2 hold if H_4, R^n, and R are replaced by $H_{4,e}$, Ω, and Ω, respectively.

<u>Theorem 2.4.</u> Let J be the operator from $L^2(R^n)$ to $L^2(\Omega)$ given by the restrictions of functions to Ω . Then the wave operators

$$W_{\pm}(H_{4,e},H_1;J) = \underset{t\to\pm\infty}{\text{s-lim}}\ e^{itH_{4,e}}Je^{-itH_1}$$

exist and coincide with $JW_{\pm}(H_4,H_1)$, where H_4 is the direct sum of $H_{4,e}$ with the operator associated with an interior problem satisfying conditions similar to (C.6'). The principle of invariance also holds.

<u>Remark 2.5.</u> Condition (C.6) is satisfied if h_4 is s-coercive (s>0) near $\partial\Omega$: $h_4[\rho u] \ge a|\rho u|^2_s - b|\rho u|^2_0$, a>0. In particular, (C.6) holds if $H_4 = H_3$. Thus, H_3 is a special case of H_4. H_3 corresponds to the direct sum of exterior and interior Dirichlet problems.

More generally, (C.6) is satisfied if H_4 is the direct sum of operators associated with exterior and interior general boundary value problems which are s-coercive. It is a more subtle question to determine whether an extension H_4 given by general boundary conditions is semibounded or s-coercive. Among recent literatures we quote Fujiwara[4], Fujiwara and Shimakura[5], and Grubb[6].

In general, h_4 may have "boundary terms" which do not separate exterior and interior domains. An example of such forms in R^1, with $K = R^1 \setminus \Omega$ being contracted to the origin, is $h_4[u] = \int_{-\infty}^{\infty} |u'(x)|^2 dx - 2|u(0)|^2$ on $H^1(R^1)$. The associated operator is $-u''(x)$ defined for $u \in H^1(R^1) \cap H^2(R^1 \setminus \{0\})$ satisfying $u'(0+) - u'(0-) = -2u(0)$. This operator has the eigenvalue -1. Similar examples can be constructed with $K = [a,b]$.

3. Lemmas

3.1. Let A_{00} be a symmetric operator in a Hilbert space H such that $A_{00} \geq 1$ and let A_0 be the Friedrichs extension of A_{00}. Let A_1 be another selfadjoint extension of A_{00} such that $A_1 \geq 1$ and let a_j, j=0,1, be the closed Hermitian form associated with A_j. Let D_j be the Hilbert space $\mathsf{D}(a_j)$ with norm $a_j[u]$, $u \in \mathsf{D}_j$. Since a_1 is an extension of a_0, D_0 can be regarded as a closed subspace of D_1. Let P be the orthogonal projection in D_1 onto D_0.

Lemma 3.1. We have

$$A_1^{-1} - A_0^{-1} = (1-P)A_1^{-1} \geq 0, \quad (A_1^{-1} - A_0^{-1})^{1/2} = (1-P)A_1^{-1/2}T,$$

where T is a partial isometry in H.

The lemma is a consequence of Theorems 1 and 12 of Kreĭn[8]; or one can prove it directly using some arguments given in [2].

3.2. By adding a suitable constant to all H_j, we may and shall assume without loss of generality that $H_j \geq 1$, $j = 1,\cdots,4$. Let $\mathbf{D}_j$ be the Hilbert space $\mathbf{D}(h_j)$ with norm $h_j[u]$. In order to apply Lemma 3.1, we take A_{00} to be $L_{|\mathcal{D}(R)}$, A_0 to be H_3, and A_2 to be either H_2 or H_4. Then, letting P_k, $k = 2,4$, be the orthogonal projection in $\mathbf{D}_k$ onto $\mathbf{D}_3$, we have

$$(3.1) \qquad V_k \equiv H_k^{-1} - H_3^{-1} = (1-P_k)H_k^{-1}, \quad k = 2,4,$$

$$(3.2) \qquad V_k^{1/2} = (1-P_k)H_k^{-1/2}T_k,$$

where T_k is a partial isometry in $L^2(R^n)$.

Lemma 3.2. For $k = 2,4$ we have

$$(1-P_k)H_k^{-1/2} \in B(L^2(R^n), L^2_{\delta/2}(R^n)) \cap B_\infty(L^2(R^n)),$$

where δ is the constant appearing in (C.4).

Remark 3.3. If h_k is m-coercive (in particular, if $k = 2$), we have a stronger assertion that $(1-P_k)H_k^{-1/2} \in B(L^2(R^n), H^m_{\delta/2}(R^n))$.

Proof of Lemma 3.2. Since H_2 is a special case of H_4, we give the proof for H_4. (Note that (C.6) for H_2 follows from Gårding's inequality in the entire space.) Let $\partial\Omega \subset \{x \mid |x|<R\}$, $R > 0$, and let $\rho \in \mathcal{D}(R^n)$ be such that $\rho(x) = 1$ if $|x| < R+2$ and $\rho(x) = 0$ if $|x| > R+3$. Put $\eta = 1-\rho$ and let T_1 (or T_2) be the operator which maps $u \in (1-P_4)\mathbf{D}_4$ to $\rho u \in L^2(R^n)$ (or $\eta u \in L^2(R^n)$). Since $H_4^{-1/2}$ is a unitary operator from $L^2(R^n)$ to $\mathbf{D}_4$, it suffices to show that

$$T_j \in B((1-P_4)\mathbf{D}_4, L^2_{\delta/2}(R^n)) \cap B_\infty((1-P_4)\mathbf{D}_4, L^2(R^n)), \ j = 1,2.$$

This is obvious for T_1 by (C.6). To prove it for T_2, we put

$$\Omega_\ell = \{x \mid |x|>R+\ell\}, \ \ell = 0,1,2,3, \quad \Omega_{2,3} = \{x \mid R+2<|x|<R+3\}.$$

1°. $u \in (1-P_4)D_4$ implies $h_4[u,\phi] = 0$ for any $\phi \in \mathcal{D}(R)$. Since $\mathcal{D}(R) \subset D(H_4)$, we have $h_4[u,\phi] = (u,H_4\phi)$ and hence $(u,L\phi) = 0$, $\phi \in \mathcal{D}(R)$. Therefore, by the interior regularity of weak solutions of the elliptic equation $Lu = 0$ it follows that $u \in H^{2m,loc}(R)$ and $Lu(x) = 0$ in R. Put $g(x) = L(\eta u)(x)$. $g(x) = 0$ if $|x| < R+2$ or $|x| > R+3$. Hence, using the interior estimate, we obtain

$$(3.3) \qquad |g|_0 = |g|_{0,\Omega_{2,3}} \le c|u|_{2m,\Omega_{2,3}} \le c|u|_0.$$

Here and in what follows we denote various constants not depending on u by the same letter c. c may depends on domains concerned etc.

2°. We can choose a countable number of open balls U_j of radius 1 in such a way that $\Omega_2 \subset \bigcup_{j=1}^{\infty} U_j \subset \Omega_1$ and that each $x \in \Omega_0$ belongs to at most n balls among the balls U_j' of radius 2 concentric to U_j. The interior estimate gives $|\eta u|_{2m,U_j} \le c_j(|g|_{0,U_j'} + |u|_{0,U_j'})$. However, since $D^\gamma a_{\alpha\beta}$ is bounded in Ω_0 by (C.1), the constants c_j can be taken independent of j. Then, it is easy to see that $\eta u \in H^{2m}(R^n)$ and that $|\eta u|_{2m} = |\eta u|_{2m,\Omega_2} \le c(|g|_0 + |u|_0) \le c|u|_0$, where we used (3.3). Using the interior estimate again we finally get

$$(3.4) \qquad |u|_{2m,\Omega_2} = |u|_{2m,\Omega_{2,3}} + |\eta u|_{2m,\Omega_3} \le c|u|_0.$$

3°. By the general Leibniz formula of Hörmander we get

$$L_0(\eta u)(x) = (\eta L_0 u)(x) + h(x) \ , \quad L_0 = P_1(D) \ ,$$

$$h(x) = \sum_{0<|\alpha|} \frac{1}{\alpha!} (P_1^{(\alpha)}(D)u)(x) D^\alpha \eta(x) \ .$$

$h(x) = 0$ unless $x \in \Omega_{2,3}$ and we have $|h|_{0,\Omega_{2,3}} \le c|u|_{2m,\Omega_{2,3}} \le c|u|_0$. Hence, $h \in L_s^2(R^n)$ for any $s > 0$ and $|h|_{(s)} \le c(s)|u|_0$. On the other hand, it follows easily from $Lu(x) = 0$ $(x \in R)$, (C.4), and

(3.4) that $\eta L_0 u \in L^2_{\delta/2}(R^n)$ and that $|\eta L_0 u|_{(\delta/2)} \le c|u|_{2m,\Omega_2} \le c|u|_0$. Thus, we have proved that $|L_0(\eta u)|_{(\delta/2)} \le c|u|_0$.

Put $v = L_0(\eta u)$. Then, the above inequality means that $|\hat{v}|_{\delta/2} \le c|u|_0$, where ^ stands for the Fourier transform. We therefore see that $|\xi^\nu(\eta u)\hat{}(\xi)|_{\delta/2} = |\xi^\nu P_1(\xi)^{-1}\hat{v}(\xi)|_{\delta/2} \le c|\hat{v}|_{\delta/2} \le c|u|_0$, if $|\nu| \le 2m$. We thus obtain

$$(3.5) \qquad |\eta u|_{2m,(\delta/2)} \le c|u|_0 \le ch_4[u]^{1/2}, \quad u \in (1-P_4)D_4 .$$

This shows in particular that $T_2 \in B((1-P_4)D_4, L^2_{\delta/2}(R^n))$. The required compactness of T_2 is obvious by (3.5). Q.E.D.

We remark that inequality (3.5) also proves the assertion mentioned in Remark 3.3. m-coercivity is needed to get the estimate $|\rho u|_{2m,(\delta/2)} \le ch_4[u]^{1/2}$.

(3.4) shows that $(1-P_4)D_4 \subset H^{2m}(R\backslash U)$, where U is a neighbourhood of $\partial\Omega$. Since $P_4 D_4 = H^m(R)$, this proves Proposition 2.1. Note that condition (C.6) was not used in the proof of (3.4).

4. Proof of theorems

4.1. Once we have Lemmas 3.1 and 3.2 the rest of the proof will be a routine and somewhat tedious verification of various assumptions introduced in (I). Thus, the rest of this note is far from self-contained and leans heavily on the account given in (I) and (II). The writer apologizes for this style of exposition.

We assume as in § 3 that $H_k \ge 1$, $k = 1,\cdots,4$, and show that H_1^{-1} and H_4^{-1} satisfy assumptions introduced in (I). (They are listed as (A.1) - (A.4) on p.226 of (II).) The following notations and conventions will be used: $\sum_{|\alpha|,|\beta|\le m}$ is abbreviated to $\sum_{\alpha,\beta}$ and $L^2(R^n)$ etc. to L^2 etc.; M is the operator of multiplication by

$(1+|x|^2)^{\delta/4}$; $c_{\alpha\beta}(x) = (1+|x|^2)^{\delta/2} a_{\alpha\beta}(x) \in L^\infty$ (cf. (C.4)); $C_{\alpha\beta}$ is the operator of multiplication by $c_{\alpha\beta}$; $[T]^a$ denotes the closure of an operator T ; $\sum_\nu \oplus X_\nu$ is the direct sum of X_ν; when A_ν is an operator from X to X_ν , the operator $\sum_\nu \oplus A_\nu$ from X to $\sum_\nu \oplus X_\nu$ is defined by $(\sum_\nu \oplus A_\nu)u = \sum_\nu \oplus A_\nu u$, $u \in D(A_\nu)$. Hereafter, formulas, theorems, etc. of (I) or (II) will be referred to as, e.g., (I.2.4), Theorem II.1.6.

Let $H_k = \int_1^\infty \lambda dE_k(\lambda)$ be the spectral resolution of H_k. The associated spectral measure is denoted by the same letter E_k. For a set $\Delta \subset (0,\infty)$ we put $\tilde{\Delta} = \{\lambda | \lambda^{-1} \in \Delta\}$. Then, the spectral measure $\tilde{E}_k$ associated with H_k^{-1} is given by $\tilde{E}_k(\Delta) = E_k(\tilde{\Delta})$.

4.2. We start from the formula

$$(4.1) \qquad H_4^{-1} - H_1^{-1} = V_1 - V_2 + V_4 ,$$

where $V_1 = H_2^{-1} - H_1^{-1}$ and V_2 and V_4 are as given by (3.1). V_2 and V_4 can be decomposed as follows:

$$(4.2) \quad V_k = A_k C_k B_k, \quad A_k = M, \quad B_k = V_k^{1/2}, \quad C_k = M^{-1} V_k^{1/2}, \quad k = 2,4.$$

Note that $C_k \in B(L^2)$ by (3.2) and Lemma 3.2.

To decompose V_1 we use the following spaces and operators:

$$K_1 = \sum_{\alpha,\beta} \oplus K_{\alpha\beta} , \quad K_{\alpha\beta} = L^2 ,$$

$$A_0 = \sum_{\alpha,\beta} \oplus MD^\alpha, \quad B_0 = \sum_{\alpha,\beta} \oplus MD^\beta, \quad C_0 = \sum_{\alpha,\beta} \oplus C_{\alpha\beta} ,$$

$$G_2 = 1 - C_0 (B_0 H_2^{-1/2}) [H_2^{-1/2} A_0^*]^a \in B(K_1) .$$

A_0 etc. are operators denoted by A etc. in (II). As was shown in (II), (I.2.4) - (I.2.8) hold. Putting $\zeta = 0$ in (I.2.4), we get

$$(4.3) \qquad V_1 = H_2^{-1} - H_1^{-2} = -[H_1^{-1} A_0^*]^a \, G_2 C_0 B_0 H_1^{-1} = A_1^* C_1 B_1 ,$$

where we put $A_1 = A_0 H_1^{-1}$, $B_1 = B_0 H_1^{-1}$, $C_1 = -G_2 C_0$. We now set

$$K = K_1 \oplus K_2 \oplus K_4 \ , \quad K_2 = K_4 = L^2 \ ,$$

$$A = A_1 \oplus A_2 \oplus A_4 \ , \quad B = B_1 \oplus B_2 \oplus B_4 \ , \quad C = C_1 \oplus (-C_2) \oplus C_4 \ .$$

Then, it follows from (4.1), (4.2), and (4.3) that $H_4^{-1} = H_1^{-1} + A^*CB$, which shows that (A.1) of (II) is satisfied.

4.3. Hereafter, we fix $I = (a,b) \subset I_1$, where I_1 is as in Theorem 2.2, and verify other assumptions in (I) for H_1^{-1} , H_4^{-1} and $\tilde{I}$.

The spectral representation of $H_1^{-1}\tilde{E}_1(\tilde{I})$ is derived from that of $H_1E_1(I)$ by the change of variable $\lambda \to \mu = \lambda^{-1}$. To be more precise, let $\Sigma = \{\xi | P_1(\xi)=c\}$, $c \in I$, and let $F : E_1(I)L^2 \to L^2(I;L^2(\Sigma))$ be the spectral representation of $H_1E_1(I)$ defined in (II). (F is the Fourier transform expressed by the coordinate $\lambda \in I$ and $\omega \in \Sigma$.) Define $\tilde{F} : \tilde{E}_1(\tilde{I})L^2 = E_1(I)L^2 \to L^2(\tilde{I};L^2(\Sigma))$ by $(\tilde{F}u)(\lambda) = \lambda^{-1}(Fu)(\lambda^{-1})$. $\tilde{F}$ gives the spectral representation of $H_1^{-1}\tilde{E}_1(\tilde{I})$ required in Assumption I.3.2.

Assumption I.3.4 requires the compactness of either AH^{-1} or BH^{-1}. In the present case B itself is compact. In fact, the compactness of $B_k = V_k^{1/2}$, $k = 2,4$, was proved in Lemma 3.2 and that of $B_1 = \sum_{\alpha,\beta} \oplus MD^{\beta}H_1^{-1}$ is obvious because $D(H_1) = H^{2m}$ and $|\beta| \le m$.

Assumption I.3.5 is met if the range of A^* is dense. This is in fact true in the present case because A is one-to-one.

4.4. We next examine the assumptions involving "trace operators". Assumption I.3.3 requires us to find $B(K,L^2(\Sigma))$-valued locally Hölder continuous functions $\tilde{T}(\lambda;A)$ and $\tilde{T}(\lambda;B)$ of $\lambda \in \tilde{I}$ such that

$$(\tilde{F}\tilde{E}_1(\tilde{I})A^*u)(\lambda) = \tilde{T}(\lambda;A)u, \quad (\tilde{F}\tilde{E}_1(\tilde{I})B^*u)(\lambda) = \tilde{T}(\lambda;B)u, \quad u \in K.$$

We know (cf.(II)) that there exists a $B(L^2,L^2(\Sigma))$-valued locally Hölder continuous function $T(\mu;M)$ of $\mu \in I$ such that

$(FE_1(I)Mu)(\mu) = T(\mu;M)u$. ($T(\mu;M)$ is essentially the operator of taking the trace on $\{P_1(\xi) = \mu\}$ of Mu.) Let $\phi(\mu,\omega)$ be as in Proposition II.2.2 and write $u \in K$ as $u = (\sum_{\alpha,\beta} \oplus u_{\alpha,\beta}) \oplus u_2 \oplus u_4$. Then it is not difficult to see that

$$\tilde{T}(\lambda;A)u = \sum_{\alpha,\beta} \phi(\lambda^{-1},\omega)^{\alpha} T(\lambda^{-1};M)u_{\alpha\beta} + \lambda^{-1}T(\lambda^{-1};M)(u_2+u_4);$$

$$\tilde{T}(\lambda;B)u = \sum_{\alpha,\beta} \phi(\lambda^{-1},\omega)^{\beta} T(\lambda^{-1};M)u_{\alpha\beta} + \lambda^{-1}T(\lambda^{-1};M)(u_2+u_4)$$

satisfy the requirement made above.

For verifying Assumptions I.5.12 and I.5.15 we define $A_{(1)} \in B(K)$ as $A_{(1)} = (\sum_{\alpha,\beta} \oplus M) \oplus M \oplus M$. $A_{(1)}$ corresponds to A_1 of § I.2.3. Then, the space K_γ determined by the norm $|A_{(1)}^{-\gamma}u|_K$ is equal to $K_{1,\gamma} \oplus K_{2,\gamma} \oplus K_{4,\gamma}$, $K_{1,\gamma} = \sum_{\alpha,\beta} \oplus L^2_{\gamma\delta/2}$ and $K_{2,\gamma} = K_{4,\gamma} = L^2_{\gamma\delta/2}$ and Assumption I.5.12, namely $A = A_{(1)}D$, is fulfilled with $D = (\sum_{\alpha,\beta} \oplus D^{\alpha}H_1^{-1}) \oplus I \oplus I : L^2 \to K_0 \subset K_{-1}$.

Assumption I.5.15 requires the Hölder continuity of specified type of $T(\lambda;B)C^*w$, where $w \in K_\gamma$, $\gamma \geq 0$. In (I), However, this condition was never used for $\gamma > 1$. So, we verify it here only for $0 \leq \gamma \leq 1$. Then, as is easily seen, it suffices to show that C_k^* maps $K_{k,\gamma}$ into itself, $k = 1,2,4$, $0 \leq \gamma \leq 1$. For $k = 2,4$, we have $C_k^* = [V_k^{1/2}M^{-1}]^a$ and hence $|C_k^*u|_{(\delta/2)} \leq |V_k^{1/2}|_{B(L^2,L^2_{\delta/2})}|u|_{(\delta/2)}$, $u \in L_{\delta/2}$. From this it follows by interpolation that C_2^* and C_4^* map $L^2_{\gamma\delta/2}$ into itself, $0 \leq \gamma \leq 1$. For $C_1^* = C_0G_2^*$ it suffices to show that $G_2^* = 1 - (A_0H_2^{-1/2})[H_2^{-1/2}B_0^*]^aC_0$ maps $K_{1,\delta/2}$ into itself. But, this follows from $|A_0H_2^{-1/2}w|^2_{K_{1,\delta/2}} = \sum_{\alpha,\beta} |MD^{\alpha}H_2^{-1/2}w|^2_{(\delta/2)} \leq c|w|_0^2$.

4.5. There still remains some technical assumptions. Assumption I.5.13 and (1) of Theorem I.6.1 are trivial, because only bounded operators are involved in the present case.

By the definition of the space $\mathfrak{H}$ and the operator A appearing in Assumption I.5.19, we see readily that $\mathfrak{Y}$ is continuously imbedded in $L^2_{-\delta/2}$ and Ay has the form $Ay = (\sum_{\alpha,\beta} \oplus z_{\alpha\beta}) \oplus M\bar{I}y \oplus M\bar{I}y$, $y \in \mathfrak{Y}$, where $\bar{I}$ is the imbedding $\mathfrak{Y} \to L^2_{-\delta/2}$. Hence, $y \in K_1$ implies $M\bar{I}y \in L^2_{\delta/2}$; but the latter is equivalent to $y \in L^2$. Furthermore $|y|_0 \le |Ay|_{K_1}$.

The space $\mathfrak{V}$ appearing in Assumption I.5.20 is the set of all $\phi \in L^2$ such that $|(H_1^{-1}\phi, v)| \le c|Av|_K$, $|(\phi, v)| \le c|v|_0$. We claim that $L^2_{\delta/2} \subset \mathfrak{V}$. In fact, this follows from the following two facts: (1) $|v|_{(-\delta/2)} \le |Av|_K$; and (2) $\phi \in L^2_{\delta/2}$ implies $H_1^{-1}\phi \in L^2_{\delta/2}$. (1) is trivial and (2) is proved easily by taking the Fourier transform. Since $(H_1^{-1}\phi, v)$ and $(H_4^{-1}\phi, v)$ are bounded forms, the possibility of approximation mentioned in Assumption I.5.20 is now obvious.

<u>4.6.</u> Proof of Theorem 2.2. All necessary assumptions having been verified, we can now apply the results of (I) to H_1^{-1} and H_4^{-1} . As a consequence, statements similar to Theorem II.1.5 hold. i) and iv) follow from this by switching to inverse operators. Theorem I.6.1 shows that the principle of limiting absorption holds for $(H_4^{-1} - \tilde{\zeta})^{-1}$ as operators from $L^2_{\delta/2}$ to $\mathfrak{Y}$, and a fortiori to $L^2_{-\delta/2}$. To obtain (2.1), it suffices to start from

$$(h_4 - \bar{\zeta})[\phi, u(\zeta)] = (\phi, v) \ , \quad \zeta = \lambda \pm i\varepsilon \ , \quad \phi \in D(R), \quad v \in L^2_{\delta/2} \ ,$$

and note that the left side converges to the left side of (2.1). v) is derived from the invariance principle applied to H_1^{-1} and H_4^{-1} .

Proof of Theorems 2.3 and 2.4. Let $H_{4,i}$ be the Friedrichs extension of $L_{|D(\Omega')}$ and put $H_4 = H_{4,e} \oplus H_{4,i}$. Then, it is clear that H_4 satisfies (C.5) and (C.6). Since $H_{4,i}$ has purely discrete spectrum, the structure of the continuous spectrum of H_4 and $H_{4,e}$ will be the same. Thus, i) and iv) for $H_{4,e}$ follow from those for H_4. Since the exterior and the interior parts are separated in H_4, the limiting absorption principle for $(H_4 - \zeta)^{-1}$ will yield that for $(H_{4,e} - \zeta)^{-1}$. A small problem here is that the exceptional set $\{\lambda_n\}$

for H_4 contains eigenvalues of $H_{4,i}$ and such an eigenvalue should be excluded from $\{\lambda_n\}$ for $H_{4,e}$. To handle this point, we argue conveniently as follows. Lower order coefficients of L can be discontinuous across $\partial\Omega$ (cf. (C.1)). In particular, if we change $a_{00}(x)$ only in Ω' by adding a large positive constant, Theorem 2.2 still remains true for $H_4 = H_{4,e} \oplus H_{4,i}$. However, the lowest eigenvalue of $H_{4,i}$ can be made as large as we wish by such a change of $a_{00}(x)$. This takes care of the problem and ii) for $H_{4,e}$ is proved.

Theorem 2.4 follows from v) of Theorem 2.2 in a routine way. Note that $H_{4,i}$ can be replaced by any operator described in Theorem 2.4.

References

[1] Belopol'skiĭ, A. L., and M. Š. Birman, The existence of wave operators in scattering theory for pairs of spaces, Izv. Akad. Nauk SSSR Ser. Mat. 32(1968), 1162-1175 (Russian), English transl. Math. USSR-Izv. 2(1968), 1117-1130.

[2] Birman, M. Š., Perturbations of continuous spectrum of a singular elliptic operator under the change of the boundary and boundary conditions, Vest. Leningrad. Univ. Ser. Mat., Meh., Astron. 1962, No. 1, 22-55 (Russian).

[3] Deĭč, V. G., An application of the method of nuclear perturbations in scattering theory for a pair of spaces, Izv. Vysš. Učebn. Zaved. Matematika 1971, no.6(109), 33-42 (Russian).

[4] Fujiwara, D., On some homogeneous boundary value problems bounded below, J. Fac. Sci. Univ. Tokyo Sect. IA 17(1970), 123-152.

[5] ——— and N. Shimakura, Sur les problèmes aux limites elliptiques stablement variationnels, J. Math. Pures Appl. 49(1970), 1-28.

[6] Grubb, G., On coerciveness and semiboundedness of general boundary problems, Israel J. Math. 10(1971), 32-95.

[7] Ikebe, T., On the eigenfuction expansion connected with the exterior problem for the Schrödinger equation, Japan. J. Math. 36(1967), 33-55.

[8] Kato, T., Scattering theory with two Hilbert spaces, J. Fuctional Anal. 1(1967), 342-369.

[9] Kreĭn, M. G., The theory of selfadjoint extensions of semibounded Hermitian operators and its applications, I, Mat. Sb. 20(62) (1947), 431-495 (Russian).

[10] Kuroda, S. T., Scattering theory for differential operators, I, operator theory, J. Math. Soc. Japan 25(1973), 75-104; II, self-adjoint elliptic operators, ibid 25(1973), 222-234.

[11] Mochizuki, K., Spectral and scattering theory for second order elliptic differential operators in an exterior domain, Lecture Notes, Univ. Utah, 1972.

[12] Shenk, N. and D. Thoe, Eigenfunction expansions and scattering theory for perturbations of $-\Delta$, J. Math. Anal. Appl. 36(1971), 313-351.

Swirling Flow

J. B. McLeod

1. Introduction

It was von Kármán [1] who first realised that the fluid motion above an infinite rotating disc which is rotating about an axis perpendicular to its plane can under suitable circumstances be reduced to the study of a pair of non-linear ordinary differential equations. Batchelor [2] extended this discussion to motion between two rotating discs rotating about a common axis perpendicular to their planes, and our object in this paper is to survey the progress that has been made in the analytical treatment of these boundary-value problems. The problem in which there is just one disc and the fluid occupies the whole space above it we shall refer to as the singular problem, and the problem with the two discs, which reduces to a boundary-value problem with finite boundaries, we shall refer to as the regular problem. If one is tempted to believe that "regular" problems are always simpler than the corresponding "singular" ones, then swirling flow provides a convenient contradiction; for so far as existence of solutions is concerned, the singular problem is reasonably well understood, while the regular problem remains unsolved except in certain special cases, and on uniqueness there are essentially no results at all for either the singular or the regular problem.

To set up the singular problem first, the equations are for functions $f(x)$, $g(x)$, where $0 \leq x < \infty$, and

$$f''' + ff'' + \tfrac{1}{2}(g^2 - f'^2) = \tfrac{1}{2}\Omega_\infty^2 , \tag{1.1}$$

$$g'' + fg' = f'g, \tag{1.2}$$

with the boundary conditions

$$f(0) = a, \quad f'(0) = 0, \quad g(0) = \Omega_0; \quad f'(\infty) = 0, \quad g(\infty) = \Omega_\infty . \tag{1.3}$$

To connect f, g, x with the physical variables, take the axis of rotation as the axis of cylindrical coordinates (r, ϕ, z), r being the distance from the axis and z the height above the disc. Let (u, v, w) be the velocity components with respect to (r, ϕ, z), and let ω_0, ω_∞ be the angular velocities of the fluid at $z = 0, \infty$ respectively. Define $\omega = \sqrt{(\omega_0^2 + \omega_\infty^2)}$. Then

$$x = (2\omega/\nu)^{\frac{1}{2}} z,$$

where ν is the (constant) kinematic viscosity of the fluid, and

$$f(x) = -w/\sqrt{(2\nu\omega)}, \quad f'(x) = u/\omega r, \quad g(x) = v/\omega r .$$

It follows that $\Omega_0 = \omega_0/\omega$, $\Omega_\infty = \omega_\infty/\omega$, so that we have the (physical) relation

$$\Omega_0^2 + \Omega_\infty^2 = 1. \tag{1.4}$$

On the other hand there is no reason analytically why we should restrict Ω_0 and Ω_∞ so as to satisfy the relation (1.4), since the boundary-value problem (1.1) - (1.3) is perfectly meaningful without it; and it will in fact be convenient not to restrict Ω_0 and Ω_∞ in this way, since in the course of the analysis we will wish to alter the value of Ω_0 without altering that of Ω_∞. The restriction (1.4) will therefore not be imposed. It is also clear from the physical interpretation that the constant a appearing in the boundary conditions (1.3) is a measure of any suction (positive values of a) or blowing (negative values of a) which is imposed at the disc.

In the case of the regular problem, it is convenient to normalise the variables slightly differently. If the lower and upper discs are rotating with angular velocities ω_0 and ω_1 respectively, then the equations can be put in the form

$$\varepsilon H^{iv} + HH''' + GG' = 0, \quad -1 \leq x \leq 1, \tag{1.5}$$

$$\varepsilon G'' + HG' - H'G = 0, \quad -1 \leq x \leq 1, \tag{1.6}$$

where x measures the distance along the axis of rotation, the discs being placed at $x = \pm 1$, and ε, H, G are given by

$$\varepsilon = \nu/2\omega_0, \quad H(x) = -w/2\omega_0, \quad G(x) = -v/\omega_0,$$

(u, v, ω) being as before the velocity components in cylindrical polar coordinates (r, ϕ, x). The boundary conditions are

$$\begin{aligned} H(-1) = H'(-1) = H(1) = H'(1) = 0, \\ G(-1) = -1, \quad G(1) = -\omega_1/\omega_0 . \end{aligned} \tag{1.7}$$

(If we were to allow suction or blowing at the discs, the values $H(-1)$, $H(1)$ would be prescribed, but no longer zero.)

It should be noted that (1.5) is immediately integrable to give

$$\varepsilon H''' + HH'' + \tfrac{1}{2}(G^2 - H'^2) = \text{constant}, \tag{1.8}$$

which is comparable with (1.1). This comparison shows that the simpler nature of the singular problem arises from the fact that the physical situation at infinity demands the prescribing of the constant of integration, so that the boundary-value problem is a fifth-order system with five boundary conditions to be satisfied. The regular problem, on the other hand, has to be regarded as either a sixth-order system with six boundary conditions, or else as a fifth-order system in which the constant in (1.8), as well as H and G, has to be determined from the six boundary conditions.

Finally, it is possible to remove the factor ε in (1.5), (1.6) by setting

$$f(x) = \varepsilon^{-1}H(x), \; g(x) = \varepsilon^{-1}G(x),$$

and the equations (1.5), (1.6) then reduce to (1.1), (1.2), with ε now making its appearance in the boundary conditions instead of in the equations. We have exhibited the forms (1.5), (1.6) because one of the most studied problems in the two-disc case (from both the physical and the numerical stand-point) has been the question of the behaviour of the flow when $\omega_0 = -\omega_1$ and both are large in modulus. In this situation, ε becomes small, and (1.5), (1.6) indicate that the problem is now one of a singular perturbation. To solve this (and its solution is one of the problems we shall discuss), it seems best to consider the problem in the form (1.5), (1.6).

The singular case is discussed first when $\Omega_\infty = 0$ (section 2) and then when $\Omega_\infty \neq 0$ (section 3), the two situations requiring quite different treatment. The regular case, particularly with $\omega_0 = -\omega_1$, is then treated in sections 4 and 5.

2. <u>The singular case</u>: $\Omega_\infty = 0$

Since the replacement of g by -g does not alter the equations (1.1) and (1.2), but only the boundary condition $g(0) = \Omega_0$, we may suppose without loss of generality that $\Omega_0 \geq 0$. The case $\Omega_0 = \Omega_\infty = 0$ is not interesting as far as existence is concerned, since $f = a$, $g = 0$ gives a trivial solution. We will therefore suppose $\Omega_0 > 0$. Existence theorems for sufficiently large suction have been proved by von Kármán and Lin [3] and by Howard [4], but a more general result is the following.

<u>Theorem 1.</u> <u>If</u> $\Omega_\infty = 0$, $\Omega_0 > 0$, <u>then the boundary-value problem</u> (1.1) - (1.3) <u>has at least one solution which has the properties that</u>

$$0 \leq f' < \Omega_0, \;\; g > 0, \;\; g' < 0,$$

f" <u>is first positive and ultimately negative, with one and only one zero</u>.

The proof of this uses a development of the "shooting technique" used by Iglisch [5] and Coppel [6] for the Falkner-Skan equation and by Ho and Wilson [7] and McLeod and Serrin [8] for pairs of differential equations arising from the study of convection and compressible boundary layer problems. We seek appropriate initial conditions for f" and g' so that the resulting solution (f, g) will have the correct behaviour at infinity. The full proof is given in [9], so that there is no need to repeat it here, but we will go a little further with the details in order to exhibit the structure of the proof more clearly.

Let (f, g) be the (unique) solution of (1.1) and (1.2) which satisfies the initial conditions

$$f(0) = a, \quad f'(0) = 0, \quad f''(0) = \alpha, \quad g(0) = \Omega_0, \quad g'(0) = \beta,$$

where, in view of the properties given in the statement of the theorem, we are interested only in $\alpha > 0$, $\beta < 0$. We can then prove the following two lemmas.

Lemma 1. *If $\alpha > 0$ is fixed, then for $|\beta|$ sufficiently small there exists some $x^+ > 0$ with $g'(x^+) > 0$, while*

$$g(x) > 0 \quad \text{for} \quad 0 \leq x \leq x^+ ,$$

i.e. g' becomes zero before g does.

(This result comes from the facts that $g'(0)$ is small, while (1.2) implies that

$$g' \exp(\int_0^x f\, dt)$$

is increasing for small x. It is then easy to conclude that g' becomes zero for some small value of x, and certainly before g does.)

Lemma 2. *If $\alpha > 0$ is fixed, then for $|\beta|$ sufficiently large there exists some $x^- > 0$ with $g(x^-) < 0$, while*

$$g'(x) < 0 \quad \text{for} \quad 0 \leq x \leq x^- ,$$

i.e. g becomes zero before g' does.

(Intuitively, $g'(0)$ is now large and negative, so that g initially decreases rapidly from the value Ω_0, and will reach zero before g' has altered significantly.)

If, given any $\alpha > 0$, a value of $\beta < 0$ is such that there exists an $x^+ > 0$ with the properties in Lemma 1, then β will be said to belong to S_α^+, while if there exists an $x^- > 0$ with the properties in Lemma 2, then β will be said to belong to S_α^-. Lemmas 1 and 2 guarantee that S_α^+ and S_α^- are non-void. Moreover, since solutions of (1.1) and (1.2) depend continuously on their initial values, it is clear that S_α^+ and S_α^- are open sets relative to the semi-axis $\beta < 0$. Finally, they are evidently disjoint. Since the semi-axis $\beta < 0$ is a connected set, it follows that there must be at least one value of β which belongs to neither S_α^+ nor S_α^-. For such a value of β, say $\beta(\alpha)$, the solution must have the property, so long as it continues to exist, that g does not vanish before g', nor g' before g, and so either g, g' vanish simultaneously (which is impossible since $g = g' = 0$ at any point implies from (1.2) that $g \equiv 0$) or neither g nor g' vanish at all, i.e. $g > 0$, $g' < 0$.

It is now a matter of carrying out a similar variation in α, which $\beta = \beta(\alpha)$, and applying a second connectedness argument to show that there is at least one value of α for which the solution exists for all x, with the properties that $f' > 0$ except at $x = 0$, $g > 0$, $g' < 0$, $f''(\infty) = 0$. It is then not difficult to deduce from the equations that the boundary conditions at infinity are satisfied, and that the other properties of f', f'' stated in the theorem are true, and the theorem is proved.

3. The singular case: $\Omega_\infty \neq 0$

In this case, it was shown in [9] that if solutions exist for (1.1) - (1.3), then they certainly cannot satisfy $f' \geq 0$, and must exhibit an oscillatory behaviour for large x. This result was achieved directly from the differential equations by a reductio ad absurdum method, without any need to obtain an explicit expression for the asymptotic expansion of a solution, but it was confirmation of the work of Rogers and Lance [10], who obtained heuristically the asymptotic expansion for large x and showed that it implied oscillatory behaviour. The asymptotic expansion was finally obtained rigorously in [11], and results for a more general equation are given by Hartman in [12].

The effect of this oscialltory behaviour is to render unlikely any successful application of the shooting technique, which depends upon the existence of simple inequalities satisfied by the solution, and some new approach to the problem becomes necessary. It is possible to give results on existence which are valid only when the suction parameter a is sufficiently large or the quantity $|\Omega_0 - \Omega_\infty|$ is sufficiently small, and the first of these is due to Watson [13], while others are given by Hartman [12] and Bushell [14]. However, the theorem for $\Omega_\infty \neq 0$ which compares with Theorem 1 for $\Omega_\infty = 0$ is the following.

Theorem 2. The boundary-value problem consisting of (1.1) - (1.3) possesses a solution for all values of the parameter a, provided that $\Omega_0 > 0$, $\Omega_\infty > 0$ (or $\Omega_0 < 0$, $\Omega_\infty < 0$). Further, the solution has the property (if $\Omega_0 > 0$) that $g > 0$ for all x.

The restriction that Ω_0 and Ω_∞ be of the same sign is not surprising in view of the difficulty which Rogers and Lance [10] and Evans [15] found in coping numerically with solutions when Ω_0 and Ω_∞ are of opposite sign, and in view too of the non-existence proof when $\Omega_0 = -\Omega_\infty$ and $a \leq 0$ which is one of the results in [16]. (See the remarks after Lemma 3 below.) It is perhaps worth mentioning that I believe, with a certain amount of evidence, that it is only when $\Omega_0 = -\Omega_\infty$ and a is not too large and positive that a solution fails to exist, but it must be emphasised that the proof of this is not yet complete.

The first proof of Theorem 2 was given in [17], but a shorter proof (and one applicable to slightly more general equations and boundary conditions) is due to Hartman [18]. Both depend on the same a priori estimates for solutions, but the method of application of them is different. I want to sketch here the original proof, not because it is the best but because the ideas involved in it lead to a basis for a degree theory for a class of operators wider than that usually considered in degree theory. We take this point up again at the end of the sketch of the proof of Theorem 2.

The framework of the proof is as follows. There certainly exists a solution when $\Omega_0 = \Omega_\infty$, i.e. the trivial solution $f = a$, $g = \Omega_0 = \Omega_\infty$; and in fact this is the only solution when $\Omega_0 = \Omega_\infty$ and either $a \leq 0$ or $g > 0$, results which are of some intrinsic interest but, curiously enough, are also essential to the existence proof when $\Omega_0 \neq \Omega_\infty$. (The proofs of these uniqueness results are indicated after Lemmas 3, 4.) If we now perturb the value of Ω_0 away from Ω_∞, we can show using Schauder's fixed point principle that a solution continues to exist at least for values of Ω_0 sufficiently close to Ω_∞. Suppose now that we consider sufficiently $\Omega_0 > \Omega_\infty$, and that a solution is thus guaranteed for $\Omega_0 < \Omega^*$. The next step is to prove that the solution must continue to exist for $\Omega_0 = \Omega^*$, and this can be made a consequence of the a priori estimates that exist for solutions. In fact, the solution is determined by the values of three parameters, which are related to its asymptotic behaviour at infinity; these parameters are bounded as $\Omega_0 \to \Omega^*$ (and the source of these bounds we indicate later), and so by the Bolzano-Weierstrass theorem they tend to limits as $\Omega_0 \to \Omega^*$, possibly though a suitable sequence of values; the limiting values of the parameters lead to a solution for $\Omega_0 = \Omega^*$.

To complete the proof, we have to show that the solution continues to exist for $\Omega_0 > \Omega^*$ (and so, by repetition, for all Ω_0). Here, by looking at the analytic dependence of the solution on the three parameters which determine it, we are able to argue that, if there is no solution for $\Omega_0 > \Omega^*$, then there must be at least a second solution for $\Omega_0 < \Omega^*$; we show in fact that if $\delta > 0$ is sufficiently small, then the total number of solutions corresponding to the two values $\Omega_0 = \Omega^* \pm \delta$ and "close to" the already known solution is even, including the already known solution itself, i.e. the total number of solutions "bifurcating" or "merging" at $\Omega_0 = \Omega^*$ is even (including the already known solution). If there is no solution for $\Omega_0 > \Omega^*$, there must therefore be a second for $\Omega_0 < \Omega^*$.

Suppose then that we cannot continue beyond $\Omega_0 = \Omega^*$, and let us trace what happens to the second solution as we reduce Ω_0. We remark that in this continuation process we may always suppose that the solution we obtain satisfies $g > 0$. For if not, then there would be a first value of Ω_0 for which $g > 0$ is false, and for which therefore g just touches the value zero, so that there is some x_0 with $g(x_0) = 0$, $g'(x_0) = 0$. But in view of (1.2), this implies $g \equiv 0$ and is impossible.

Since $g > 0$, there is as we have already pointed out only one solution for $\Omega_0 = \Omega_\infty$, and it is also possible to prove that bifurcation cannot occur at $\Omega_0 = \Omega_\infty$. As we reduce Ω_0, therefore, the second solution that we are now following must cease as a second distinct solution before $\Omega_0 = \Omega_\infty$; and this can happen, by a repetition of previous arguments, only if the second solution reaches a further bifurcation point, $\Omega_0 = \Omega^{**}$, say, where it merges either with a third solution, which we can then trace, or with the first. If it merges with the first solution, we now have three solutions in a neighbourhood of Ω^{**}, the first for $\Omega_0 < \Omega^{**}$ and the first and second for $\Omega_0 > \Omega^{**}$, and so there must be yet another solution which we can trace.

If then we suppose for contradiction that a solution cannot be found for Ω_0 beyond Ω^*, the tracing process we have been considering can be continued indefinitely, with Ω_0 always lying between Ω_∞ and Ω^*, and with always uniform bounds on the solutions. This forces us to the conclusion that there must be a bifurcation point with infinitely many solutions bifurcating from it, and this is easily refuted, leading to the required contradiction. We must be able, therefore, to continue beyond $\Omega_0 = \Omega^*$, and the theorem is proved.

It remains to indicate the source of the a priori bounds, and also of uniqueness for $\Omega_0 = \Omega_\infty$. These are based on the following two lemmas, which are proved by straightforward manipulations on the equations.

Lemma 3. For any solution (f, g) of (1.1), (1.2), we have that

$$\frac{d}{dx}(f''^2 + g'^2) \qquad (3.1)$$

either is identically zero or has at most zero, being in the second case negative before the zero (if it exists) and positive after.

If (f, g) satisfies the boundary conditions (1.3) at infinity, then it is possible to argue from the asymptotics of solutions that $f''^2(\infty) + g'^2(\infty) = 0$, so that, from Lemma 3, $f''^2 + g'^2$ is either a strictly decreasing function or else identically zero. The identically zero case leads, under the boundary conditions (1.3), to the solution $f = a$, $g = \Omega_0$, and is possible only when $\Omega_0 = \Omega_\infty$. If $f''^2 + g'^2$ is strictly decreasing, we evaluate (3.1) at $x = 0$ and obtain

$$-2a\{f''^2(0) + g'^2(0)\} - f''(0)\{\Omega_0^2 - \Omega_\infty^2\} < 0, \qquad (3.2)$$

and if $a < 0$, this gives one source of bounds on $f''(0)$, $g'(0)$, and so $f''(x)$, $g'(x)$ for all x. Also, if $\Omega_0 = \pm\,\Omega_\infty$ and $a \leq 0$, the inequality (3.2) is impossible, and so there can be no solution to (1.1) - (1.3) for $\Omega_0 = -\Omega_\infty$ and $a \leq 0$, and only the trivial one for $\Omega_0 = \Omega_\infty$ and $a \leq 0$.

We have another result of a similar character.

Lemma 4. For any solution (f, g) of (1.1), (1.2) for which $g > 0$, we have that

$$\frac{d}{dx}\left(\frac{f'^2 + g^2 + \Omega_\infty^2}{g}\right) \qquad (3.3)$$

either is identically zero or has at most one zero, being in the second case negative before the zero (if it exists) and positive after.

This result implies that $(f'^2 + g^2 + \Omega_\infty^2)/g$ must have the property that it does not exceed the maximum of its values at $x = 0, \infty$. But these two values are known from the boundary conditions (1.3), and so $(f'^2 + g^2 + \Omega_\infty^2)/g$ is uniformly bounded for Ω_0 in any bounded interval $0 < K_1 \leq \Omega_0 \leq K_2$, say. This in turn implies uniform bounds for f', g, and from these it is possible to deduce all the other bounds that are required. Also, by evaluating (3.3) at $x = 0$, we can, as from Lemma 3, obtain a second uniqueness result, this time without restrictions on a but with the proviso that $g > 0$.

Finally, we return to the remark made earlier that the continuation process employed in this proof leads to wider applications. The essence of the argument is that an even number of solutions bifurcating from any bifurcation point ensures that the parity of the total number of solutions which can be reached by the continuation process is independent if Ω_0. Since the parity is odd when $\Omega_0 = \Omega_\infty$, it must be odd for all Ω_0, and so there exists at least one solution for all Ω_0. This maintenance of parity is the essential feature of Leray-Schauder degree theory, or of any degree theory, and shows that we can aim at the development of a degree theory (and apply it, as here, to the existence of solutions) for any non-linear operator $F(\cdot, \lambda)$ depending on some parameter λ and possessing the property that any value of λ which is a bifurcation point of the equation $F(\cdot, \lambda) = 0$ yields an even number of bifurcating branches. But if we have a non-linear operator whose Fréchet derivative is a Fredholm operator, then the Lyapunov-Schmidt process, (as, for example, in [19]), shows that in general there are indeed an even number of bifurcating solutions. By "in general" we mean that the result is true provided that certain quantities in the analysis of the operator do not take exceptional values, but even if they do, we can avoid the difficulty by approximating to such an operator by operators for which the exceptional values are not taken, defining the degree for these approximating operators, and then in the limit establishing the degree for the original operator. That it is possible to develop a degree theory for operators whose Fréchet derivative is Fredholm seems to have been realised first by Smale [20], and then taken up by Elworthy and Tromba [21], although their approach is quite different from that outlined above.

4. The regular case: existence theory

For the regular case, existence theory is currently limited to three papers. The first, by Hastings [22], proves existence provided that the angular velocities of both discs are sufficiently small. Elcrat [23] has also used what is essentially a perturbation approach, but it is carried through with sufficient precision that definite numerical estimates can be given of the extent of the allowable perturbation; further, Elcrat perturbs both about the rest state, with $\omega_0 = \omega_1 = 0$, and about the rigid body rotation, with $\omega_0 = \omega_1$.

The third paper [24] looks at the case where $\omega_0 = -\omega_1$, and the object is to prove the existence and discuss the behaviour of anti-symmetric solutions, so that H and G in (1.5) and (1.6) are odd functions of x. So far as existence is concerned, the fact that ε is small in (1.5) and (1.6) is irrelevant, since we would hope (and are able) to prove existence regardless of the size of ε, and so we make the substitution

$$f(x) = \varepsilon^{-1} H(x), \quad g(x) = \varepsilon^{-1} G(x),$$

and use anti-symmetry to reduce the problem to one for $x \in [0, 1]$. The boundary-value problem then becomes

$$f^{iv} + ff''' + gg' = 0, \quad 0 \leq x \leq 1, \tag{4.1}$$

$$g'' + fg' - f'g = 0, \quad 0 \leq x \leq 1, \tag{4.2}$$

with the boundary conditions

$$\begin{aligned} &f(0) = f''(0) = f'(1) = f(1) = 0, \\ &g(0) = 0, \qquad g(1) = R > 0\,, \end{aligned} \tag{4.3}$$

and we can prove the following theorem.

<u>Theorem 3.</u> <u>There is a pair of functions</u> (f, g) <u>which satisfy equations</u> (4.1) - (4.3). <u>Moreover,</u>

$$f \leq 0$$

<u>and there are three distinguished points</u> x_1, x_2, x_3, <u>with</u> $0 < x_1 < x_2 < 1$, $0 < x_3 < x_2 < 1$, <u>such that</u>

$$f'(x) < 0 \ \underline{\text{for}}\ 0 \leq x < x_1, \quad f'(x) > 0 \ \underline{\text{for}}\ x_1 < x < 1, \tag{4.4}$$

$$f''(x) > 0 \ \underline{\text{for}}\ 0 < x < x_2, \quad f''(x) < 0 \ \underline{\text{for}}\ x_2 < x \leq 1, \tag{4.5}$$

$$f'''(x) > 0 \ \underline{\text{for}}\ 0 \leq x < x_3, \quad f'''(x) < 0 \ \underline{\text{for}}\ x_3 < x \leq 1, \tag{4.6}$$

<u>while</u>

$$\begin{aligned} g(x) &> 0 \qquad \text{for} \quad 0 < x \leq 1, \\ g'(x) &> 0 \qquad \text{for} \quad 0 \leq x \leq 1, \\ g''(x) &\geq 0 \qquad \text{for} \quad 0 \leq x \leq 1. \end{aligned}$$

(The theorem also gives bounds on f, g and their derivatives in terms of the constant R appearing in the boundary conditions, but we will not specify these here.)

The proof of Theorem 3 depends on two lemmas.

Lemma 5. Let $\bar{g} \in C^1[0,1]$ and satisfy

$$\bar{g}(0) = 0, \quad \bar{g}(1) = R, \quad \bar{g}'(x) \geq 0.$$

Let $\bar{f} \in C^1[0, 1]$, with $\bar{f}(x) \leq 0$. Then there is a unique $\tilde{f} \in C^4[0, 1]$ such that

$$\tilde{f}^{iv} + \bar{f}\tilde{f}''' = -\bar{g}\bar{g}', \quad 0 \leq x \leq 1,$$

with

$$\tilde{f}(0) = \tilde{f}''(0) = \tilde{f}(1) = \tilde{f}'(1) = 0 .$$

Moreover

$$-\frac{1}{2}R^2 \leq \tilde{f}(x) \leq 0,$$

and there exist points $\tilde{x}_1, \tilde{x}_2$ such that (4.4), (4.5) hold with $\tilde{f}, \tilde{x}_i$ replacing f, x_i $(i = 1, 2)$. Finally, if $\bar{g}'(x) > 0$, then there exists also $\tilde{x}_3$ such that $\tilde{f}, \tilde{x}_3$ satisfy (4.6).

Given the linearity of the problem in $\tilde{f}$, this lemma is not difficult to establish. To prove existence and uniqueness, we have merely to show that the homogeneous problem has only the trivial solution, and this, and the resultant properties of $\tilde{f}$, arise from straightforward manipulations with the equations.

Lemma 6. Let $\tilde{f} \in C^2[0, 1]$ and satisfy $\tilde{f}(0) = 0$, $\tilde{f}(x) \leq 0$ and the properties (4.4), (4.5) for some points $\tilde{x}_1, \tilde{x}_2$. Then there exists a unique $\tilde{g} \in C^2[0, 1]$ such that

$$\tilde{g}'' + \tilde{f}\tilde{g}' - \tilde{f}'\tilde{g} = 0, \quad 0 \leq x \leq 1,$$

with

$$\tilde{g}(0) = 0, \quad \tilde{g}(1) = R.$$

Moreover,

$$\tilde{g}'(x) > 0, \quad \tilde{g}''(x) \geq 0.$$

Again, being a linear problem this is not difficult to establish. Then Lemmas 5, 6 together imply that we have a map $(\bar{f}, \bar{g}) \rightarrow (\tilde{f}, \tilde{g})$ defined on the set $F \otimes G$ where

$$F = \{\bar{f} \in C^1[0, 1] : -\frac{1}{2}R^2 \leq \bar{f} \leq 0, \quad \bar{f}(0) = \bar{f}(1) = 0\},$$

$$G = \{\bar{g} \in C^1[0, 1] : \bar{g}(0) = 0, \quad \bar{g}(1) = R, \quad \bar{g}' \geq 0\} .$$

The analysis behind Lemmas 5, 6 provides even more information on bounds for $(\tilde{f}, \tilde{g})$ then we have explicitly exhibited here, and these bounds make it clear that the map $(\bar{f}, \bar{g}) \rightarrow (\tilde{f}, \tilde{g})$ maps the convex set $F \otimes G$ into a compact subset of itself. The Schauder fixed-point theorem then guarantees a fixed point for the map, and this fixed point is of course a solution to the boundary-value problem.

5. The regular case: behaviour as $\varepsilon \downarrow 0$

We return now to the equations in the form (1.5), (1.6) and are interested, as in section 4, in anti-symmetric solutions. Betchelor [2] conjectured that, in the limit of large Reynolds number (small ε), the main body of the fluid is separated into two parts, rotating with opposite angular velocities with a narrow central transition layer through which the fluid adjusts from one rate of rotation to the other. On the other hand, Stewartson [25] conjectured that the main body of the fluid is only slightly distrubed at large Reynolds number.

Numerical computations have been carried out by Lance and Rogers [26], Pearson [27] and Greenspan [28], but the evidence given by these is conflicting. Tam [29] has applied the method of matched asymptotic expansions to suggest the non-uniqueness of the solution. Serrin [30] has commented on the computational results and the mathematical difficulty of the problem.

It is proved in [24] that for any odd solution of the boundary-value problem which also satisfies the condition $G' \geq 0$, and so for the particular solution whose existence is guaranteed by Theorem 3, we can obtain precise estimates on the size and behaviour of the solution as $\varepsilon \downarrow 0$. Some of these results are given in detail in Theorem 4 below, but we remark that the behaviour so found is consistent with Stewartson's predictions and not with Betchelor's. At the same time, the absence of any uniqueness proof amongst the results means that a solution of Batchelor's type is not completely ruled out, although our investigations of the equations enable us to say that certain behaviours are just not consistent with the equations, and that in particular the solution obtained numerically by Greenspan [28] is impossible. The point here is that the function H obtained by Greenspan satisfies (when scaled by a factor ε) the conditions satisfied by $\tilde{f}$ in Lemma 6, but his function G, which must by Lemma 6 satisfy $G' > 0$, fails to do so.

Before stating Theorem 4, we remark that it is part of the proof of Lemma 5 that a solution with $G' > 0$ will satisfy $H \leq 0$ and possess the three points x_1, x_2, x_3 associated with (4.4) - (4.6).

Theorem 4. Any odd solution of (1.5) - (1.7) which has $G' \geq 0$ has the following behaviour as $\varepsilon \downarrow 0$.

(i) $x_1, x_2, x_3 \to 1$ with $1 - x_2$, $1 - x_3$ precisely of order $\varepsilon^{\frac{1}{2}}$, while $(1 - x_1)/\varepsilon^{\frac{1}{2}} \to \infty$, $1 - x_1 = O(\varepsilon^{\frac{1}{2}} \log \varepsilon)$.

(ii) $\sup_{0 \leq x \leq 1} |H(x)|$ is precisely of order $\varepsilon^{\frac{1}{2}}$.

(iii) $-H'(0)$ is precisely of order $\varepsilon^{\frac{1}{2}}$, while $H'(x_2)$, which gives the maximum of H', is precisely of order 1.

(iv) There exists a constant $K > 0$ such that, uniformly in x,

$$H''(x) = O(\varepsilon^{-\frac{1}{2}} \exp\{-K\varepsilon^{-\frac{1}{2}}(1-x)\}),$$

$$H'''(x) = O(\varepsilon^{-1} \exp\{-K\varepsilon^{-\frac{1}{2}}(1-x)\}),$$

while

$$-H''(1) \text{ is precisely of order } \varepsilon^{-\frac{1}{2}},$$

$$-H'''(1) \text{ is precisely of order } \varepsilon^{-1}.$$

(v) Similarly, G, G', G'' are exponentially small except near $x = 1$, and $G'(1)$ is precisely of order $\varepsilon^{-\frac{1}{2}}$.

(vi) In the boundary layer, the solution behaves like a suitably scaled version of a solution of the von Kármán singular problem possessing the monotonicity properties that the solution obtained in Theorem 1 possesses.

The result (vi) is the result that a formal application of matched asymptotic expansions would give. For if we make the change of variables

$$1 - x = \varepsilon^{\frac{1}{2}}\xi, \quad -\varepsilon^{\frac{1}{2}}H(x) = \phi(\xi), \quad G(x) = \psi(\xi),$$

and if the constant in (1.8) is in some sense negligible, then the equation (1.8) takes the form (1.1) with $\Omega_\infty = 0$. At the same time, since there is no uniqueness result for solutions of the von Kármán problem, we cannot say that the solution which arises in the boundary layer is necessarily the same as that in Theorem 1.

The proof of Theorem 4 is contained in a long series of lemmas, each at least reasonably straightforward, through which one moves steadily closer to the final required result. Lemmas 3, 4 are important steps in the process.

References

1. T. von Kármán, "Über laminare und turbulente Reibung," Z. Angew. Math. Mech. 1, 232-252 (1921).
2. G. K. Batchelor, "Note on a class of solutions of the Navier-Stokes equations representing steady rotationally-symmetric flow," Quart. J. Mech. Appl. Math. 4, 29-41 (1951).
3. T. von Kármán and C. C. Lin, "On the existence of an exact solution of the equations of Navier-Stokes," Comm. Pure Appl. Math. 14, 645-655 (1961).
4. L. N. Howard, "A note on the existence of certain viscous flows," J. Math. and Phys. 40, 172-176 (1961).

5. R. Iglisch, "Elementarer Existenzbeweis für die Strömung in der laminaren Grenzschicht zur Potential strömung $V = u_1 x^m$ mit $m > 0$ bei Absaugen und Ausblasen," Z. Angew. Math. Mech. 33, 143-147 (1953).

6. W. A. Coppel, "On a differential equation of boundary-layer theory," Phil. Trans. Roy. Soc. A 253, 101-136 (1960).

7. D. Ho and H. K. Wilson, "On the existence of a similarity solution for a compressible boundary layer," Arch. Rational Mech. Anal. 27, 165-174 (1967).

8. J. B. McLeod and J. Serrin, "The existence of similar solutions for some boundary layer problems", Arch. Rational Mech. Anal. 31, 288-303 (1968).

9. J. B. McLeod, "Von Kármán's swirling flow problem," Arch. Rational Mech. 33, 91-102 (1969).

10. M. H. Rogers and G. N. Lance, "The rotationally symmetric flow of a viscous fluid in the presence of an infinite rotating disc," J. Fluid Mech. 7, 617-631 (1960).

11. J. B. McLeod, "The asymptotic form of solutions of von Kármán's swirling flow problem," Quart. J. Math. (Oxford) (2) 20, 483-496 (1969).

12. P. Hartman, "The swirling flow problem in boundary layer theory," Arch. Rational Mech. Anal. 42, 137-156 (1971).

13. J. Watson, "On the existence of solutions for a class of rotating disc flows and the convergence of a successive approximation scheme," J. Inst. Math. Appl. 1, 348-371 (1966).

14. P. J. Bushell, "On von Kármán's equations of swirling flow," J. London Math. Soc. (2) 4, 701-710 (1972).

15. D. J. Evans, "The rotationally symmetric flow of a viscous fluid in the presence of an infinite rotating disc with uniform suction," Quart. J. Mech. Appl. Math. 22, 467-485 (1969).

16. J. B. McLeod, "A note on rotationally symmetric flow above an infinite rotating disc," Mathematika 17, 243-249 (1970).

17. J. B. McLeod, "The existence of axially symmetric flow above a rotating disk," Proc. Roy. Soc. A. 324, 391-414 (1971).

18. P. Hartman, "On the swirling flow problem," Indiana Univ. Math. J. 21, 849-855 (1972).

19. J. B. McLeod and D. H. Sattinger, "Loss of stability and bifurcation at a double eigenvalue," J. Funct. Anal. 14, 62-84 (1973).

20. S. Smale, "An infinite dimensional version of Sard's theorem," Amer. J. Math. 87, 861-866 (1965).

21. K. D. Elworthy and A. J. Tromba, "Degree theory on Banach manifolds," Proc. Symp. Pure Maths. Amer. Math. Soc. 18, Pt. 1, 86-94 (1970).

22. S. P. Hastings, "On existence theorems for some problems from boundary layer theory," Arch. Rational Mech. Anal. 38, 308-316 (1970).

23. A. R. Elcrat, "On the swirling flow between rotating coaxial disks," J. Diff. Equ., to appear.

24. J. B. McLeod and S. V. Parter, "On the flow between two counter-rotating infinite plane disks," Arch. Rational Mech. Anal., to appear.

25. K. Stewartson, "On the flow between two rotating coaxial disks," Proc. Cambridge Philos. Soc. 49, 333-341 (1953).

26. G. N. Lance and M. H. Rogers, "The axially symmetric flow of a viscous fluid between two infinite rotating disks," Proc. Roy. Soc. A 266, 109-121 (1962).

27. C. E. Pearson, "Numerical solutions for the time-dependent viscous flow between two rotating coaxial disks," J. Fluid Mech. 21, 623-633 (1965).

28. D. Greenspan, "Numerical studies of flow between rotating coaxial disks," J. Inst. Math. Appl. 9, 370-377 (1972).

29. K. K. Tam, "A note on the asymptotic solution of the flow between two oppositely rotating infinite plane disks," SIAM J. Appl. Math. 17, 1305-1310 (1969).

30. J. Serrin, "Existence theorems for some compressible boundary layer problems," Studies in Applied Math. 5, (SIAM) Symposium held at Madison, Wisconsin, Summer 1969. Edited by J. Nohel (1969).

A SURVEY OF SPECTRAL THEORY FOR PAIRS OF ORDINARY DIFFERENTIAL OPERATORS

Åke Pleijel

INTRODUCTION. The spectral theory of formally symmetric differential equations $Su = \lambda u$ was initiated in 1910 by H. Weyl for real second order equations. Of special importance for Weyl's theory is the use of his well-known contracting circles. The theory works equally well for $Su = \lambda ru$ provided $r(x) > 0$ for all x under consideration. If symmetric boundary conditions are introduced, it leads to spectral theorems in a Hilbert space with metric determined by $\int r|u|^2$. In 1950 K. Kodaira extended Weyl's method to real operators S of arbitrary even order. A different method was simultaneously obtained by I.M. Glasmann. General operators S were treated in 1965 by T. Kimura and M. Takahasi who thereby completed Kodaira's work.

An extension to equations $Su = \lambda Tu$ with two operators S and T, S of higher order than T, is due to Weyl when S is real and of second order, while T is multiplication by a function r which is now allowed to take both positive and negative values. A Dirichlet integral $(u,u)_S = \int(p|u'|^2 + q|u|^2)$ is assumed to be positive and can replace the non-definite $\int r|u|^2$. Also in the theory of more general operators there are two similarly treatable situations, a T-positive one in which a Dirichlet integral $(\cdot,\cdot)_T$ belonging to T is positive, and a S-positive in which a Dirichlet integral $(\cdot,\cdot)_S$ of S has this property. In both cases symmetric boundary conditions Z are defined. Since $Su = Tv$ is a relation, spaces $H_\infty(Z)$ depending upon Z and formally corresponding to an eigenvalue $\lambda = \infty$ must be taken into account in spectral theorems. Such spaces for instance enter the theory of $Su = \lambda ru$ when $r(x)$ vanishes on a subinterval (Pleijel, Ark. Mat. Astr. Fys. 30 A : 21, 1944, for the corresponding p.d.e. problem). They also appear in T-positive theories.

Disregarding several less systematic contributions, an extension of the classical theory has been given for higher order operators by Fred Brauer [6] in 1958. Under the assumption $(u,u)_T \geq c \int |u|^2$, c constant > 0, for functions with compact supports, Brauer considers $T^{-1}S$, where T^{-1} is the invers of the Friedrichs extension of T. This means a restriction of the final symmetric boundary conditions. The results of the classical theory are transferred to the new situation, but the underlying condition is rather heavy, compare Section 9. The treatment is clearly confined to T-positive cases, but also to special boundary conditions.

It should also be mentioned that in recent years F.W. Schäfke, H.-D. Niessen and A. Schneider have established a general theory for systems of first order equations, see for instance [12], which under certain conditions can be reduced to $Su = \lambda Tu$.

A direct study of $Su = \lambda Tu$ along the lines originated by Weyl was started by the author in 1968, [13], [14]. The present survey concerns this study of S- and T-positive cases. It contains the definition of general symmetric boundary conditions and the corresponding spaces $H_\infty(Z)$. The survey begins with a presentation of the S-positive theory.

1. A GREEN'S FORMULA AND A HERMITEAN FORM. The operator S shall have a representation

$$S = \sum_{j=0}^{m} \sum_{k=0}^{m} D^j a_{jk} D^k , \quad D = id/dx ,$$

on an interval I containing none or one (or both) of its endpoints a and b, $a < b$. The functions a_{jk} shall be "sufficiently regular" on I and enjoy hermitean symmetry, $\overline{a}_{jk} = a_{kj}$, in which case

S is formally symmetric. The Dirichlet integral

$$(u,v)_{S,J} = \int_J \Sigma\Sigma\, a_{jk} D^k u\, \overline{D^j v}$$

is obtained by partial integrations of $Su\cdot\overline{v}$ over a compact subinterval J of I.

The operator T shall be given by a similar sum as S, again with regular hermitean coefficients b_{jk}, $0 \leq j, k \leq n$. In the S-positive theory it is assumed that $a_{mm}(x) \neq 0$ on I. The operator T shall have lower order than S. Then $n \leq m$.

It is natural to consider functions u and $\dot{u}$ which are regular on I and related by $Su = T\dot{u}$. The set of such pairs $U = (u,\dot{u})$ form the linear space $E(I)$. By partial integrations the Green formula

$$i^{-1}((\dot{u},v)_{S,J} - (u,\dot{v})_{S,J}) = [q_x(U,V)]_J \tag{1.1}$$

is obtained when $U = (u,\dot{u})$, $V = (v,\dot{v}) \in E(I)$ and J is a compact subinterval of I. The out-integrated part on the right hand side contains a hermitean form $q_x(U,V)$. By computation it can be seen that

$$q_x(U,U) = \sum_1^m |\cdot|^2 - \sum_1^m |\cdot|^2 ,$$

where the dots indicate linear forms. This proves the double inequality

$$\operatorname{sig} q_x \leq (m,m) \tag{1.2}$$

for the signature of q_x. On the solution space

$$E_\lambda(I) = \{U = (u,\lambda u) \in E(I)\}$$

related to $Su = \lambda Tu$, it can be seen that

$$\operatorname{sig} q_x = (m,m) \quad \text{on } E_\lambda(I),\ x \in I, \tag{1.3}$$

provided $\lambda \neq 0$. The dimension of $E_\lambda(I)$ is $M = 2m$.

2. POSITIVE AND FINITE DIRICHLET INTEGRALS. In the S-positive theory it is assumed that $(u,u)_S \geq 0$ (subscript J) is non-negative and also increasing with J, provided that the compact subinterval J of I contains a certain interval J_0. It is furthermore assumed that for such intervals J

$$\underset{J}{\|u\|_S} = \underset{J}{(u,u)_S^{1/2}}$$

defines a norm on $E_\lambda(I)$ for every non-real λ.

Square brackets shall indicate finiteness of Dirichlet integrals over I. Thus

$$E[I] = \{U = (u,\dot{u}) \in E(I) : \underset{I}{(u,u)_S} < \infty,\ \underset{I}{(\dot{u},\dot{u})_S} < \infty\},$$

$$E_\lambda[I] = \{U = (u,\lambda u) \in E_\lambda(I) : \underset{I}{(u,u)_S} < \infty\}.$$

The right hand side of (1.1) is denoted by $\underset{J}{Q}$ so that

$$\underset{J}{Q}(U,V) = q_\beta(U,V) - q_\alpha(U,V) \tag{2.1}$$

when α,β are the endpoints of J. Because of (1.1)

$$\underset{J}{Q}(U,V) = i^{-1}(\underset{J}{(\dot{u},v)_S} - \underset{J}{(u,\dot{v})_S}) \tag{2.2}$$

when $U,V \in E(I)$. On $E_\lambda(I)$ and with $V = U$ this formula is reduced to

$$\underset{J}{Q}(U,U) = c(\lambda)\ \underset{J}{(u,u)_S},\quad c(\lambda) = i^{-1}(\lambda - \bar{\lambda}),$$

which shows that $c(\lambda)\ \underset{J}{Q}$ is positive definite on $E_\lambda(I)$ when $\mathrm{Im}(\lambda) \neq 0$ and $J \supset J_0$. From (2.1) and (1.2) it follows that

$$\mathrm{sig}\ \underset{J}{Q} \leq (M,M). \tag{2.3}$$

This shows that $E_\lambda(I)$ is maximal in $E(I)$, i.e. $E_\lambda(I)$ has no proper extension (dimension $> M$) in $E(I)$ on which $c(\lambda)\ \underset{J}{Q}$ remains positive definite.

If $U,V \in E[I]$, one can let J tend to I in (2.2), hence in

(2.1), obtaining in this way

$$\underset{I}{Q}(U,V) = i^{-1}((\dot{u},v)_{\underset{I}{S}} - (u,\dot{v})_{\underset{I}{S}}), \tag{2.4}$$

$$\underset{I}{Q}(U,V) = q_b(U,V) - q_a(U,V), \tag{2.5}$$

where $q_b(U,V) = \lim_{\beta\to b} q_\beta(U,V)$, $q_a(U,V) = \lim_{\alpha\to a} q_\alpha(U,V)$. When U,V are in $E_\lambda[I]$ the formula (2.4) can be reduced to

$$\underset{I}{Q}(U,V) = c(\lambda)\,(u,v)_{\underset{I}{S}}, \quad c(\lambda) = i^{-1}(\lambda - \bar{\lambda}),$$

and $c(\lambda)\underset{I}{Q}$ is positive definite on $E_\lambda[I]$.

3. AN IDENTITY. Weyl developed his theory for a half-closed interval $I = \{x:\ a \leq x < b\}$ and with a boundary condition at $x = a$. Another approach to the spectral theory is obtained by a generalization of Weyl's contracting circles to the case of an arbitrary interval without imposing any restriction of the final symmetric boundary condition before-hand. The basis for such a generalization is the following identity. Let $U = (u,\dot{u})$ belong to $E(I)$, put $f = \dot{u} - \lambda u$ and assume that

$$(f,f)_{\underset{I}{S}} < \infty. \tag{3.1}$$

For any $V = (v,\lambda v)$ in $E_\lambda(I)$ one obtains from Green's formula that

$$(c(\lambda)\underset{J'}{Q}(U-V,\ U-V) - (f,f)_{\underset{I-J'}{S}}) - (c(\lambda)\underset{J}{Q}(U-V,\ U-V) - (f,f)_{\underset{I-J}{S}}) =$$

$$= (|c(\lambda)|\ \|u-v\|_{\underset{J'-J}{S}} - \|f\|_{\underset{J'-J}{S}})^2 +$$

$$+ |c(\lambda)|((\|u-v\|_{\underset{J'-J}{S}} + \|f\|_{\underset{J'-J}{S}})^2 - \|u-v+\frac{ic(\lambda)}{|c(\lambda)|}f\|^2_{\underset{J'-J}{}}) \tag{3.2}$$

when $J_0 \subset J \subset J'$ (compact) $\subset I$. The identity is deduced by inserting $\dot{u} = \lambda u + f$ in the expressions for $\underset{J}{Q}(U-V,\ U-V)$, $\underset{J'}{Q}(U-V,\ U-V)$ according to (2.2). According to the assumptions in Section 2, the expression $\|\cdot\|_{\underset{J'-J}{S}}$ is a seminorm and the right hand side of (3.2) is non-negative.

In connection with (3.2) the sets

$$\Sigma_J : V \in E_\lambda(I),\ c(\lambda)\underset{J}{Q}(U-V,\ U-V) - \underset{I-J}{(f,f)_S} \leq 0$$

are considered for $U \in E(I)$, with (3.1) satisfied. Since $c(\lambda)\underset{J}{Q}$ is positive definite on $E_\lambda(I)$, the set Σ_J is the interior of an ellipsoid when $J \supset J_0$. The center $U(J)$ of Σ_J is determined by $\underset{J}{Q}(U-U(J),\ E_\lambda(I)) = 0$, $U(J) \in E_\lambda(I)$. Due to the maximality of $E_\lambda(I)$ in $E(I)$, stated after (2.3), the inequality

$$c(\lambda)\underset{J}{Q}(U-U(J),\ U-U(J)) \leq 0$$

holds true and shows that Σ_J contains $U(J)$. From (3.2) it follows that the non-empty and compact ellipsoids Σ_J, $J \to I$, contract to a limit ellipsoid containing at least its center $U(I)$. This leads to the following

<u>COMPENSATION THEOREM</u>. <u>If</u> $U = (u,\dot{u}) \in E(I)$, <u>and</u>

$$\underset{I}{(\dot{u}-\lambda u,\ \dot{u}-\lambda u)_S} < \infty \tag{3.3}$$

<u>holds for a non-real value</u> λ, <u>there is a unique</u> $U(I)$ <u>in</u> $E_\lambda(I)$ <u>such that</u> $U - U(I) \in E[I]$, <u>and</u> $\underset{I}{Q}(U-U(I),\ E_\lambda[I]) = 0$. <u>For this</u> $U(I)$

$$c(\lambda)\underset{I}{Q}(U-U(I),\ U-U(I)) \leq 0. \tag{3.4}$$

<u>4. MAXIMALITY</u>. The condition (3.3) of the compensation theorem is fulfilled if $U \in E[I]$. Then $U(I)$ belongs to $E_\lambda[I]$ and (3.4) shows that $c(\lambda)\underset{I}{Q}$ cannot be positive definite on the linear hull $\{U,\ E_\lambda[I]\} = \{U-U(I),\ E_\lambda[I]\}$ if the $E[I]$-element U does not belong to $E_\lambda[I]$. Thus the space $E_\lambda[I]$ on which $c(\lambda)\underset{I}{Q}$ is positive definite, is maximal in $E[I]$ with this property.

Since $c(\overline{\lambda}) = -c(\lambda)$ the form $c(\lambda)\underset{I}{Q}$ is negative definite on

$E_{\bar\lambda}[I]$, and $E_{\bar\lambda}[I]$ is also maximal in $E[I]$. Consequently Q_I is non-degenerate on the direct sum

$$E_\lambda[I] \dotplus E_{\bar\lambda}[I], \tag{4.1}$$

and the sum is maximal in $E[I]$ with this property.

From the intrinsic nature of the maximal properties it follows that <u>the deficiency pair</u>

$$(\dim E_\lambda[I], \dim E_{\bar\lambda}[I]), \ \mathrm{Im}(\lambda) > 0 ,$$

<u>is constant, independent of</u> λ.

<u>5. SYMMETRIC BOUNDARY CONDITIONS</u>. To link the pair S,T to a symmetric operator or relation in a Hilbert space, linear subspaces of $E[I]$ are considered on which $(\dot u,v)_{S_I} = (u,\dot v)_{S_I}$. According to (2.4) such spaces are nullspaces for Q_I. From the maximality of $E_\lambda[I] \dotplus E_{\bar\lambda}[I]$ it follows that every nullspace can be extended to a maximal one which has the form of a direct sum

$$Z = E[I]^{\perp} \dotplus Z'. \tag{5.1}$$

Here

$$E[I]^{\perp} = \{U \in E[I] : Q_I(U,E[I]) = 0\}$$

which is a nullspace. The space Z' is a maximal nullspace in $E_\lambda[I] \dotplus E_{\bar\lambda}[I]$. <u>Maximal nullspaces</u> (5.1) <u>are symmetric boundary conditions</u>. They can all be listed by spaces Z' in a finite dimensional space $E_\lambda[I] \dotplus E_{\bar\lambda}[I]$, $\mathrm{Im}(\lambda) \neq 0$. <u>A symmetric boundary condition</u> $(u,\dot u) \in Z$ <u>is a condition for the pair</u> $(u,\dot u)$, $Su = T\dot u$, <u>and not a condition to be satisfied by a single function</u>. For a solution of $Su = \lambda Tu$, a symmetric boundary condition in general contains the eigenvalue parameter λ. Similarly symmetric boundary conditions are defined in a T-positive theory. If T is the identity they can be reduced to conditions for a function only, and then coincide with the classical ones.

6. RESOLVENT OPERATOR. Let λ belong to the halfplane, $\mathrm{Im}(\lambda) > 0$ or $\mathrm{Im}(\lambda) < 0$, in which

$$\dim E_\lambda[I] \geq \dim E_{\bar{\lambda}}[I], \tag{6.1}$$

and let $Z = E[I]^{\perp} \dot{+} Z'$ be a symmetric boundary condition. Due to (6.1) the space $E_\lambda[I] \dot{+} E_{\bar{\lambda}}[I]$ coincides with $E_\lambda[I] \dot{+} Z'$ and every element in this space can be compensated by an $E_\lambda[I]$-element so that the difference belongs to Z'. By the compensation theorem of Section 3 (and integration of $Sw = Tv$) it can be proved that to any regular function v with $(v,v)_{\underset{I}{S}} < \infty$, there exists a unique u such that $(u, \lambda u + v) \in Z$ which defines the operator $R(\lambda)$, $u = R(\lambda)v$. Let L be the set of all such functions v. Its closure $H = \overline{L}$ with respect to $(\cdot,\cdot)_{\underset{I}{S}}$ is the Hilbert space basic for the theory.

It can be seen that $R(\lambda)$ maps L onto the domain $\mathcal{D}(Z)$ of the relation Z. The operator is bounded and its closure $\overline{R(\lambda)}$ maps H onto the domain $\mathcal{D}(\overline{Z})$ of the closure $\overline{Z}$ of Z in $H \oplus H$. The set $\mathcal{D}(\overline{Z})$ is a subset of the subhilbertspace $H(Z) = \overline{\mathcal{D}(Z)}$.

7. EXTENSIONS TO MAXIMAL SYMMETRIC RELATIONS. In contrast to the case when T is the identity, the resolvent operator $\overline{R(\lambda)}$ is in general not one-to-one. It can be seen that $H_\infty(Z) = \{v \in H : \overline{R(\lambda)}v = 0\}$ coincides with $\{v \in H : (0,v) \in \overline{Z}\}$. $H(Z)$ and $H_\infty(Z)$ are orthogonal, and $H = H(Z) \oplus H_\infty(Z)$. In the sequel, let $R(\lambda)$ and $\overline{R(\lambda)}$ denote the restrictions to $H(Z)$ of these operators. These restrictions are one-to-one and $\overline{R(\lambda)}$ maps $H(Z)$ onto $\mathcal{D}(\overline{Z}) \subset H(Z) = \overline{\mathcal{D}(Z)}$. It is easy to see that the domain $\{u : (u,\dot{u}) \in Z$ for a $\dot{u} \in H(Z)\}$ defines an operator $\dot{u} = Pu$ in $H(Z)$ which is clearly symmetric. A compari-

son of the restriction of $R(\lambda)$ and the operator P on account of $(u, \lambda u+v) \in Z$, $(u,\dot{u}) \in Z$ gives $(P-\lambda)R(\lambda) = 1$. For the closures one obtains $(\overline{P}-\lambda)\overline{R(\lambda)} = 1$. But here $\overline{R(\lambda)}$ is one-to-one because of the restriction to $H(Z)$. $\overline{P}-\lambda$ has the range $H(Z)$ and $\overline{P}$ is a maximal symmetric extension of P. This extension is selfadjoint if and only if

$$\dim E_{\lambda}[I] = \dim E_{\overline{\lambda}}[I].$$

The result agrees with the theory of symmetric relations on a Hilbert space due to R. Arens [1]. In [3], [4], [5] Bennewitz has presented this theory with the application to differential operators in mind, and with a thorough discussion of the extension of such relations to maximal ones. The existence of a space H_{∞} is characteristic for the theory of symmetric relations compared to the theory of symmetric operators.

In the selfadjoint case when $\dim E_{\lambda}[I] = \dim E_{\overline{\lambda}}[I]$, the spectral theorem of $\overline{P}$ assigns eigenspaces $\mathcal{E}(\Delta)$ to real sets Δ. In addition there is the eigenspace $\mathcal{E}(\infty) = H_{\infty}(Z)$ belonging to $\lambda = \infty$. The spaces $\mathcal{E}(\Delta)$ have the usual properties. They span $H(Z)$, and together with $\mathcal{E}(\infty)$ the entire space H.

It remains to show that $\mathcal{E}(\Delta)$, when Δ is bounded, consists of regular functions to which S and T can be applied. If $u \in \mathcal{E}(\Delta)$, Δ bounded, then also $\overline{P}^{k}u = v$ belongs to $\mathcal{E}(\Delta)$ for an arbitrary positive integer k. In a setting due to Bennewitz [5], Weyl's lemma tells that in a weak solution pair of $Su = T\dot{u}$,

$$\int u\overline{S\varphi} = \int \dot{u}\,\overline{T\varphi} \quad \text{for all } \varphi \in C_0^{\infty}(I),$$

the function u possesses more derivatives than $\dot{u}$. From $\overline{P}^{k}u = v \in \mathcal{E}(\Delta)$ the desired result follows because of the regularity of v as an element of H.

8. THE T-POSITIVE CASE. In this case a Dirichlet integral $(\cdot,\cdot)_T$ of T is subject to the same conditions which were previously imposed upon $\underset{J}{(\cdot,\cdot)}_S$. The operator T has the same even order for all x while the order of S, still constant for all x, may be even or odd.

The previous discussion only involved Green's formula and the signature properties of q_x in Section 1. There is a similar Green's formula containing $\underset{J}{(\cdot,\cdot)}_T$ instead of $\underset{J}{(\cdot,\cdot)}_S$ and with a new form q_x. For this new q_x statements similar to (1.2) and (1.3) are valid (if $M = 2m-1$, (1.2) is replaced by $\operatorname{sig} q_x \leq (m,m-1)$ or $\leq (m-1,m)$). Therefore the preceeding discussion is valid also in the T-positive theory and gives corresponding results.

9. THE SPECTRAL THEOREM UNDER MORE GENERAL CONDITIONS. Recently Bennewitz [4], [5] deduced a spectral theory for $Su = \lambda Tu$ on the basis of Weyl's lemma and the theory of symmetric Hilbert space relations. This was done under the only condition that an inequality

$$\int_J |u|^2 \leq C(J)\,\underset{I}{(u,u)}$$

holds true with a finite constant $C(J)$ for every compact subinterval J. The expression on the right hand side is the Dirichlet integral of S or T extended over I. The theory covers the preceeding deduction of a spectral theory and Bennewitz could also treat certain cases when $\underset{I}{(\cdot,\cdot)}$ is semi-definite as well as problems for partial differential equations.

This reduces the interest of the previous generalization of Weyl's deduction of the spectral theory. However, this generalization is easily adapted to other questions which are concerned with parts of I and therefore less accessible to Bennewitz' theory. An example is Weyl's limit circle and limit point classification of $Su = \lambda Tu$ at an endpoint of I.

<u>10. LIMIT TYPE AT AN ENDPOINT</u>. In the study of the solution set of $Su = \lambda Tu$ near an endpoint which is not contained in I, it is no restriction to assume I half-closed, $I = \{x : a \leq x < b\}$. The identity (3.2) is then used with $J = [a,x]$ and $J' = [a,x']$, $a<x<x'<b$. Expressions $c(\lambda)q_a(U-V, U-V)$ cancel so that

$$(c(\lambda)q_{x'}(U-V, U-V) - \overset{b}{\underset{x'}{(f,f)}}) - (c(\lambda)q_x(U-V, U-V) - \overset{b}{\underset{x}{(f,f)}}) \qquad (10.1)$$

equals the non-negative expression in (3.2). Here $\overset{\beta}{\underset{\alpha}{(\cdot,\cdot)}}$ denotes the considered Dirichlet integral. In contrast to $c(\lambda)Q_J$ the form $c(\lambda)q_x$ is not positive definite on $E_\lambda(I)$. We therefore replace $E_\lambda(I)$ by a subspace $b_\lambda(I)$ on which $c(\lambda)q_x$ is positive definite when x is near b. The space $b_\lambda(I)$ shall be maximal i.e. have the dimension m if for instance $M = 2m$, $\operatorname{sig} q_x = (m,m)$ on $E_\lambda(I)$. Weyl ellipsoids connected with the identity for (10.1) are

$$\Sigma_x : V \in b_\lambda(I),\ c(\lambda)q_x(U-V, U-V) - \overset{b}{\underset{x}{(f,f)}} \leq 0 .$$

One puts $b_\lambda[I] = b_\lambda(I) \cap E[I]$, and for a similar subspace $b_{\bar\lambda}(I)$ of $E_{\bar\lambda}(I)$ also $b_{\bar\lambda}[I] = b_{\bar\lambda}(I) \cap E[I]$. As in Section 3 one obtains a compensation theorem and it follows that the pair

$$(\dim b_\lambda[I],\ \dim b_{\bar\lambda}[I]),\ \operatorname{Im}(\lambda) > 0 ,$$

is independent of λ and also of the choice of $b_\lambda(I)$ and $b_{\bar\lambda}(I)$. With a slight change of the usual convention, the pair is called <u>the limit type of</u> S <u>and</u> T <u>at</u> b. The limit point case at b occurs when the pair equals $(0,0)$.

The compensation theorem states the existence of an element $U(b)$ in $b_\lambda(I)$ for which $c(\lambda)q_b(U-U(b), U-U(b)) \leq 0$. If $U \in E[I]$, the element $U(b)$ belongs to $b_\lambda[I]$, and if $\dim b_\lambda[I] = 0$ it follows that $c(\lambda)q_b(U,U) \leq 0$ for all U in $E[I]$. The converse is also true. If $\dim b_{\bar\lambda}[I] = 0$, the inequality $c(\lambda)q_b(U,U) \geq 0$ holds on $E[I]$ since $c(\bar\lambda) = -c(\lambda)$. As a consequence <u>the condition</u>

$q_b = 0$ on $E[I]$ is necessary and sufficient for the limit point case at b.

The inequality $c(\lambda)q_b(U - U(b), U - U(b)) \leq 0$ also shows that $b_\lambda[I]$ is maximal in $E[I]$ as a subspace on which $c(\lambda)q_b$ is positive definite. Hence the maximal signature of q_b in $E[I]$ equals the limit type of S,T at b. This value is assumed on the direct sum $b_\lambda[I] \dotplus b_{\bar\lambda}[I]$.

The compensation theorem connected with the identity for (10.1) applies when $U \in E_\lambda(I)$ since $f = \dot{u} - \lambda u$ is then identically 0. The range of the mapping $U \to U - U(b)$ from $E_\lambda(I)$ is a space $a_\lambda[I] \subset E_\lambda[I]$ on which $c(\lambda)q_b$ is non-positive because of $c(\lambda)q_b(U - U(b), U - U(b)) \leq 0$. Green's formula on $E_\lambda(I)$ then shows that $c(\lambda)q_x$ is negative definite on $a_\lambda[I]$ when x is near a. The space also has maximal dimension (m if $M = 2m$) so that $E_\lambda(I) = a_\lambda[I] \dotplus b_\lambda(I)$. Intersection by $E[I]$ gives $E_\lambda[I] = a_\lambda[I] \dotplus b_\lambda[I]$. Thus $\dim b_\lambda[I] = \dim E_\lambda[I] - m$ which again shows the λ-independence of the limit type. It also shows that $\dim E_\lambda[I] \geq m$ which gives a well-known lower bound for the number of linearly independent solutions of $Su = \lambda Tu$ which are "integrable square" near an endpoint with respect to the considered Dirichlet integral.

All considerations above are valid in the T-positive theory even when the order of S is odd, $M = 2m - 1$. In certain statements m must be replaced by $m - 1$ in this case.

11. SIGNATURES OF THE BOUNDARY FORMS.

The maximal signature of q_b on $E[I]$ equals the limit type at b. Similarly, and for an interval I of arbitrary type, the maximal signature of q_a on $E[I]$ agrees with a limit type at the lower (negative) endpoint a. If Z_a and Z_b are (maximal) nullspaces in $E[I]$ for q_a and q_b respectively, the intersection $Z_a \cap Z_b$ is a nullspace for $\underset{I}{Q} = q_b - q_a$ i.e. a symmetric boundary condition provided it is maximal. It follows for instance that separated boundary conditions exist in the selfadjoint case only for even order equations.

12. INTERRUPTED INTERVAL. The identity (3.2) is also valid when $U = (u,\dot{u})$ belongs to the space $E(I_a \cup I_b)$ of pairs defined only on intervals I_a, I_b neighbouring a and b, and there satisfying $Su = T\dot{u}$. The form Q_J is defined as in (2.1) when $J = [\alpha,\beta]$ and $\alpha \in I_a$, $\beta \in I_b$. The inequality (2.3) remains valid. The condition (3.1) is replaced by

$$(f,f)_{I_a \cup I_b} < \infty, \quad f = \dot{u} - \lambda u .$$

The space $E(I)$ can be considered as a subspace of $E(I_a \cup I_b)$ and the extended identity is applied for elements V in $E_\lambda(I)$. In the same way as in Section 3 and 4 one obtains that the previous spaces $E_\lambda[I]$, $E_{\bar{\lambda}}[I]$ and $E_\lambda[I] \dotplus E_{\bar{\lambda}}[I]$ have the same maximality properties as earlier, but now as subspaces of the larger space $E[I_a \cup I_b]$ which is defined in an obvious way.

The space $E[I_a \cup I_b]^\perp = \{U \in E[I_a \cup I_b] : Q_I(U, E[I_a \cup I_b]) = 0\}$ is a nullspace for $Q_I = q_b - q_a$ but clearly also for q_b and q_a separately. Because of the maximality of $E_\lambda[I] \dotplus E_{\bar{\lambda}}[I]$ the orthogonality condition in the definition of $E[I_a \cup I_b]^\perp$ can be replaced by $Q_I(U, E_\lambda[I] \dotplus E_{\bar{\lambda}}[I]) = 0$ which in case $U \in E[I]$ **holds if** $U \in E[I]^\perp$. Thus $E[I]^\perp \subset E[I_a \cup I_b]^\perp$. Consequently q_b and q_a vanish separately also on $E[I]^\perp$.

The previous reasonings lead to an enlarged definition of symmetric boundary conditions depending only on values taken near the end points. They also allow the introduction of a Green's function $g(x,y,\lambda)$ belonging to a given symmetric boundary condition Z and defined in the classical way with usual discontinuity properties at x

etc. The function exists in the halfplane where $\dim E_\lambda[I] \geq \dim E_{\bar\lambda}[I]$, and in the selfadjoint case for all non-real λ. In the T-positive theory it has all expected properties. In the S-positive theory there are certain deviations. Thus the equation $Su - \lambda Tu = Tv$ together with a symmetric boundary condition is in the selfadjoint case solved by

$$\lambda u(x) + v(x) = \underset{I}{(v(\cdot), \overline{g(\cdot,x,\lambda)})_S},$$

and the resolvent identity takes the form

$$\lambda g(x,y,\lambda) - \mu g(x,y,\mu) = (\lambda - \mu) \underset{I}{(g(x,\cdot,\lambda), \overline{g(\cdot,y,\mu)})_S}.$$

For the introduction of Green's functions see Pleijel [16] and for applications Karlsson [10].

13. MISCELLANEOUS STATEMENTS. A well-known theorem in the classical theory of a formally symmetric equation $Su = \lambda u$ states, that _if_ $\dim E_\lambda[I]$ _and_ $\dim E_{\bar\lambda}[I]$ _both coincide with the order_ M _of_ S, _then the spectrum belonging to any symmetric boundary condition is discreet_. Bert Karlsson [10] has proved that the statement is valid in the T-positive theory of a pair S,T. In a forthcoming paper [11] he also studies the S-positive theory and deduces corresponding results.

A theorem by W.N. Everitt [7] for a formally symmetric equation $Su = \lambda u$ of even order M tells that _if_ $\dim E_\lambda[I] = M$, _then also_ $\dim E_{\bar\lambda}[I] = M$. In [9] this theorem is extended by Bert Karlsson to the T-positive theory of a pair S,T. In [11] the same question is treated in the S-positive theory. The results agree with Everitt's but for the case when the difference $M - N$ of the orders of S and T equals 1. In this case Karlsson gives a necessary and sufficient condition for the statement to hold true. This settles a discussion in

Everitt [8] about the case when M is odd. Bert Karlsson has also given examples showing the existence of exceptional pairs with $M - N = 1$, M arbitrary.

For operators S and T on a half-closed interval $I = \{x : a \leq x < b\}$ one can consider the possibility of a S-positive theory with respect to a form

$$\overset{x}{\underset{a}{(u,v)_S}} + \sum_{j=0}^{m-1} \sum_{k=0}^{m-1} A_{jk} D^k u(a) \overline{D^j v(a)} \tag{13.1}$$

instead of $\overset{x}{\underset{a}{(u,v)_S}}$. The order of S is $M = 2m$ and the coefficients A_{jk} shall be hermitean. For $m = 1$ and when the pair S,T is in the limit point case at $b = +\infty$, the question has recently been studied by Atkinson, Everitt and Ong [2] with the construction of Titchmarsh's m-coefficient in mind. Some comments on their paper are found in Pleijel [17]. If the additional sum in (13.1) is not positive definite, a finite number of non-real eigenvalues can occur. The problem is related to Krein - Iohvidov's theory for symmetric operators on a generalized Hilbert space with non-definite metric.

REFERENCES

[1] R. ARENS, Operational calculus of linear relations. Pacific J. Math. 11, 1961, 9-23.

[2] F.V. ATKINSON, W.N. EVERITT and K.S. ONG, On the m-coefficient of Weyl for a differential equation with an indefinite weight function. To appear in Proc. London Math. Soc.

[3] CHR. BENNEWITZ, Symmetric relations on a Hilbert space. Proc. Conference on the Theory of Ordinary and Partial Differential Equations, Dundee, Scotland, March 1972, Springer Lecture Notes 280, 212-218.

[4] CHR. BENNEWITZ, Spectral theory for pairs of partial differential expressions. Uppsala University, Department of Mathematics, Report 1974:3, 1-4.

[5] CHR. BENNEWITZ, Remarks on the spectral theory for pairs of ordinary differential operators. Uppsala University, Department of Mathematics, Report 1974:4, 1-20.

[6] FRED BRAUER, Spectral theory for the differential equation $Lu = \lambda Mu$. Canad. J. Math. 10, 1958, 431-446.

[7] W.N. EVERITT, Singular Differential Equations I: The Even Order Case. Math. Ann. 156, 1964, 9-24.

[8] W.N. EVERITT, Integrable-Square, Analytic Solutions of Odd-Order, Formally Symmetric, Ordinary Differential Equations. Proc. London Math. Soc. Third Series, XXV, July 1972, 156-182.

[9] BERT KARLSSON, Generalization of a theorem of Everitt. To appear in Proc. London Math. Soc.

[10] BERT KARLSSON, A compactness theorem for pairs of formally self-adjoint ordinary differential operators. Uppsala University, Mathematics Department, Report 1974:2, 1-10.

[11] BERT KARLSSON, On the limit circle case for pairs of ordinary symmetric differential operators in a left-positive theory. To appear in Uppsala University, Mathematics Department Report.

[12] H.-D. NIESSEN and A. SCHNEIDER, Integraltransformationen zu singulären S-hermiteschen Rand-Eigenwertproblemen. Manuscripta Math. 5, 1971, 133-145.

[13] ÅKE PLEIJEL, Some remarks about the limit point and limit circle theory. Ark. Mat. 7:21, 1968, 543-550.

[14] ÅKE PLEIJEL, Complementary remarks about the limit point and limit circle theory. Ark. Mat. 8:6, 1969, 45-47.

[15] ÅKE PLEIJEL, Spectral theory for pairs of ordinary formally self-adjoint differential operators. Journal Indian Math. Soc. 34, 1970, 259-268.

[16] ÅKE PLEIJEL, Green's functions for pairs of formally selfadjoint ordinary differential operators. Conference on the Theory of Ordinary and Partial Differential Equations, Dundee, Scotland, March 1972, Springer Lecture Notes 280, 131-146.

[17] ÅKE PLEIJEL, Generalized Weyl circles. Conference on the Theory of Ordinary and Partial Differential Equations, Dundee, Scotland, March 1974, Springer Lecture Notes. To appear.

Deficiency indices and properties of spectrum of some classes of differential operators

F. S. Rofe-Beketov

This lecture surveys recent results obtained by the author and his co-workers A. G. Brusentsev, A. M. Holkin, V. I. Hrabustovskii, V I Kogan. A more detailed exposition, full proofs and bibliography can be found in the original papers (see refs.).

1. Ordinary differential equations and systems.

Deficiency indices (see [1], [2]).

Consider the differential equation of an arbitrary order $r \geq 1$ with the $n \times n$ matrix coefficients:

$$\ell[y] \overset{\text{def}}{=} \sum_{k=0}^{r} i^k \ell_k[y] = \lambda W(t) y, \quad 0 \leq t < \infty, \tag{1.1}$$

where

$$\ell_{2j} = D^j p_j(t) D^j, \qquad p_j^*(t) = p_j(t)$$

$$\ell_{2j+1} = \tfrac{1}{2} D^j \{ D q_j(t) + q_j^*(t) D \} D^j; \quad D = \frac{d}{dt} .$$

The highest coefficient in $\ell[y]$ in (1.1), i.e. $p_m(t)$ when $r = 2m$, or $\mathrm{Re}[q_m(t)] = \frac{1}{2}(q_m + q_m^*)$ when $r = 2m+1$, is non-degenerate at $t \in [0, \infty]$. The matrix weight is $W(t) \geq 0$ (in the sense of quadratic forms). The case $\det W(t) \equiv 0$ is not excluded. When the coefficients are not smooth enough, (1.1) has to be regarded as a quasi-differential equation ([1], [3]). $N(\lambda)$ is the number of linearly independent and W-square-integrable solutions $y(t)$ of (1.1) such that

$$\int_0^\infty y^*(t) W(t) y(t) dt < \infty. \tag{1.2}$$

Theorem 1.1 $N(\lambda) = \text{const.}$ in $\mathrm{Im}\,\lambda > 0$, and $N(\lambda) = \text{const.}$ in $\mathrm{Im}\,\lambda < 0$.

We remark that the fact that $W(t)$ can be identically degenerate makes it much more difficult to extend the general theory of symmetric operators to (1.1). Therefore, a mere reference to the theory of deficiency indices of symmetric operators is not sufficient to prove Theorem 1.1. This theorem is related to results announced by S. A. Orlov in 1971 without proofs.

Theorem 1.2. The formal deficiency indices of (1.1), viz. $N_{\pm} \overset{\text{def}}{=\!=} N(\lambda)$, $(\operatorname{Im}\lambda \gtrless 0)$, have the following estimates

$$mn \leq N_{\pm} \leq rn, \qquad r = 2m, \tag{1.3}$$

$$\left.\begin{aligned} mn + \nu_{-}[\operatorname{Re} q_m] &\leq N_{+} \leq rn, \\ mn + \nu_{+}[\operatorname{Re} q_m] &\leq N_{-} \leq rn \end{aligned}\right\} r = 2m + 1 \tag{1.4}$$

(Here, $\nu_{\pm}[Q]$ denotes the number of positive or negative eigenvalues of the Hermitian matrix Q. Since $\det[\operatorname{Re} q_m(t)] \neq 0$ and $q_m(t)$ is continuous, $\nu_{\pm}[\operatorname{Re} q_m(t)]$ is independent of t when $r = 2m + 1$).

Theorem 1.3. If $r > 1$ and if $N(\lambda_0) = rn$ for some λ_0 in $\mathbb{C}$, then $N(\lambda) = rn$ for all $\lambda \in \mathbb{C}$. In particular,

$$N_{+} = rn \text{ if and only if } N_{-} = rn, \quad (r > 1). \tag{1.5}$$

This theorem gives a positive solution of Professor W. N. Everitt's problem [4, p. 181]. Note that W. N. Everitt estimated $N_{\pm}$ for the scalar equation $\ell[y] = \lambda y$ in the general case and established (1.5) when $r = 2m$. For a system like (1.1) but of first order $(r = 1)$:

$$\begin{gathered} \frac{i}{2}\{[q(t)y]' + q^*(t)y'\} = [\lambda W(t) + p(t)]y, \\ q(t) \in AC_{loc}[0,\infty); \quad W, p \in \mathcal{L}^1_{loc}[0,\infty), \end{gathered} \tag{1.6}$$

the following theorem is valid.

Theorem 1.4. If all the solutions of (1.6) are W-square-integrable (1.2) for some $\lambda = \lambda_0 \in \mathbb{C}$ if

$$\inf_{0 \leq t < \infty} \{(\operatorname{sgn} \operatorname{Im} \lambda_0) \int_0^t \operatorname{Sp} [(\operatorname{Re} q(\tau))^{-1} W(\tau)] d\tau\} > -\infty, \tag{1.7}$$

then all the solutions of (1.6) are W-square-integrable for any $\lambda \in \mathbb{C}$. (If $\operatorname{Im} \lambda_0 = 0$, then $\operatorname{sgn} \operatorname{Im} \lambda_0 = 0$ and (1.7) are fulfilled automatically.) This is the generalization of Theorem 9.11.2 in F. V. Atkinson's book [5].

In proving Theorems 1.1 - 1.3 we reduced (1.1) to the canonical first-order system with $rn \times rn$ matrix coefficients

$$\mathcal{F} \frac{dx}{dt} = \lambda \mathcal{H}(t)x, \quad t \in [0, \infty),$$

$$\mathcal{F}^* = -\mathcal{F} = \text{Const}, \ \mathcal{F}^2 = -I_{rn}, \quad 0 \leq \mathcal{H}(t) \in \mathcal{L}^1_{loc} \tag{1.8}$$

(I_s is the $s \times s$ unit matrix.) $N(\lambda)$ is invariant under this reduction.

$$\nu_+[\mathcal{F}/i] = \nu_-[\mathcal{F}/i] = mn, \quad r = 2m, \tag{1.9}$$

$$\left.\begin{aligned} \nu_+[\mathcal{F}/i] &= mn + \nu_+[\operatorname{Re} q_m], \\ \nu_-[\mathcal{F}/i] &= mn + \nu_-[\operatorname{Re} q_m] \end{aligned}\right\} \quad r = 2m + 1. \tag{1.10}$$

The proofs of the above results are given in [1].

It is interesting to find out which of the pairs $\{N_+, N_-\}$ consistent with Theorems 1.2 and 1.3 are realizable for differential equations, or systems, of particular types, on the semi-axis. For example, for the scalar differential equations with $r = 2m$ and with real coefficients, we always have $m \leq N_+ = N_- \leq 2m$: and as I. M. Glazman showed, the $N_\pm$ can take any values between the limits indicated. The equality $N_+ = N_-$ need not necessarily hold when the coefficient in symmetric equations or systems are complex. According to G. A. Kalyabin's results [6], for the system $-y'' + Q(t)y = \lambda y$ on the semi-axis, with $Q^* = Q$, any values of $\{N_+, N_-\}$ allowed by Theorems 1.2 and 1.3 are possible. Our paper [2] described a new class of differential equations whose asymptotic solutions as $x \to \infty$ (and so also their deficiency indices) are determined explicitly by the method of undetermined coefficients. This class is intermediate between the equations with asymptotically constant coefficients and the equations asymptotically similar to Euler's equation.

Theorem 1.5. Let the coefficients in the equation

$$a_r(x)y^{(r)} + a_{r-1}(x)y^{(r-1)} + \dots + a_0(x)y = 0 \tag{1.11}$$

be holomorphic in a certain open sector $S \supset R^+$ (with angle $< \pi$) of the right-hand half-plane, and suppose they have the following asymptotic expansions as $x \to \infty$:

$$a_j(x) \sim x^j\{a_{j,0} + \sum_{k=1}^{\infty} a_{j,k}x^{-k}\}, \quad a_{r-\ell,0} \neq 0, \quad r \geq \ell \geq 0, \; j = 0, \dots, r-\ell$$

$$a_j(x) \sim x^{r-\ell}\{a_{j,0} + \sum_{k=1}^{\infty} a_{j,k}x^{-k}\}, \; a_{r,0} \neq 0, \; j = r-\ell+1, \dots r.$$

If the roots μ_j of

$$a_{r,0}\mu^{\ell} + a_{r-1,0}\mu^{\ell-1} + \dots + a_{r-\ell,0} = 0$$

are simple, then (1.11) has a fundamental system of solutions $y_1, \dots, y_r$ such that ℓ of the solutions have in S as $x \to \infty$ an asymptotic expansion

$$y_j(x) \sim e^{\mu_j x} x^{z_j} (1 + \sum_{k=1}^{\infty} b_{j,k}x^{-k}), \quad j = 1, \dots \ell,$$

and $(r - \ell)$ of the solutions have the following asymptotic expansions

$$y_{\ell+j}(x) \sim x^{z_{\ell+j}}(\ln x)^{\sigma_{\ell+j}}(1 + \sum_{k=1}^{\infty} b_{\ell+j,k}x^{-k}) \quad j = 1, \dots r-\ell,$$

where $z_{\ell+j}$ are the roots of

$$R(z) \stackrel{\text{def}}{=\!=} \sum_{j=1}^{r-\ell} a_{j,0} \prod_{k=0}^{j-1} (z-k) + a_{0,0} = 0.$$

Here the $\sigma_{\ell+j}$ are certain integers such that $0 \leq \sigma_{\ell+j} \leq r - \ell - 1$, and all of them vanish if $R(z)$ has no multiple roots or roots differing by an integer.

In both the cases $r = 2m$ and $r = 2m + 1$ the above class contains the equations with any deficiency indices $\{N_+, N_-\}$ within the limits allowed by Theorems 1.2 and 1.3, with the additional restrictions

$$|N_+ - N_-| \leq 1.$$

Important results on the points mentioned are to be found in the books [5], [7]-[10].

2. Elliptic equations and systems. (Essential self-adjointness, discreteness of spectrum, etc.) (See [11]-[14]).

In this section we suppose the coefficients of differential expressions to be sufficiently smooth functions of $x \in R^n$.

We introduce a special representation of an arbitrary partial differential operator as a superposition of ordinary operators. Let

$$L = \sum_{|\alpha| \leq \sigma} a_\alpha(x)\,\partial^\alpha, \quad \partial^\alpha = \partial_1^{\alpha_1} \dots \partial_n^{\alpha_n}, \quad \partial_j = \frac{\partial}{\partial x_j}. \tag{2.1}$$

The $a_\alpha(x)$ are $r \times r$ matrices. Let $\eta \in R^n$,

$$\partial_\eta^k = (\eta, \nabla_x)^k = \Big(\sum_{j=1}^n \eta_j \partial_j\Big)^k; \quad \omega = \{\eta : |\eta| = 1\},$$

$P_\alpha(\eta)$ is a homogeneous polynomial in η of degree $|\alpha|$ which is determined by the system of equalities

$$\int_\omega P_\alpha(\eta)\,\eta^\beta\, d\omega = \frac{\alpha!}{|\alpha|!}\,\delta_{\alpha,\beta} \quad \text{for all multi-indices } \beta \text{ such that } |\beta| = |\alpha|. \tag{2.2}$$

Lemma 2.1. Any differential expression L (2.1) can be represented in the form

$$(Lu)(x) = \int_\omega \sum_{\nu=0}^{\sigma} a_\nu(x, \eta)\, \partial_\eta^k\, u(x)\, \mu(d\omega_\eta) \tag{2.3}$$

with a certain non-negative measure μ on ω. If L is formally self-adjoint, then

$$L = \sum_{\nu=0}^{\sigma} i^\nu \ell_\nu , \tag{2.4}$$

where

$$\ell_{2j}[u] = \int_\omega (\eta, \nabla_x)^j \{p_{2j}(x, \eta)\, \partial_\eta^j u(x)\}\, \mu\, (d\omega_\eta), \tag{2.4'}$$

$$\ell_{2j+1}[u] = \tfrac{1}{2}\int_\omega (\eta, \nabla_x)^j \{(\eta, \nabla_x) p_{2j+1}(x, \eta) +$$

$$+ p_{2j+1}(x, \eta)\, \partial_\eta\}\, \partial_\eta^j\, u(x)\, \mu(d\omega_\eta), \tag{2.4''}$$

$p_\nu(x, \eta)$ are the Hermitian matrices-functions which can be chosen to be continously dependent on η, and $p_o(x, \eta)$ are those which are completely independent of η: $p_o(x, \eta) \equiv p_o(x)$.

The Lebesgue measure $d\omega$ may always be used as the measure μ and then we can put

$$a_\nu(x, \eta) = \sum_{|\alpha| = \nu} a_\alpha(x) P_\alpha(\eta), \quad \nu = 0, \ldots \sigma, \tag{2.5}$$

$$p_\sigma(x, \eta) = i^{-\sigma} \sum_{|\alpha| = \sigma} a_\alpha(x) P_\alpha(\eta). \tag{2.6}$$

In particular,

$$\Delta^m u(x) = \frac{(2m + n - 2)!!}{(2m - 1)!![(n - 2)!!]s_n} \int_\omega \partial_\eta^{2m}\, u(x)\, d\omega_\eta , \tag{2.7}$$

where Δ is the Laplacian and s_n the hyper-area of ω.

Finally, the measure μ can always be concentrated at a finite number of points (a number which, in general, exceeds the dimension n of the space, but in the absence of mixed derivatives, at the end-points of the coordinate unit vectors).

We distinguish three classes of essentially self-adjoint in $\mathcal{L}^2_r(R^n)$ strongly elliptic operators L of order $\sigma = 2m$:

$$(-1)^m \sum_{|\alpha|=2m} a_\alpha(x)\xi^\alpha \geq \epsilon(x)\,|\xi|^{2m}\, I_r, \tag{2.8}$$

where $\epsilon(x) > 0$, $\xi \in R^n$, $\mathcal{L}^2_r(R^n)$ is the space of square-integrable vector-functions with r components. We shall use the notation $\langle\cdot,\cdot\rangle$ and $(\cdot,\cdot)$ for the scalar product in $\mathcal{L}^2_r(R^n)$ and that in the unitary finite-dimensional space E, respectively. The sufficient conditions for the essential self-adjointness of the formally symmetric operator L are formulated as an inequality between the quadratic forms (on the finite functions) of L and of a standard comparison operator L_c under certain additional constraints on the coefficients of L. For L_c we shall use the following expression

$$L_c f(x) = \sum_{s=0}^{m} (-1)^s c_s \int_\omega (\eta, \nabla_x)^s [V_s(x,\eta)\partial^s_\eta f\,]\,\mu\,(d\omega_\eta), \tag{2.9}$$

where the c_s are real constants, and the matrices are such that

$$V_s(x,\eta) \geq 0,\ (s \geq 0);\ c_0V_0(x,\eta) \equiv c_0V_0\,(x) \geq -\,CQ(x)I_r; \tag{2.9'}$$

here $Q(x)$ is a measurable function with $1 \leq Q(x) \leq \infty$; the possibility that $Q(x) = \infty$ on a set of positive measure is not excluded.

<u>Theorem A.</u> If the coefficients of L (2.4-4") satisfy

$$|(p_\nu(x,\eta)h_1, h_2)| \leq C\,|V^{\frac{1}{2}}_{[\frac{\nu+1}{2}]}(x,\eta)h_1|\,|V^{\frac{1}{2}}_{[\frac{\nu}{2}]}(x,\eta)h_2| \tag{2.10}$$

where $\nu > 0$, for all $h_1, h_2 \in E_r$ and if

$$|\langle Lf, f\rangle| \geq \langle L_c f, f\rangle, \text{ for all } f \in C_0^\infty(R^n) \text{ which map } R^n \text{ to } E_r, \tag{2.11}$$

then L is essentially self-adjoint and for all $u \in C^{2m}(R^n)$ such that $Lu \in \mathcal{L}_r^2(R^n)$ the integral converges:

$$\int_{R^n \times \omega} Q^{-1}(x)(V_s(x, \eta)\partial_\eta^s u, \partial_\eta^s u)\, \mu(d\omega_\eta)dx < \infty; \quad s = 1, \ldots, m, \tag{2.12}$$

provided that L_c belongs to one of the classes described in Theorems 2.1-2.3. (The conditions for $V_s(x, \eta)$ are also preserved when some of the constants c_s in (2.9) are zero.)

From now on, $P(x)$ stands for a function in $C^{2m}(R^n)$ such that

$$0 < P(x) \to \infty \quad \text{as} \quad |x| \to \infty \tag{2.13}$$

and the domains $G^\tau = \{x \in R^n : P(x) < \tau\}$ are simply connected and have piecewise smooth boundaries.

<u>Theorem 2.1</u> If the coefficients of the comparison operator L_c are

$$V_s(x, \eta) = a^{2s}(x, \eta)q^{2m-2s}(x)I_r, \quad s = 0, \ldots, m \tag{2.14}$$

where the scalar functions $1 \leq q(x) \leq \infty$, $a(x, \eta) > 0$ satisfy the conditions

$$a(x, \eta) \in C^m(R^n), \; q^{-m}(x) \in C^{2m}(R^n), \tag{2.15}$$

$$|\partial_\eta[a(x, \eta)q^{-1}(x)]| \leq C \qquad (\text{only if } m > 1) \tag{2.16}$$

$$|\partial_\eta^j q^{-m}(x)| \leq C a^{-j}(x, \eta)q^{-(m-j)}(x), \quad j = 1, \ldots, m \tag{2.17}$$

$$|\partial_\eta^j P(x)| \leq CP(x)a^{-j}(x, \eta)q^{-(2m-j)}(x), \quad j = 1, \ldots, m \tag{2.18}$$

and

$c_m > 0$, $c_0 < 0$, then Theorem A is true for any $c_s > -\infty$, $(s = 0, \ldots, m - 1)$ even when the validity of (2.11) is assumed only for those f for which $\operatorname{supp} f(x) \subset \{x : q(x) \neq \infty\}$.

<u>Theorem 2.2</u>. If

$$V_s(x, \eta) = h^{2s}(x, \eta)g^{2m-2s}(x)I_r \tag{2.19}$$

where the functions $h(x, \eta)$, $g(x) > 0$; $h(x, \eta)$, $g(x) \in C^m(R^n)$ satisfy the conditions

$$|(\eta, \nabla_x)[g^{m-s}(x)h^s(x, \eta)]| \leqslant Cg^{m-s+1}(x)h^{s-1}(x, \eta), \quad s = 1, \ldots, m-1 \tag{2.20}$$

((2.20) is required at $m > 1$ only), and if there is

$$P(x) \text{ as in (2.13), and } \epsilon > 0 \text{ such that } P(x) \geqslant g^{\epsilon}(x), \tag{2.21}$$

$$|\partial^j_\eta P(x)| = o\,\{P(x)[g(x)h^{-1}(x, \eta)]^j\}, \quad j = 1, \ldots, m \tag{2.22}$$

then Theorem A is valid at c_0, $c_m > 0$, $c_s = 0$, $s = 1, \ldots, m - 1$.

<u>Theorem 2.3</u>. If in (2.9) $c_s > 0$, $V_s(x, \eta) \geqslant \delta I_r$, $s = 0, \ldots, m$ and the conditions*)

$$\gamma^2(x)(V_{s+1}(x, \eta)h, h) \leqslant C(V_s(x, \eta)h, h)^{1-\frac{1}{m-s}} |h|^{\frac{2}{m-s}} \tag{2.23}$$

hold, where $h \in E_r$, $s = 0, \ldots, m - 1$, $|x| \geqslant N$; and $\gamma(x)$ is such a function that

$$0 \leqslant \gamma(x) \leqslant 1, \quad |\partial^j_\eta P(x)| \leqslant C\,\gamma^j(x), \qquad j = 1, \ldots, m, \tag{2.24}$$

then Theorem A holds.

Note that Theorem 2.1 embraces the class of non-semi-bounded operators including the operators whose self-adjointness is determined by the well-known Titchmarsh and Sears theorem [15. Nos. 22.15-16] (see also [14], [16]). The

---

* (2.23) is certainly satisfied if

$$\gamma^2(x)V_{s+1}(x, \eta) \leqslant C[V_s(x, \eta)]^{1-\frac{1}{m-s}}, \quad s = 0, \ldots, m - 1$$

operators considered in Theorems 2.2 and 2.3 are semi-bounded. From Theorems 2.1 - 2.3 one may deduce a number of criteria for essential self-adjointness of the operation L which are formulated directly as conditions on the coefficients of L (see in particular [11]). An additional requirement in Theorem 2.2 on the growth of the lower coefficient, viz.,

$$g(x) \to \infty, \qquad |x| \to \infty,$$

guarantees the discreteness of spectrum of L. Theorem 2.2 is related to some results of A. G. Kostyuchenko and V. V. Grushin.

Here, as examples, we derive some corollaries of Theorems 2.1-2.3 for ordinary scalar operators $(n = r = 1)$. Some of these results for the fourth-order operators are due to W. N. Everitt and A. Devinatz. The proofs are given in [12].

<u>Corollary 2.1</u>. Let the coefficients of the differential operator L (2.4-4"), $(n = r = 1)$ satisfy

$$\epsilon a^{2m}(x) \leq p_{2m}(x) \leq C a^{2m}(x); \quad |p_\nu(x)| \leq C a^\nu(x)\, q^{2m-\nu}(x), \quad \nu = 1, \ldots, 2m-1 \tag{2.25}$$

$$p_0(x) \geq -C q^{2m}(x), \tag{2.26}$$

where $C, \epsilon > 0$, and the functions $a(x) > 0$, $1 \leq q(x) \leq \infty$ of (2.15) satisfy the conditions

$$\left| \frac{d^j}{dx^j} [a(x)q^{-1}(x)] \right| [a(x)q^{-1}(x)]^j \leq C, \quad j = 0, \ldots, m-1 \tag{2.27}$$

$$\left| \frac{d^j}{dx^j} q^{-m}(x) \right| \leq C a^{-j}(x)\, q^{-(m-j)}(x), \quad j = 0, \ldots, m \tag{2.28}$$

$$\int^{+\infty} a^{-1}(x) q^{-2m+1}(x)\,dx = \infty; \quad \int_{-\infty} a^{-1}(x) q^{-2m+1}(x)\,dx = \infty. \tag{2.29}$$

Then L is essentially self-adjoint in $\mathcal{L}^2(-\infty, \infty)$.

Corollary 2.2. Let h(x) and g(x) be respectively equal to $|x|^{\rho_1}$, $|x|^{\rho_2}$ or $e^{\alpha_1 |x|}$, $e^{\alpha_2 |x|}$ and let the following conditions hold:

$$|p_\nu(x)| \leqslant C h^\nu(x)\, g^{2m-\nu}(x), \quad \nu > 0; \quad p_{2m}(x) > 0, \tag{2.30}$$

$$\sum_{\nu=0}^{2m} p_\nu(x)\xi^\nu \geqslant \epsilon[h^{2m}(x)\xi^{2m} + g^{2m}(x)], \quad x, \xi \in (-\infty, \infty), \tag{2.31}$$

$$\left|\frac{d}{dx} p_\nu(x)\right| = o(g^{2m-\nu+1}(x) h^{\nu-1}(x)), \qquad \nu = 0, \dots, 2m. \tag{2.32}$$

Then L is essentially self-adjoint if

$$\rho_1 \leqslant \rho_2; \qquad \alpha_1 < \alpha_2 \tag{2.33}$$

Corollary 2.3. Let L (2.4-4") have real coefficients such that $p_{2m}(x) \geqslant \epsilon > 0$, $p_{2\nu}(x) \geqslant -C$ and when $|x| \geqslant 1$

$$p_{2(\nu+1)}(x) \leqslant C|x|^2 \,(1 + |p_{2\nu}(x)|^{1-\frac{1}{m-\nu}}), \quad \nu = 0, \dots, m-1. \tag{2.34}$$

Then L is essentially self-adjoint.

A. G. Brusentsev [13] considered a number of topics in the qualitative spectral analysis of non-self-adjoint elliptic systems. Here the principle of splitting (I. M. Glazman [17]) is established for arbitrary elliptic systems which are not assumed to be strongly elliptic or formally self-adjoint. In particular, by means of this principle, sufficient conditions for the absence of an essential spectrum were found in terms of the coefficients of an elliptic system. It was shown, for example, that the spectrum of the scalar operator $(-\Delta)^m + Q(x)$ with a complex Q(x) is purely discrete if, when $|x| \to \infty$, $Q(x) \to \infty$ and remains outside an arbitrary sector containing the negative semi-axis, under certain conditions on the regularity of behaviour of Q(x) at infinity. Moreover, in [13] a number of conditions were obtained for the coincidence of the minimum and maximum operators generated by an arbitrary elliptic system. In particular these conditions include the results of P. Hess and F. E. Browder. For the scalar non-self-adjoint elliptic expression of the second order the conditions for coincidence of the minimum and maximum operators were obtained in [14]. They are presented below, though not in the most general form.

Theorem 2.4. Suppose that for the elliptic expression

$$Mu = -\nabla(B(x)\nabla u) + (\nabla u, \overline{c}(x)) + p(x)u, \tag{2.35}$$

with $\det B(x) \neq 0$, $c(x) = \{c_j(x)\}$, $\overline{c}(x) = \{\overline{c}_j(x)\}$ for $j = 1, 2, \ldots, n$

given $\epsilon > 0$, $K \geq 0$, the following inequality holds for all $u(x) \in C_0^\infty(R^n)$:

$$|\langle Mu, u\rangle| \geq \langle L_{\epsilon K} u, u\rangle,$$

where

$$L_{\epsilon K} u = -\epsilon\nabla(A(x)\nabla u) - KQ(x)u,$$

$$A(x) = \{B^*(x)B(x) + B(x)B^*(x)\}^{\frac{1}{2}} > 0.$$

Suppose there are piecewise-smooth functions $Q^{-\frac{1}{2}}(x)$, and $P(x)$ as in (2.13) and a sequence of domains G^τ (2.13'), $1 \leq Q(x) \leq \infty$ such that

$$|A^{\frac{1}{2}}(x)\nabla Q^{-\frac{1}{2}}(x)| \leq \text{const}, \quad (\text{a.e.}),$$

$$|A^{-\frac{1}{2}}(x)c(x)| + |A^{-\frac{1}{2}}(x)\overline{c}(x)| \leq KQ^{\frac{1}{2}}(x),$$

$$|A^{\frac{1}{2}}(x)\nabla P(x)| = O(\tau)Q^{-\frac{1}{2}}(x), \quad x \in G^\tau, \quad \tau \to \infty.$$

Then the minimum and maximum operators generated by M in (2.35) coincide in $\mathcal{L}^2(R^n)$.

The results on the points mentioned are to be found in works [15], [17]-[19].

3. Spectrum perturbation and stability of solutions for Hill's equation. (See [20]-[27]).

As shown in [20], in each of the sufficiently remote spectral gaps of the perturbed Hill's operator

$$-y'' + [q(x) + p(x)]y = \lambda y, \quad (-\infty < x < \infty), \tag{3.1}$$

under the condition

$$(1 + |x|)p(x) \in \mathcal{L}^1(-\infty, \infty) \tag{3.2}$$

there may appear not more than two eigenvalues $(p(x), q(x) \in R^1$, $q(x + 1) = q(x) \in \mathcal{L}^1_{loc})$.

Theorem 3.1. [21]. If under the above conditions

$$\int_{-\infty}^{\infty} p(x)dx \neq 0, \tag{3.3}$$

then in each of the sufficiently remote spectral gaps of the perturbed Hill's operator there is exactly one eigenvalue.

Proof. Let $\varphi(x, \lambda)$, $\theta(x, \lambda)$, $e_j(x, \lambda)$ be the solutions of the non-perturbed equation

$$-y'' + q(x)y = \lambda y,$$

$$\theta(0, \lambda) = \varphi'(0, \lambda) = 1, \quad \theta'(0, \lambda) = \varphi(0, \lambda) = 0,$$

$$e_j(x, \lambda) = \varphi(1, \lambda)\theta(x, \lambda) + [\rho_j(\lambda) - \theta(1, \lambda)]\varphi(x, \lambda), \quad j = 1, 2;$$

$$e_j(x + 1, \lambda) = \rho_j(\lambda)e_j(x, \lambda),$$

where

$$\rho^2 - [\theta(1, \lambda) + \varphi'(1, \lambda)]\rho + 1 = 0,$$

$|\rho_1(\lambda)| \leq 1$, $|\rho_2(\lambda)| \geq 1$. At the end-points of the gaps $\rho_j = \pm 1$, and $e_1(x, \lambda) = e_2(x, \lambda)$ are periodic.

Lemma 3.1. [20]. Eq. (3.1) has solutions of the following form

$$E_1(x, \lambda) = e_1(x, \lambda) - \int_x^{\infty} K(x, t, \lambda)p(t)e_1(t, \lambda)dt,$$

$$E_2(x, \lambda) = e_2(x, \lambda) + \int_{-\infty}^{x} K(x, t, \lambda)p(t)e_2(t, \lambda)dt,$$

where $K(x, t, \lambda)$ satisfies (3.1) with respect to x,

$$K(t, t, \lambda) = 0, \quad K'_x(x, t, \lambda)\,|_{x=t} = 1,$$

$$|K(x, t, \lambda)| \leqslant C\, \frac{1 + |x - t|}{1 + |\sqrt{\lambda}|}\, \exp\{\alpha(\lambda)\,|x - t|\},$$

$$\alpha(\lambda) = \ln|\rho_2(\lambda)| \geqslant 0. \tag{3.4}$$

The eigenvalues of (3.1) are the roots of the Wronskian

$$W(\lambda) = E_1(x, \lambda)E'_2(x, \lambda) - E'_1(x, \lambda)E_2(x, \lambda),$$

and when $\varphi(1, \lambda) = 0$, one of the non-perturbed solutions $e_j(x, \lambda)$ and $E_j(x, \lambda)$ is identically zero, and then the normalization of such a solution has to be changed. Therefore the eigenvalues of (3.1) in the gap $(\lambda'_k, \lambda''_k)$ $k = 0, 1, 2, \ldots$ are exactly those λ for which

$$\widetilde{W}(\lambda) \overset{\text{def}}{=\!=} \frac{1}{\varphi(1, \lambda)}\, W(\lambda) = 0, \quad \lambda \in (\lambda'_k, \lambda''_k). \tag{3.5}$$

We shall put $\varphi(1, \lambda_k) \neq 0$, where λ_k is any of the points λ'_k, λ''_k, $k = 0, 1, 2, \ldots$, since this can always be achieved by a suitable choice of origin on the x-axis. Then (See [28], ch. VIII)

$$\operatorname{sgn} \varphi(1, \lambda'_k) = -\operatorname{sgn} \varphi(1, \lambda''_k). \tag{3.6}$$

According to Lemma 3.1

$$K'_x(x, t, \lambda)E_j(x, \lambda) - K(x, t, \lambda)E'_j(x, \lambda) = E_j(t, \lambda), \quad j = 1, 2; \tag{3.7}$$

$$(e_1E'_2 - e'_1 E_2)(x, \lambda) = w(\lambda) + \int_{-\infty}^{x} p(t)e_1(t, \lambda)E_2(t, \lambda)dt, \tag{3.8}$$

where

$$w(\lambda) = (e_1e'_2 - e'_1e_2)(x, \lambda); \quad w(\lambda'_k) = w(\lambda''_k) = 0. \tag{3.9}$$

Hence

$$W(\lambda) = (e_1E_2' - e_1'E_2)(x, \lambda) - \int_x^{\infty} [K(x, t, \lambda)E_2'(x, \lambda)$$

$$- K_x'(x, t, \lambda) E_2(x, \lambda)]p(t)e_1(t, \lambda)dt \tag{3.10}$$

$$= w(\lambda) + \int_{-\infty}^{\infty} p(t)e_1(t, \lambda)E_2(t, \lambda)dt \equiv w(\lambda) + \mathcal{T}(\lambda),$$

where

$$\mathcal{T}(\lambda) = \int_{-\infty}^{\infty} p(x)e_1(x, \lambda)e_2(x, \lambda)dx$$

$$+ \int_{-\infty}^{\infty} p(x)e_1(x, \lambda)dx \int_{-\infty}^{x} K(x, t, \lambda)p(t)e_2(t, \lambda)dt \equiv \mathcal{T}_1(\lambda) + \mathcal{T}_2(\lambda). \tag{3.11}$$

Employing the asymptotic behaviour of the solutions when $\lambda_k \to \infty$, Riemann-Lebesgue's theorem, and the estimate for (3.4), we get

$$\mathcal{T}_1(\lambda_k) = \left\{ \max_x |e_1(x, \lambda_k)|^2 \right\} \left\{ \tfrac{1}{2} \int_{-\infty}^{\infty} p(x)dx + o(1) \right\},$$

$$|\mathcal{T}_2(\lambda_k)| \le \frac{C}{\sqrt{|\lambda_k|}} \left\{ \max_x |e_1(x, \lambda_k)|^2 \right\} \int_{-\infty}^{\infty} |p(t)| dt \int_{-\infty}^{\infty} (1 + |x|) |p(x)| dx.$$

Therefore, due to (3.3), (3.9-3.11) and (3.5-3.6)

$$\operatorname{sgn} \widetilde{W}(\lambda_k') = - \operatorname{sgn} \widetilde{W}(\lambda_k''), \quad (\lambda_k \to \infty). \tag{3.12}$$

Thus, in each of the remote gaps there are eigenvalues but not more than two (See [20]). Let us show that $\widetilde{W}(\lambda)$, regular in each gap, can have only simple roots there. This together with (3.12), will imply that a remote gap contains one eigenvalue only. By the standard method it is easy to show that

$$E_1E_2 = \frac{d}{dx} (E_1'\dot{E}_2 - E_1\dot{E}_2') = \frac{d}{dx} (E_2'\dot{E}_1 - E_2 \dot{E}_1'),$$

where the point denotes differentiation with respect to λ. Hence if $\widetilde{W}(\tilde{\lambda}) = 0$, i.e., if $\tilde{\lambda}$ is an eigenvalue for which $E_1(x, \tilde{\lambda})$ and $E_2(x, \tilde{\lambda})$ are linearly dependent, then one obtains

$$\int_{-\infty}^{\infty} + \int_{0}^{\infty} (E_1E_2)(x, \tilde{\lambda})dx = (E_1'\dot{E}_2 - E_1\dot{E}_2')(0, \tilde{\lambda}) - (E_2'\dot{E}_1 - E_2\dot{E}_1')(0, \tilde{\lambda}),$$

i.e.

$$\int_{-\infty}^{\infty} (E_1E_2)^{\cdot}(x, \tilde{\lambda})dx = -\dot{W}(\tilde{\lambda}),$$

or by changing the normalization:

$$\dot{\tilde{W}}(\tilde{\lambda}) = -\int_{-\infty}^{\infty} \left\{\frac{(E_1E_2)(x, \lambda)}{\varphi(1, \lambda)}\right\}_{\lambda=\tilde{\lambda}} dx \neq 0.$$

Therefore the roots $\tilde{W}(\lambda)$ are simple. Q.E.D.

(In connexion with this theorem see also V. A. Zheludev's work [29])

For the perturbed Hill's operator (3.1) with the Hermitian $n \times n$ matrix coefficients $q(x)$, $p(x)$ V. I. Khrabustovsky [22], [23] found sufficient conditions under which perturbation gives not more than a finite number of discrete levels in each spectral gap. In this situation a phenomenon having no analogues in the scalar case has been discovered: if at the end-point λ_0 of the semi-infinite spectral gap $(-\infty, \lambda_0)$ of Hill's operator the spectrum multiplicity is not maximum $(< 2n)$, then it is possible under rather broad assumptions to separate the part of perturbation "responsible" for displacing the lower spectrum boundary and for the eigenvalues appearing in the gap $(-\infty, \lambda_0)$ from the other part which does not effect the above factors. In the scalar case the discrete spectrum in the semi-infinite gap was studied by M. Sh. Birman (See [17, No. 56]).

The problems of the spectral analysis for Hill's equation are closely connected with the stability of its solutions. (See [24].) We formulate below the results proved in [25]-[27].

Let $\mathcal{P}$ denote the class of operator-valued functions $P(t + 1) = P(t) = P^*(t)$ in the separable Hilbert space H, which are weakly measurable, $\| P(t) \| \in \mathcal{L}^1_{loc}$

The boundary-value problem

$$y'' + \lambda P(t)y = 0, \qquad (3.13)$$

$$y(0) + y(1) = y'(0) + y'(1) = 0 \tag{3.13'}$$

has a spectrum-free neighbourhood of zero. Let $(\lambda_{-1}, \lambda_1)$ be the maximum neighbourhood: $\lambda_{-1} < 0 < \lambda_1$.

Theorem 3.2. With $\lambda \in (0, \lambda_1)$ all the solutions for (3.13) $P(t) \in \mathcal{P}$, are stably bounded on the t-axis (i.e., along with all the solutions for $y'' + \lambda\tilde{P}(t)y = 0$, $\tilde{P}(t) \in \mathcal{P}$, when $\int_0^1 \|P(t) - \tilde{P}(t)\|dt$ is sufficiently small) in each of the following cases.

Either 1°: $P(t)$ is integrable, according to Bochner, in the uniform operator topology (B-integrable),

$$P_{av} \stackrel{def}{=} \int_0^1 P(t)dt \geq 0, \tag{3.14}$$

and there is a $\delta > 0$ such that, for all $f \in H$,

$$\int_0^1 \|P(t)f\|dt > \delta\|f\|, \quad (f \neq 0); \tag{3.15}$$

or 2°: $\qquad P_{av} \gg 0,$

i.e. there is a $\delta > 0$ such that, for all $f \in H$,

$$(P_{av}f, f) > \delta\|f\|^2, \quad (f \neq 0); \tag{3.16}$$

or 3°: $\qquad P_{av} \geq 0$

and $$(Q^2)_{av} \gg 0, \quad Q(t) \stackrel{def}{=} \int_0^t (P(s) - P_{av})ds. \tag{3.17}$$

Corollary 3.1. If in cases 1° or 3° $P_{av} = 0$, then all the solutions for (3.13) are stably bounded at $0 \neq \lambda \in (\lambda_{-1}, \lambda_1)$.

Note that case 1° of Theorem 3.2 in the scalar case transforms into A. M. Lyapunov's result, and when $\dim H < \infty$ into M. G. Krein's result. Under the additional requirement that $P(t)$ should be B-integrable, case 3° is described in the book [30, p.344] and is a particular case of 1°. If $\dim H = \infty$, B-integrability in 1° must not be ignored, and, unlike the finite-dimension case, conditions (3.15), (3.16), (3.17) cannot be weakened, by putting $\delta = 0$ in them. This follows from the theorem mentioned below, where $P(t)$ is not supposed to be either self-adjoint, or periodic, or B-integrable.

Theorem 3.3. Let $P(t)$ be weakly measurable $\|P(t)\| \in \mathcal{L}^1_{loc}$. Then if there exists a sequence of the normalized vectors such that

$$\| \int_0^t (t - s) sP(s) x_k ds \| \to 0$$

locally uniformly over t as $k \to \infty$, then (3.13) has a solution which is unbounded on the t-axis.

REFERENCES

[1] V. I. Kogan and F. S. Rofe-Beketov, On square-integrable solutions of symmetric systems of differential equations of arbitrary order (to appear in the Proc. Roy. Soc. Edinb.).

[2] V. I. Kogan and F. S. Rofe-Beketov, On the question of the deficiency indices of differential operators with complex coefficients, Matem. Fizika i funktsional Anal., vyp.2 (Kharkov, 1971), 45-60 (Engl. Transl. to appear in the Proc. Roy. Soc. Edinb.).

[3] F. S. Rofe-Beketov, Self-adjoint extensions of differential operators in a space of vector functions, DAN SSSR, 184, No. 5 (1969), 1034-1037.

[4] W. N. Everitt, Integrable-square, analytic solutions of odd-order, formally symmetric, ordinary differential equations, Proc. Lond. Math. Soc., (3), 25 (1972), 156-182.

[5] F. V. Atkinson, Discrete and continuous boundary problems, Acad. Press, N.Y., 1964.

[6] G. A. Kalyabin, The number of solutions in $L_2(0,\infty)$ of a self-adjoint system of second order differential equations, Funktsional Anal. i Prilozhen., 6 No. 3 (1972), 74-76.

[7] J. C. Gokhberg and M. G. Krein, Theory of Volterra operators in Hilbert space and its applications, "Nauka", Moscow, 1967.

[8] B. M. Levitan and I. S. Sargsyan, Introduction to spectral theory, "Nauka", Moscow, 1970.

[9] M. A. Naimark, Linear differential operators, 2nd edn, "Nauka", Moscow, 1969.

[10] W. Wasow, Asymptotic expansions for ordinary differential equations, John Wiley and Sons, N.Y., 1965.

[11] A. G. Brusentsev and F. S. Rofe-Beketov, On the self-adjointness of high-order elliptic operators, Funktsional Anal. i Prilozhen., 7 No. 4 (1973), 78-79.

[12] A. G. Brusentsev and F. S. Rofe-Beketov, Conditions for the self-adjointness strongly elliptic systems of arbitrary-order, (to appear in Matem. Sbornik).

[13] A. G. Brusentsev, Certain problems of the qualitative spectral analysis of arbitrary-order elliptic systems, Matem. Fizika i Funktsional. Anal., vyp.4 (Kharkov, 1973), 93-116.

[14] F. S. Rofe-Beketov and A. M. Holkin, Conditions for the self-adjointness of second-order elliptic operators of the general type, Theor. Funktsiy Funktsional. Anal. i Prilozh., vyp. 17 (1973), 41-51.

[15] E. C. Titchmarsh, Eigenfunctions expansions associated with second-order differential equations, Part II, Oxford, at the Claredon Press, 1958.

[16] F. S. Rofe-Beketov, Conditions for the self-adjointness of the Schrödinger operator, Mat. Zametki, 8, No. 6 (1970), 741-751.

[17] I. M. Glazman, Direct methods of qualitative spectral analysis, "Fizmatgiz", Moscow, 1963. Eng. transl.

[18] I. M. Gelfand and G. E. Shilov, Certain questions of differential equations theory, "Fizmatgiz", Moscow, 1958.

[19] Ju. M. Berezanskiy, Expansions in eigenfunctions of self-adjoint operators, Am. Math. Soc., Monograph. Transl., vol. 17, 1968.

[20] F. S. Rofe-Beketov, A test for the finiteness of the number of discrete levels introduced into the gaps of a continuous spectrum by perturbations of a periodic potential, DAN SSSR, 156, No. 3 (1964), 515-518.

[21] F. S. Rofe-Beketov, Hill's operator perturbation, which has a first moment and a non-vanishing integral, introduces one discrete level into each distant spectral gap, Matem. Fizika i Funktsional. Anal., vyp. 4, (Kharkov, 1973), 158-159.

[22] V. I. Khrabustovskiy, On perturbations of the spectrum of self-adjoint differential operators with periodic matrix coefficients, ibid. 117-138.

[23] V. I Khrabustovskiy, On perturbations of the spectrum of arbitrary order self-adjoint differential operators with periodic matrix coefficients, ibid, vyp.5 (to appear).

[24] F. S. Rofe-Beketov, On the spectrum of non-self-adjoint differential operators with periodic coefficients, DAN SSSR, 152, No. 6 (1963), 1312-1315.

[25] F. S. Rofe-Beketov and V. I. Khrabustovskiy, The stability of the solutions of Hill's equation with an operator coefficient, Teor. Funktsiy Funktsional. Anal. i Prilozh., vyp. 13 (1971), 140-147.

[26] F. S. Rofe-Beketov and V. I. Khrabustovskiy, The stability of the solutions of Hill's equation with an operator coefficient that has a non-negative mean value, ibid, vyp. 14 (1971), 101-105.

[27] F. S. Rofe-Beketov and V. I. Khrabustovskiy, Letters to the editors, ibid, 195.

[28] E. A. Coddington and N. Levinson, Theory of ordinary differential equations, McGraw-Hill, N.Y. and London, 1955.

[29] V. A. Želudev, The perturbation of the spectrum of the one-dimensional selfadjoint Schrödinger operator with periodic potential, Problemy Matem. Fiziki, No. 4 (Leningrad, 1970), 61-82.

[30] Ju. L. Daleckiĭ and M. G. Kreĭn, Stability of the solutions of differential equations in Banach space, "Nauka", Moscow, 1970.

Eigenvalue Problems for Nonlinear Second Order Differential Equations

Klaus Schmitt

§ 1. In this paper we are concerned with eigenvalue problems for nonlinear differential equations of the form

$$(1.1) \qquad x'' + \lambda(a(t)x + f(t,x,x')) = 0 \ , \quad 0 \le t \le 1 \ ,$$

$$(1.2) \qquad x(0) = 0 = x(1) \ ,$$

where a: $[0,1] \longrightarrow (0,\infty)$ and f: $[0,1] \times \mathbb{R} \times \mathbb{R} \longrightarrow \mathbb{R}$ are continuous and λ is a parameter. Throughout most of the paper it will be assumed that

$$(1.3) \qquad |f(t,x,y)| = o(|x| + |y|), \text{ as } |x| + |y| \to 0.$$

The last requirement implies that $x \equiv 0$ is a solution of (1.1), (1.2). The question of interest therefore is to provide conditions on f which guarantee the existence of nontrivial solutions (λ,x) of (1.1), (1.2).

Problem (1.1), (1.2) may be rewritten as an equivalent integral equation

$$(1.4) \qquad x(t) = \lambda \int_0^1 G(t,s)(a(s)x(s) + f(s,x(s),x'(s)))\,ds$$

or, for short, as an operator equation in the Banach space E,

$$(1.5) \qquad x = \lambda G(ax + f(x)),$$

$E = C^1([0,1], \mathbb{R})$, with norm $\|\cdot\|$ defined by

$$\|x\| = \max_{[0,1]} |x(t)| + \max_{[0,1]} |x'(t)|.$$

The hypotheses imposed on f imply that the operator defined by the right hand side of (1.4) is Fréchet differentiable at $x = 0 \in E$, the Fréchet derivative being a compact linear operator. The linearized equation (at $x = 0$) has the form

(1.6) $$x = \lambda G(ax),$$

equivalent to the linear Sturm-Liouville problem

(1.7) $$x'' + \lambda a(t)x = 0$$
$$x(0) = 0 = x(1) .$$

This problem has an infinite sequence of eigenvalues

$$0 < \lambda_0 < \lambda_1 < \dots < \lambda_n < \dots , \lim_{n\to\infty} \lambda_n = +\infty ,$$

the eigenspace associated with each eigenvalue being one-dimensional and each element associated with λ_i has i nodes in (0,1) (see [6]), i = 0,1,... .

A result of Leray-Schauder (see [7]) thus implies that each eigenvalue of the linear problem (1.7) is a bifurcation point of (1.1), (1.2), i.e. every neighbourhood of $(\lambda_i, 0)$ in $\mathbb{R} \times E$ contains a solution (λ, x) of (1.1), (1.2) with $\|x\| \neq 0$. (An elementary verification of this result by means of Prüfer transformation techniques is contained in [8], see also [1], [3]).

We shall impose additional requirements upon f which will imply a global result, i.e. we shall show that under suitable hypotheses every $\lambda > 0$ is an eigenvalue of (1.1), (1.2) and establish a classification of associated eigensolutions x in terms of their nodal properties in (0,1). The hypotheses and methods of proof have been motivated by the papers [2], [4], [5], [9], [10-12], [13], [16,17]. In fact much of what is to follow may already be found in these papers.

The requirements are the following:

(1.8) $$\frac{f(t,x,y)}{x} \to +\infty \quad \text{as } |x| \to +\infty ,$$

uniformly with respect to (t,y), $0 \leq t \leq 1$, $y \in \mathbb{R}$,

or

(1.9) $\frac{f(t,x,y)}{x} \to -\infty$ as $|x| \to +\infty$,

uniformly with respect to (t,y), $0 \leq t \leq 1$, $y \in \mathbb{R}$,

and

(1.10) for each bounded x-set K, there exists a monotone increasing continuous function

$\phi_K: [0,\infty) \to (0,\infty)$ such that

$\int^{\infty} \frac{sds}{\phi_K(s)} = +\infty$ and

$|f(t,x,y)| \leq \phi_K(|y|)$, $0 \leq t \leq 1$, $x \in K$.

This latter condition is the Nagumo condition as given by Hartman, it may in particular be verified in case

(1.11) $\frac{f(t,x,y)}{y^2} \to 0$ as $|y| \to \infty$

uniformly on bounded (t,x) sets.

A consequence of (1.10) (and thus (1.11)) is that bounded families of solutions of (1.1) for bounded λ are precompact in E. (If one is dealing with vector equations (1.10) no longer will suffice to yield precompactness in E of bounded families, on the other hand (1.11), interpreted appropriately, is still sufficient (see [13]).)

It should be noted here that the results to be discussed are valid in the more general setting where x'' is replaced by $(p(t)x')' + q(t)x$, $p(t) > 0$, q, continuous, and the boundary conditions (1.2) by $\alpha x(0) + \beta x'(0) = 0 = \gamma x(1) + \delta x'(1)$, $(\alpha^2+\beta^2)(\gamma^2+\delta^2) \neq 0$.

We adopt the following notation:

(1.12) $S_n = \{x \in E$: x satisfies (1.2), x has exactly n nodes in (0,1) and $x'(0)x'(1) \neq 0\}$, $n = 0,1,2,\ldots$.

$B_r = \{x \in E: \|x\| < r\}$.

We shall establish the following results.

Theorem A. Let f satisfy (1.3), (1.8), (1.10). Then for every $\lambda \in (0,\lambda_n)$, there exists a nontrivial solution x of (1.1), (1.2) with $x \in S_n$, $n = 0,1,2,\ldots$.

Theorem B. Let f satisfy (1.3), (1.9), (1.10). Then for every $\lambda \in (\lambda_n,\infty)$, there exists a solution $x \in S_n$ of (1.1), (1.2), $n = 0,1,2,\ldots$.

In fact much more can be concluded: In the case of Theorem A, one may establish the existence of a connected branch of solutions (λ,x) eminating from $(\lambda_n,0)$, $x \in S_n$, and extending to the left whereas in the case of Theorem B a connected branch eminates from $(\lambda_n,0)$ and extends to the right.

That these results pertain may be verified using the proofs of the above results (to follow) together with the homotopy invariance theorem of Leray-Schauder degree. Such arguments may be found in various places (see e.g. [10], [15]).

In addition to these results we also shall show how results similar to the above may be obtained for the boundary value problem

$$x'' + f(t,x,x') = 0, \quad x(0) = 0 = x(1),$$

where f satisfies (1.8) and (1.10) but not necessarily (1.3).

§ 2. Throughout this section we shall assume that f satisfies conditions (1.3), (1.8), (1.10).

Lemma 2.1. Let $[\xi,\eta] \subset (0,\infty)$ be a compact interval with $\lambda_n \notin [\xi,\eta]$. Then there exists $\delta > 0$ (depending only on $[\xi,\eta]$) such that: If (λ,x) is a solution of

$$(2.1) \qquad x = \lambda G(ax + (1-\tau)f(x) + \tau x^3), \quad 0 \leq \tau \leq 1 ,$$

$x \in S_n$, $\xi \leq \lambda \leq \eta$, then $||x|| > \delta$.

Proof. If the conclusion is false, there exist sequences $\{\tau_m\} \subseteq [0,1]$, $\{\lambda^m\} \subseteq [\xi,\eta]$, $\{x_m\} \subseteq S_n$ such that $\|x_m\| \leq \frac{1}{m}$, $m = 1,2,\ldots$, and

$$x_m = \lambda^m G(ax_m + (1-\tau_m)f(x_m) + \tau_m x_m^3).$$

Thus letting $y_m = \frac{x_m}{\|x_m\|}$, we obtain

$$y_m = \lambda^m G(ay_m + \frac{(1-\tau_m)f(x_m) + \tau_m x_m^3}{\|x_m\|}) .$$

Using the complete continuity of G and (1.3) we may assume, without loss in generality that $\lim_{m\to\infty} y_m = y$, $\lim_{m\to\infty} \tau_m = \tau_o$, $\lim_{m\to\infty} \lambda^m = \lambda^o$ and obtain

$$y = \lambda^o G(ay).$$

On the other hand $y \in \bar{S}_n$. Thus since $y \neq 0$, $y \in S_n$, implying that $\lambda^o = \lambda_n$, a contradiction.

Remark. Conditions (1.8) and (1.10), of course, are not needed in the proof of Lemma 2.1. Furthermore for $\eta > 0$, sufficiently small condition (1.3) may be replaced by

(1.3)' $\quad |f(t,x,y)| \leq L(|x|+|y|)$, as $|x|+|y| \to 0$.

This together with the representation (1.4) shows that Lemma 2.1 remains valid, provided only that η is sufficiently small, but independent of n.

Lemma 2.2. Let $[\xi,\eta] \subset (0,\infty)$ be a compact interval. Then there exists $M > 0$ (depending only on $[\xi,\eta]$) such that: If (λ,x) is a solution of (2.1), $x \in S_n$, $\xi \leq \lambda \leq \eta$, then $\|x\| < M$.

Proof. (a) We first prove the following auxiliary result: If $\{x(t)\}$ is a family of solutions of (2.1) such that $|x(t)| > 0$, $0 \leq c < t < d \leq 1$, $\xi \leq \lambda \leq \eta$, $0 \leq \tau \leq 1$, then there exists a constant $D = D(d-c)$ such that $|x(t)| \leq D$.

If this were not the case we choose an interval $[r,s] \subset (c,d)$ and $k > 0$ such that any solution u of $u'' + \lambda(\varepsilon+k)u = 0$ has at least two zeros in (r,s) for $\xi \leq \lambda \leq \eta$, where $0 < \varepsilon \leq \min_{[0,1]} a(t)$. Then choose $m > 0$ so large that the condition $|x(t)| \geq m$ for at least one $t \in [c,d]$ implies $|x(t)| \geq m_1$, $r \leq t \leq s$, where $m_1 > 0$ has been chosen that

$$\frac{(1-\tau)f(t,x,y) + \tau x^3}{x} \geq k, \quad |x| \geq m_1.$$

Thus, if x is a solution with $|x(t)| \geq m$ for some $t \in [c,d]$ and $|x(t)| > 0$, $c < t < d$, must thus satisfy

$$x'' + \lambda(\varepsilon+k)x \leq 0, \; r \leq t \leq s, \text{ if } x(t) > 0, \; c < t < d,$$

or

$$x'' + \lambda(\varepsilon+k)x \geq 0, \; r \leq t \leq s, \text{ if } x(t) < 0, \; c < t < d.$$

In either case we may conclude that $u''+\lambda(\varepsilon+k)u = o$ is disconjugate on $[r,s]$, contradicting the choice of k.

(b) Condition (1.10) implies, for every $Q > 0$, the existence of a constant $N = N(Q,d-c)$ such that: If x is a solution of (2.1), $\xi \leq \lambda \leq \eta$, $0 \leq \tau \leq 1$, with $|x(t)| \leq Q$, then $|x'(t)| \leq N$, $0 \leq c \leq t \leq d \leq 1$. Thus if the conclusion of the lemma were false, there exist sequences $\{\lambda^m\} \subseteq [\xi,\eta]$, $\{\tau_m\} \subseteq [0,1]$ and $\{x_m\} \subseteq S_n$ such that $\max_{[0,1]} |x_m(t)| \geq m$ and

$$x_m = \lambda^m G(ax_m + (1-\tau_m)f(x_m) + \tau_m x_m^3), \; m = 1,2,\ldots .$$

Let $t_1^m, \ldots, t_n^m$ denote the zeros of x_m in $(0,1)$, $m = 1,2,\ldots$. Let t_i be an accumulation point of $\{t_i^m\}$, $i = 1, \ldots, n$, and choose a subsequence such that

$$\lim_{j\to\infty} t_i^{m_j} = t_i, \quad i = 1,\ldots,n.$$

We note that the case $0 < t_1 < t_2 < \ldots < t_n < 1$ cannot occur, since otherwise (here we use part (a))

$$|x_{m_j}(t)| \leq \max_{0\leq i\leq n+1} D(d_i-c_i), \; j \text{ sufficiently large},$$

where $[c_i,d_i] \subset (t_i,t_{i+1})$, $i = 1,\ldots,n-1$, $[c_o,d_o] \subset (0,t_1)$,

$[c_{n+1}, d_{n+1}] \subset (t_n, 1)$, are intervals of positive measure, contradicting that $\{x_m(t)\}$ is unbounded. Thus there exists at least one i, $1 \leq i < n$, such that $t_i = t_{i+1}$. There are several cases to consider, and since the argument in all of these is similar, we consider one such case, namely the case $t_1 = t_n$ and $1 > t_1 \geq 0$. Choose c such that $t_1 < c < 1$. Then for j sufficiently large $|x_{m_j}(t)| > 0$, $c < t < 1$. Therefore

$$\max_{t_n^{m_j} \leq t \leq 1} |x_{m_j}(t)| \leq D(1-c) \qquad \text{(by part (a)).} \tag{2.2}$$

On the other hand $\lim_{j \to \infty} |x'_{m_j}(t_n^{m_j})| = +\infty$, which cannot happen because of (2.2) and (1.10). This final contradiction establishes the Lemma.

Let us denote by

$$f_\tau(t,x,y) = (1-\tau)f(t,x,y) + \tau x^3.$$

Lemmas 2.1 and 2.2 therefore imply:

Corollary 2.3. Let $[\xi, \eta] \subset (0, \lambda_n)$. Then

$$d(I-\lambda G(a+f_\tau),\ B_M \setminus \bar{B}_\delta \cap S_n, 0) = \text{const.} \tag{2.3}$$

for $\xi \leq \lambda \leq \eta$, $0 \leq \tau \leq 1$, where M and δ are determined by Lemmas 2.2 and 2.1, respectively, and $d(\cdot,\cdot,\cdot)$ denotes the Leray-Schauder degree.

Proof. One first notes that because of Lemmas 2.1 and (2.1) the left side of (2.3) is defined. That it is constant, follows from the homotopy invariance property of Leray-Schauder degree (see [14]).

Once we are able to show that the constant defined by (2.3) is unequal to zero, the proof of Theorem A is complete. Corollary 2.3, on the other hand, implies that it suffices to evaluate

$$d(I-\eta G(a+f_1),\ B_M \setminus \bar{B}_\delta \cap S_n, 0). \tag{2.4}$$

It is the purpose of the following lemmas to evaluate the constant (2.4), i.e. we may restrict ourselves to the simpler problem

$$(2.5)\qquad \begin{aligned} &x'' + \lambda(a(t)x + x^3) = 0 \\ &x(0) = 0 = x(1). \end{aligned}$$

Lemma 2.4. Let (λ,x) be a solution of (2.5) with $x \in S_n$, then $\lambda < \lambda_n$.

Proof. This follows easily from Sturm's Comparison Theorem, see e.g. [6].

Lemma 2.5. Let $\tilde{\lambda} < \lambda_n < \tilde{\tilde{\lambda}} < \lambda_{n+1}$ and let (λ,x) be a solution of (2.5), $\tilde{\lambda} \le \lambda \le \tilde{\tilde{\lambda}}$. Then there exists $\varepsilon > 0$ such that $x \in S_k$, $k > n$, implies $\|x\| > \varepsilon$.

Proof. If the Lemma is false there exists a sequence $\{\lambda^m\} \subseteq [\tilde{\lambda},\tilde{\tilde{\lambda}}]$ and $\{x_m\} \subseteq \bigcup\limits_{k>n} S_k$ such that $\|x_m\| \le \frac{1}{m}$ and

$$x_m = \lambda^m G(ax_m + x_m^3) ,$$

again, letting $y_m = \frac{x_m}{\|x_m\|}$, we obtain

$$y_m = \lambda^m G(ay_m + \frac{x_m^3}{\|x_m\|}) .$$

Passing to subsequences, if necessary, and relabeling, we obtain $\lim\limits_{m\to\infty} y_m = y$, $\lim\limits_{m\to\infty} \lambda^m = \lambda^o$ and

$$y = \lambda^o G(ay).$$

Further $y \in \overline{\bigcup\limits_{k>n} S_k}$ and $y \neq 0$. Thus $y \in S_k$ for some $k > n$, contradicting that $\lambda^o < \lambda_{n+1}$.

Lemma 2.6. Let $\lambda \in [\tilde{\lambda},\lambda_n)$. Then for $\lambda_n - \tilde{\lambda}_n$ sufficiently small a solution $x \in S_n$ of (2.5) satisfies $\|x\| < \varepsilon$, where $\varepsilon > 0$ is given by Lemma 2.5.

Proof. If the lemma were false there exists $\{\lambda^m\}$, $\lambda^m < \lambda_n$, $\lim_{m\to\infty} \lambda^m = \lambda_n$ and $\{x_m\} \subseteq S_n$ such that $\|x_m\| \geq \varepsilon$ and

$$x_m = \lambda^m G(ax_m + x_m^3).$$

Since $\{x_m\}$ is bounded (viz. Lemma 2.2) it is precompact in E. Thus again, passing to subsequences, if necessary, and relabeling, we obtain $\lim_{m\to\infty} x_m = x$, where x satisfies

$$x'' + \lambda_n(a(t)x + x^3) = 0 ,$$

$$x(0) = 0 = x(1).$$

On the other hand $x \in S_n$, contradicting Lemma 2.4.

Lemma 2.7. Let $\eta = \tilde{\lambda}$ satisfy the conditions of Lemma 2.6. Then

(2.6) $d(I-\eta G(a+f_1),\ B_M \setminus \bar{B}_\delta \cap S_n, 0) = d(I-\eta G(a+f_1),\ B_\varepsilon \setminus \bar{B}_\delta \cap S_n, 0)$.

Proof. Apply the excision principle of Leray-Schauder degree (see [14]).

Lemma 2.8. Let η be as above, then

(2.7) $d(I-\eta G(a+f_1), B_\varepsilon, 0) = d(I-\eta G(a+f_1), B_\delta, 0) + (d(I-\eta G(a+f_1), B_\varepsilon \setminus \bar{B}_\delta \cap S_n, 0)$.

Proof. By the set additivity of Leray-Schauder degree (see [14])

$$d(I-\eta G(a+f_1), B_\varepsilon, 0) = d(I-\eta G(a+f_1), B_\delta, 0) + d(I-\eta G(a+f_1), B_\varepsilon \setminus \bar{B}_\delta, 0).$$

On the other hand, the excision principle together with lemmas 2.4 and 2.5 implies

$$d(I-\eta G(a+f_1),B_\varepsilon\setminus\bar{B}_\delta,O) = d(I-\eta G(a+f_1),B_\varepsilon\setminus\bar{B}_\delta\cap S_n,O).$$

Lemma 2.9. Let η be as above and $\lambda_n < \tilde{\tilde{\lambda}} < \lambda_{n+1}$, then

$$(2.8) \qquad d(I-\lambda G(a+f_1),B_\varepsilon,O) = \text{const.} = (-1)^{n+1}$$

for $\eta \leq \lambda \leq \tilde{\tilde{\lambda}}$.

Proof. Lemmas 2.4 - 2.5 together with the homotopy invariance theorem imply that the above degree is defined and constant, hence equal

$$d(I-\tilde{\tilde{\lambda}}G(a+f_1),B_\varepsilon,O).$$

On the other hand the only solution x of

$$x'' + \tilde{\tilde{\lambda}}(a(t)x+x^3) = O\ ,$$

$$x(O) = O = x(1)$$

with $||x|| < \varepsilon$ is $x \equiv O$. A computation as in [7, p. 136] yields

$$d(I-\tilde{\tilde{\lambda}}G(a+f_1),B_\varepsilon,O) = d(I-\tilde{\tilde{\lambda}}G(a),B_\varepsilon,O) = (-1)^{n+1}.$$

Lemma 2.10. $d(I-\eta G(a+f_1),B_\delta,O) = (-1)^n$.

Proof. See [7,p. 136].

Lemma 2.11. $d(I-\eta G(a+f_1),B_\varepsilon\setminus\bar{B}_\delta\cap S_n,O) = 2(-1)^{n+1}$.

Proof. Use Lemmas 2.8 - 2.10.

This sequence of results completes the proof of Theorem A. Further using Lemmas 2.2 through 2.11 and the remark following the proof of Lemma 2.1 we have also established.

Theorem C. Let f satisfy (1.3)', (1.8), (1.10). Then there exists $\eta > 0$ such that for every λ, $0 < \lambda \le \eta$, (1.1), (1.2) has a solution $x \in S_n$ for $n = 1,2,\ldots$.

A further consequence of these considerations is the following result.

Theorem D. Let f satisfy (1.3)', (1.8), (1.10), where $L > 0$ is such that

$$L\left(\int_0^1 |G(t,s)|ds + \int_0^1 |G_t(t,s)|ds\right) < 1.$$

Then for every $n = 0,1,2,\ldots$ the boundary value problem

$$x'' + f(t,x,x') = 0, \quad x(0) = 0 = x(1),$$

has a solution $x \in S_n$.

Proof. Let $\varepsilon > 0$ be a positive number to be chosen such that the first eigenvalue of $x'' + \lambda\varepsilon x = 0$, $x(0) = 0 = x(1)$, is bigger than 1, and

$$(\varepsilon + L)\left(\int_0^1 |G(t,s)|ds + \int_0^1 |G_t(t,s)|ds\right) < 1.$$

This condition implies a lemma like Lemma 2.1 for the differential equation

$$x'' + \mu\varepsilon x + f(t,x,x') = 0, \quad 0 \le \mu \le 1.$$

Now proceed as in Lemma 2.2 - 2.11 using μ as an additional homotopy parameter.

§ 3. In order to establish Theorem B, one considers the equation

$$x = \lambda G(ax + (1-\tau)f(x) - \tau x^3), \quad 0 \le \tau \le 1 ,$$

and establishes a sequence of lemmas analogous to those in § 2, making appropriate modifications. The major difference occurs in proving a result similar to Lemma 2.2, the proof in this case being rather straight forward.

References.

[1] F. Brauer, Nonlinear perturbations of Sturm-Liouville boundary value problems, J. Math. Anal. Appl. 22 (1968), 591-598.

[2] M. Crandall and P. Rabinowitz, Nonlinear Sturm-Liouville eigenvalue problems and topological degree, J. Math. Mech. 19 (1970), 1083-1102.

[3] M. Eastabrooks and J. Macki, A nonlinear Sturm-Liouville problem, J. Diff. Equ. 10 (1971), 181-187.

[4] G. Gustafson and K. Schmitt, Nonzero solutions of boundary value problems for second order ordinary and delay differential equations, ibid. 12 (1972), 129-147.

[5] G. Gustafson, Nonzero solutions of boundary value problems for damped nonlinear differential systems, to appear.

[6] P. Hartman, Ordinary Differential Equations, Wiley, New York, 1964.

[7] M. Krasnosel'skii, Topological Methods in the Theory of Nonlinear Integral Equations, Pergamon, New York, 1964.

[8] J. Macki and P. Waltman, A nonlinear Sturm-Liouville problem, Ind. Univ. Math. J. 22 (1972), 217-225.

[9] G.H. Pimbley, A superlinear Sturm-Liouville problem, Trans. Amer. Math. Soc. 103 (1962), 229-248.

[10] P. Rabinowitz, Nonlinear Sturm-Liouville problems for second order ordinary differential equations, Comm. Pure Appl. Math. 23 (1970), 939-961.

[11] P. Rabinowitz, Some global results for nonlinear eigenvalue problems, J. Funct. Anal. 7 (1971), 487-513.

[12] P. Rabinowitz, On bifurcation from infinity, J. Diff. Equ. 14 (1973), 462-475.

[13] K. Schmitt and R. Thompson, Boundary value problems for infinite systems of second order differential equations, ibid., to appear.

[14] J. Schwartz, Nonlinear Functional Analysis, Gordon and Breach, New York, 1969.

[15] C. Stuart, Concave solutions of singular nonlinear differential equations, Math. Z. 136 (1974), 117-135.

[16] R. Turner, Nonlinear Sturm-Liouville problems, J. Diff. Equ. 10 (1971), 141-146.

[17] R. Turner, Superlinear Sturm-Liouville problems, ibid. 13 (1973), 157-171.

Left-Definite Multiparameter Eigenvalue Problems

B.D. Sleeman

§1 The problem

This lecture is concerned largely with eigenvalue value problems associated with the following: consider the finite system of ordinary, second order, linear, formally self-adjoint differential equations in the k-parameters $\lambda_1, \lambda_2, \ldots, \lambda_k$, $k \geq 2$,

$$\frac{d^2 y_r}{dx_r^2} + \{\sum_{s=1}^{k} a_{rs}(x_r)\lambda_s - q_r(x_r)\} y_r = 0, \qquad (1)$$

$0 \leq x_r \leq 1$, $r = 1, 2, \ldots, k$, with $a_{rs}(x_r)$, $q_r(x_r)$ continuous and real valued functions defined on the interval $0 \leq x_r \leq 1$. By writing $\underset{\sim}{\lambda}$ for $(\lambda_1, \lambda_2, \ldots, \lambda_k)$ we may formulate an eigenvalue problem for (1) by demanding that $\underset{\sim}{\lambda}$ be chosen so that all the equations of (1) have non-trivial solutions with each satisfying the homogeneous boundary conditions

$$\cos\alpha_r\, y_r(0) - \sin\alpha_r \frac{dy_r(0)}{dx_r} = 0, \qquad 0 \leq \alpha_r < \pi,$$

$$\cos\beta_r\, y_r(1) - \sin\beta_r \frac{dy_r(1)}{dx_r} = 0, \qquad 0 < \beta_r \leq \pi, \qquad (2)$$

$r = 1, 2, \ldots, k$.

If $\underset{\sim}{\lambda}$ can be so chosen, then we shall refer to $\underset{\sim}{\lambda}$ as an eigenvalue of the system (1) (2); if $\{y_r(x_r, \underset{\sim}{\lambda})\}_{r=1}^{k}$ is a corresponding set of simultaneous solutions of (1) (2) then the product $\prod_{r=1}^{k} y_r(x_r, \underset{\sim}{\lambda})$ will be called an eigenfunction of this system corresponding to the eigenvalue $\underset{\sim}{\lambda}$.

Before we proceed, it is instructive to recall some fundamental notions in the one parameter case $(k = 1)$. Here we have the classical Sturm-Liouville problem defined by

$$-\frac{d^2y}{dx^2} + q(x)y = \lambda\, p(x)y, \qquad (3)$$

$0 \leq x \leq 1$, with $p(x)$, $q(x)$ continuous and real valued functions defined on the interval $0 \leq x \leq 1$, and we seek solutions satisfying the homogeneous conditions

$$\cos\alpha\, y(0) - \sin\alpha \frac{dy(0)}{dx} = 0, \qquad 0 \le \alpha < \pi,$$

$$\cos\beta\, y(1) - \sin\beta \frac{dy(1)}{dx} = 0, \qquad 0 < \beta \le \pi, \tag{4}$$

In order to treat the problem (3) (4), particularly as regards questions of completeness of eigenfunctions and the development of a spectral theory, it is desirable to interpret it in terms of linear operators in Hilbert space. Such a Hilbert space structure may be realised in one of two ways. Firstly, if we assume $p(x)$ is positive on $[0, 1]$ then we take our Hilbert space to be $L_p^2[0, 1]$. With this condition on the coefficient $p(x)$ we are led to the study of so-called "right-definite" problems for (3) (4). On the other hand if $p(x)$ changes sign in $[0, 1]$ but $q(x)$ is positive and if we further assume, for simplicity, that $\alpha \in (0, \pi/2]$, $\beta \in [\pi/2, \pi)$ then a positive definite Dirichlet integral may be associated with (3) (4) and a theory may be developed in the Hilbert space which is the completion of $C'[0, 1]$ with respect to the inner product

$$(u, v) = \int_0^1 \left(\frac{du}{dx}\frac{d\bar{v}}{dx} + q(x)\, u\bar{v}\right) dx + \cot\alpha\, u\bar{v}(0) - \cot\beta\, u\bar{v}(1). \tag{5}$$

This leads to the study of so-called "left-definite" problems.

For the multiparameter eigenvalue problem (1) (2) the appropriate generalisations of the above assumptions appear to be

(A) $$\Delta_k = \det\{a_{rs}(x_r)\}_{r,s=1}^{k} > 0 \tag{6}$$

for all $x = (x_1, x_2, \ldots, x_k) \in I^k$ (the cartesian product of the k intervals $0 \le x_r \le 1$, $r = 1, 2, \ldots, k$).

and

(B) $$\begin{vmatrix} \mu_1 & \cdots & \mu_k \\ a_{21} & \cdots & a_{2k} \\ \vdots & & \vdots \\ a_{k1} & \cdots & a_{kk} \end{vmatrix} > 0, \ \ldots\ldots, \ \begin{vmatrix} a_{11} & \cdots & a_{1k} \\ \vdots & & \vdots \\ a_{r-1,1} & \cdots & a_{r-1,k} \\ \mu_1 & \cdots & \mu_k \\ a_{r+1,1} & \cdots & a_{r+1,k} \\ \vdots & & \vdots \\ a_{k1} & \cdots & a_{kk} \end{vmatrix} > 0,$$

$$\ldots\ldots, \quad \begin{vmatrix} a_{11} & \cdots\cdots & a_{1k} \\ \cdot & & \cdot \\ \cdot & & \cdot \\ \cdot & & \cdot \\ \cdot & & \cdot \\ a_{k-1,1} & \cdots & a_{k-1,k} \\ \mu_1 & \cdots\cdots & \mu_k \end{vmatrix} > 0, \tag{7}$$

for some non-trivial k-tuple of real numbers $\mu_1, \mu_2, \ldots \mu_k$. The inequalities holding for all $\underset{\sim}{x} \varepsilon I^k$.

Henceforth the problem defined by (1) (2) and condition (A) will be called the "right-definite" multiparameter eigenvalue problem, whilst the problem defined by (1) (2) and condition (B) will be seen to lead to what may be called the "left-definite" multiparameter eigenvalue problem. We note, for further reference, that condition (B) may be expressed in the more convenient form

$$(B^*) \qquad h_s = \sum_{r=1}^{k} \mu_r a^*_{sr} > 0 \tag{8}$$

for all $\underset{\sim}{x} \varepsilon I^k$, where a^*_{sr} is the co-factor of a_{sr} in the determinant Δ_k.

§2 The Conditions (A) and (B)

In the case k = 2 there is a strong connection between conditions (A) and (B). It is easily proved, using a theorem of Atkinson [1, p. 151 Theorem 9.4.1], that (A) implies (B). However the converse is not true as may be seen from the following example.

$$\begin{aligned} -y_1'' + y_1 &= (\lambda p(x_1) - \mu)y_1, \qquad 0 \leq x_1 \leq 1, \\ -y_2'' &= (\lambda q(x_2) + \mu)y_2, \qquad 0 \leq x_2 \leq 1, \end{aligned} \tag{9}$$

together with Sturm-Liouville boundary conditions for both equations. Condition (A) demands, in this case,

$$\begin{vmatrix} p(x_1) & -1 \\ q(x_2) & 1 \end{vmatrix} = p(x_1) + q(x_2) > 0, \text{ for all } x_1, x_2 \ \varepsilon [0, 1]. \tag{10}$$

This is obviously not true except for special choices of p and q. On the other hand condition (B) corresponds to the existence of two real number numbers α and β such that

$$\begin{vmatrix} \alpha & \beta \\ q(x_2) & 1 \end{vmatrix} > 0 \qquad \begin{vmatrix} p(x_1) & -1 \\ \alpha & \beta \end{vmatrix} > 0. \tag{11}$$

If we take $\alpha = 1$ and $\beta = 0$ then (11) holds for any choice of p and q.

When $k \geq 3$, there is no relation between conditions (A) and (B). Indeed in the following example due to B. Karlsson shows that (A) may hold but (B) is violated. Consider the determinant

$$\Delta_3 = \begin{vmatrix} 1 & \cos x_1 & \sin x_1 \\ 1 & \cos x_2 & \sin x_2 \\ 1 & \cos x_3 & \sin x_3 \end{vmatrix} \tag{12}$$

defined on $I^3 = [0, \pi/3] \times [2\pi/3, \pi] \times [4\pi/3, 5\pi/3]$.

Then

$$\Delta_3 = 4 \sin\left(\frac{x_1 - x_2}{2}\right) \sin\left(\frac{x_3 - x_1}{2}\right) \sin\left(\frac{x_2 - x_3}{2}\right) > 0$$

for all $\underset{\sim}{x} = (x_1, x_2, x_3) \varepsilon I^3$. The matrix of co-factors is

$$\begin{pmatrix} \sin(x_3 - x_2) & \sin(x_1 - x_3) & \sin(x_2 - x_1) \\ \sin x_2 - \sin x_3 & \sin x_3 - \sin x_1 & \sin x_1 - \sin x_2 \\ \cos x_3 - \cos x_2 & \cos x_1 - \cos x_3 & \cos x_2 - \cos x_1 \end{pmatrix} \tag{13}$$

For condition (B) suppose there exist real numbers μ_1, μ_2, μ_3 such that

$$h_1 = \mu_1 \sin(x_3 - x_2) + \mu_2(\sin x_2 - \sin x_3) + \mu_3(\cos x_3 - \cos x_2) > 0,$$

$$h_2 = \mu_1 \sin(x_1 - x_3) + \mu_2(\sin x_3 - \sin x_1) + \mu_3(\cos x_1 - \cos x_3) > 0,$$

$$h_3 = \mu_1 \sin(x_2 - x_1) + \mu_2(\sin x_1 - \sin x_2) + \mu_3(\cos x_2 - \cos x_1) > 0.$$

Then for $x_2 = 2\pi/3$, $x_3 = 5\pi/3$ we have

$$h_1 = \mu_2\sqrt{3} + \mu_3 > 0$$

and for $x_1 = \pi/3$, $x_3 = 4\pi/3$

$$h_2 = -\mu_2\sqrt{3} + \mu_3 > 0.$$

From this it follows that $\mu_3 > 0$, but for $x_1 = 0$, $x_2 = \pi$ we have

$$h_3 = -2\mu_3 > 0$$

which gives a contradiction. Hence there are no numbers μ_1, μ_2, μ_3 such that h_1, h_2, h_3 are all positive. Conversely we may have condition (B) holding but not condition (A). This may be shown by the following example due to F.V. Atkinson. Consider

$$\Delta_3 = \begin{vmatrix} 2 & -1 & -1 \\ -1 & 2 & -1 \\ -1 & -1 & 2 \end{vmatrix} = 0. \tag{14}$$

Here (A) fails to hold, but for (B) we have

$$h_1 = h_2 = h_3 = 3(\mu_1 + \mu_2 + \mu_3) > 0$$

for all real μ_i, $i = 1, 2, 3$ such that $\mu_1 + \mu_2 + \mu_3 > 0$.

§3 The right-definite problem

In this section we state two fundamental results which are known for the problem (1) (2) under condition (A)

Theorem 1 [5] (Klein oscillation theorem)

The eigenvalues of the system (1) (2) and (A) form a countably infinite discrete set, lying in E^k (Euclidean k-space). In particular if $(p_1, \ldots, p_k)$ is a k-tuple of non-negative integers, then there is precisely one eigenvalue of this set, say $\lambda^* \in E^k$, such that if $\{y_r(x_r, \lambda^*)\}_{r=1}^k$ is a corresponding set of simultaneous solutions of (1) (2) and (A) then $y_r(x_r, \lambda^*)$ has precisely p_r zeros in $0 < x_r < 1$, $r = 1, 2, \ldots, k$.

Theorem 2 [1, 2, 3, 12]

The eigenfunctions of the system (1) (2) and (A) form a complete orthonormal set in the space of functions square integrable on I^k with weight function $\det\{a_{rs}(x_r)\}_{r,s=1}^k$.

§4 Oscillation theory under condition B

We now come to the main subject of this lecture, that is the study of the system (1) (2) when condition (B) is assumed to hold. In particular we wish to discuss the analogues of Theorems 1 and 2 in this case. Much of what we shall say has been treated in some depth by the author and A. Källström in the series of papers [6, 7, 8] and so we shall endeavour to give the flavour of the ideas and arguments involved and refer the reader to these papers for a more comprehensive treatment.

To begin with we consider the analogue of Theorem 1. Firstly we reformulate the system (1) via the non-singular transformation

$$\begin{pmatrix} \nu_1 \\ \cdot \\ \cdot \\ \cdot \\ \cdot \\ \nu_k \end{pmatrix} = \begin{pmatrix} 1 & 0 & \cdots & & & 0 \\ 0 & 1 & 0 & \cdots & & 0 \\ \cdot & & & & & \\ \cdot & & & & & \\ \cdot & & & & & \\ \cdots & & & 0 & 1 & 0 \\ \mu_1 & \mu_2 & \cdots & & \mu_{k-1} & \mu_k \end{pmatrix} \begin{pmatrix} \lambda_1 \\ \cdot \\ \cdot \\ \cdot \\ \cdot \\ \lambda_k \end{pmatrix} \qquad (15)$$

to obtain the new system

$$\frac{d^2y_r}{dx_r^2} + \{\sum_{s=1}^{k-1} (a_{rs} - a_{rk}\frac{\mu_s}{\mu_k})\nu_s + a_{rk}\frac{\nu_k}{\mu_k} - q_r\}y_r = 0, \tag{16}$$

$r = 1, 2, \ldots, k$ together with the conditions (2) and (B), where without loss of generality, μ_k is assumed positive. Let ν_k be real and fixed and consider the first $(k - 1)$ members of the system (16). In particular, using B*, we find

$$\Delta_{k-1} = \det\{a_{rs} - a_{rk}\frac{\mu_s}{\mu_k}\}_{r,s=1}^{k-1}$$

$$= a^*_{kk} + \sum_{s=1}^{k-1}\frac{\mu_s}{\mu_k}a^*_{ks} = \frac{1}{\mu_k}h_k > 0, \tag{17}$$

for all $\underset{\sim}{x} = (x_1, \ldots, x_{k-1})\epsilon I^{k-1}$. Thus for all real $\nu_k\epsilon(-\infty, \infty)$ the first $(k - 1)$ members of the system (16) together with the first $(k - 1)$ Sturm-Liouville conditions (2) constitutes a multiparameter eigenvalue problem for which the equivalent of condition (A) is satisfied. Hence for this system the Klein oscillation Theorem 1 holds. That is, for each $\nu_k\epsilon(-\infty, \infty)$, there exists precisely one eigenvalue $\underset{\sim}{\nu}^* = (\nu_1^*, \ldots, \nu^*_{k-1})$ such that if $\{y_r(x_r, \underset{\sim}{\nu}^*)\}_{r=1}^{k-1}$ is a corresponding set of simultaneous solutions of (16) (2) then $y_r(x_r, \underset{\sim}{\nu}^*)$ has precisely p_r zeros in $0 < x_r < 1$, $r = 1, \ldots, k - 1$.

Substituting for $\underset{\sim}{\nu}^*$ in the last member of (16) we are led to the one parameter eigenvalue problem

$$\frac{d^2y_k}{dx_k^2} + \{\sum_{s=1}^{k-1}(a_{ks} - a_{kk}\frac{\mu_s}{\mu_k})\nu^*_s + a_{kk}\frac{\nu_k}{\mu_k} - q_k\}y_k = 0, \tag{18}$$

$0 \le x_k \le 1$,

$$\cos\alpha_k\, y_k(0) - \sin\alpha_k\frac{dy_k(0)}{dx_k} = 0, \qquad 0 \le \alpha_k < \pi,$$

$$\cos\beta_k\, y_k(1) - \sin\beta_k\frac{dy_k(1)}{dx_k} = 0, \qquad 0 < \beta_k \le \pi. \tag{19}$$

Thus our given problem has been reduced to one of seeking whether ν_k can be chosen so that the system (18) (19) has a non-trivial solution $y_k(x_k, \underset{\sim}{\nu}^*, \nu_k)$ having precisely p_k zeros in $0 < x_k < 1$.

Rather than treat the problem (18) (19) directly we consider instead a related problem defined by

$$\frac{d^2\omega}{dx_k^2} + \{ \sum_{s=1}^{k-1} (a_{ks} - a_{kk} \frac{\mu_s}{\mu_k})\nu_s^* + a_{kk} \frac{\nu_k}{\mu_k} + Q\Omega - (q_k + Q)\} \omega = 0, \qquad (20)$$

$0 \leq x_k \leq 1$, together with the Sturm-Liouville conditions (19), where Ω is a real parameter and Q is a positive constant to be suitably chosen. Observe that when $\Omega = 1$, equation (20) reduces to (18). A real tuple (ν_k^*, Ω^*) is an eigenvalue of the system (20) (19) if for $\nu_k = \nu_k^*$, $\Omega = \Omega^*$, (20) has a non-trivial solution. Problems of the form (20) (19) have been studied in [4, 10, 11] ; indeed we have the result.

Theorem 3 [4, 10]

The totality of the real eigenvalues of the system (20) (19) is the union of a countably infinite number of closed, unbounded, disjoint subsets S_{p_k}, $p_k = 0, 1, \ldots$, of E^2. If, for each p_k, we consider S_{p_k} as a topological space in itself with the topology induced by E^2, then S_{p_k} is a connected one dimensional manifold. Moreover, for each p_k, S_{p_k} is an analytic manifold and has the further property that if $(\nu_k^*, \Omega^*) \varepsilon S_{p_k}$ then $\omega(x_k, \nu_k^*, \Omega^*)$ has precisely p_k zeros in $0 < x_k < 1$.

Also, from well known Sturm-Liouville theory we know that for each $\nu_k \varepsilon (-\infty, \infty)$, the totality of values of Ω for which (20) (19) is non-trivially solvable form a countably infinite set of real numbers $\{\Omega_{p_k}(\nu_k)\}_{p_k=0}^{\infty}$ which may be ordered as

$$\Omega_0(\nu_k) < \Omega_1(\nu_k) < \ldots\ldots\ldots$$

where $\lim_{p_k \to \infty} \Omega_{p_k}(\nu_k) = +\infty$. Furthermore $\omega(x_k, \nu_k, \Omega_{p_k}(\nu_k))$ has precisely p_k zeros in $0 < x_k < 1$. We may also prove that

$$S_{p_k} = \{(\nu_k, \Omega_{p_k}(\nu_k)) | \nu_k \varepsilon (-\infty, \infty)\},$$

where $\Omega_{p_k}(\nu_k)$ is a single valued analytic function in $-\infty < \nu_k < \infty$.

The next thing is to study the eigenvalue curve S_{p_k} ; in particular we are interested in the points of intersection of S_{p_k} with the line $\Omega = 1$. Such points, if any, will be those eigenvalues ν_k^* so that (18) has a solution $y_k(x_k, \underset{\sim}{\nu}^*, \nu_k^*)$ having precisely p_k zeros in $0 < x_k < 1$. The slope of the curve S_{p_k} at any point (ν_k, Ω_{p_k}) is easily shown to be given by

$$\frac{d\Omega_{p_k}(\nu_k)}{d\nu_k} = - \frac{\int_{I^k} \det \{a_{rs}\}_{r,s=1}^{k} (y_1 \; y_2 \cdot \ldots \cdot y_{k-1} \; \omega)^2 d\underset{\sim}{x}}{Q\int_{I^k} h_k \, (y_1 \cdot \ldots \cdot y_{k-1} \; \omega)^2 d\underset{\sim}{x}} \tag{21}$$

where I^k and h_k are as defined in section 1 above, $y_r \equiv y_r(x_r, \underset{\sim}{\nu}^*, \nu_k)$, $r = 1, \ldots, k-1$, and $\omega \equiv \omega(x_k, \nu_k, \Omega_{p_k}(\nu_k))$.

Lemma 1

If $\hat{\Omega} \leq 0$, then the line $\Omega = \hat{\Omega}$ intersects each curve S_{p_k} in precisely two points $(\nu^+_{k,p_k}(\hat{\Omega}), \hat{\Omega})$, $(\nu^-_{k,p_k}(\hat{\Omega}), \hat{\Omega})$ say, $p_k = 0, 1, \ldots,$ and

$\ldots < \nu^-_{k,1}(\hat{\Omega}) < \nu^-_{k,0}(\hat{\Omega}) < 0 < \nu^+_{k,0}(\hat{\Omega}) < \nu^+_{k,1}(\hat{\Omega}) < \ldots,$

where $\lim_{p_k \to \infty} \nu^+_{k,p_k}(\hat{\Omega}) = +\infty$, $\lim_{p_k \to \infty} \nu^-_{k,p_k}(\hat{\Omega}) = -\infty$.

This result is proved in the following way. For fixed $p_k \geq 0$ we have [4, Theorem (4.3)] $\Omega_{p_k}(0) > 0$, $\lim_{\nu_k \to \infty} \Omega_{p_k}(\nu_k) = \lim_{\nu_k \to -\infty} \Omega_{p_k}(\nu_k) = -\infty$ and so S_{p_k} intersects $\Omega = \hat{\Omega}$ in at least one point with positive abscissa and in at least one point with negative abscissa and from the analytic nature of $\Omega_{p_k}(\nu_k)$ we conclude that there is at most a finite number of points of intersection and with each such point having non-zero abscissa. Now if $(\hat{\nu}_k, \hat{\Omega})$ is a point of intersection we see, from (1) (2) with $r = 1, \ldots, k-1$ and (20) (19), that

$$\begin{aligned}
&\hat{\nu}_k \int_{I^k} \det \{a_{rs}\} (\hat{y}_1 \cdot \ldots \cdot \hat{y}_{k-1} \; \hat{\omega})^2 d\underset{\sim}{x} \\
&= \int_{I^k} \Big\{ \sum_{s=1}^{k} h_s \left[\frac{\partial(\hat{y}_1 \cdot \ldots \cdot \hat{y}_{k-1} \; \hat{\omega})}{\partial x_s} \right]^2 + \sum_{s=1}^{k-1} h_s \, q_s \, (\hat{y}_1 \cdot \ldots \cdot \hat{y}_{k-1} \; \hat{\omega})^2 \\
&\quad + h_k \, (q_k + Q - \hat{\Omega} Q) \, (\hat{y}_1 \cdot \ldots \cdot \hat{y}_{k-1} \; \hat{\omega})^2 \Big\} d\underset{\sim}{x} \\
&\quad + \sum_{s=1}^{k} \int_{I_s^k} h_s \, (\cot \alpha_s (\hat{y}_1 \cdot \ldots \cdot \hat{y}_{k-1} \; \hat{\omega})^2_{x_s = 0} - \cot \beta_s (\hat{y}_1 \cdot \ldots \cdot \hat{y}_{k-1} \; \hat{\omega})^2_{x_s = 1}) d\underset{\sim}{x}_s
\end{aligned} \tag{22}$$

where $I_s^k = \prod_{\substack{r=1 \\ r \neq s}}^{k} \times [0, 1]$, $d\chi_s = dx_1 \ldots\ldots dx_{s-1} \; dx_{s+1} \ldots\ldots dx_k$,

and where $\hat{y}_r \equiv y_r\,(x_r, \chi^*, \hat{\nu}_k)$ $r = 1, \ldots, k-1$, $\hat{\omega} \equiv \omega(x_k, \hat{\nu}_k, \hat{\Omega})$.

From the definition of h_s (see B* in section 1) we see that the right handside of (22) is positive for Q sufficiently large and $\hat{\Omega} \leq 0$. Thus from (21) and (22) we conclude that if $\hat{\nu}_k > 0$ then $d\Omega_{p_k}(\hat{\nu}_k)/d\hat{\nu}_k < 0$ and if $\hat{\nu}_k < 0$ then

$d\Omega_{p_k}(\hat{\nu}_k)/d\hat{\nu}_k > 0$. Hence $\Omega = \hat{\Omega}$ is cut by S_{p_k} in precisely two points, which we

denote by $(\nu^+_{k,p_k}(\hat{\Omega}), \hat{\Omega})$ and $(\nu^-_{k,p_k}(\hat{\Omega}), \hat{\Omega})$ where $\nu^-_{k,p_k}(\hat{\Omega}) < 0 < \nu^+_{k,p_k}(\hat{\Omega})$.

The remaining statements of the lemma follow from (21) and the fact that $\Omega_{p_k}(\nu_k) > \hat{\Omega}$ in $\nu^-_{k,p_k}(\hat{\Omega}) < \nu_k < \nu^+_{k,p_k}(\hat{\Omega})$ and that in any bounded interval of

the ν_k-axis there is at most a finite number of points of the sets $\{\nu^+_{k,p_k}(\hat{\Omega})\}^{\infty}_{p_k=0}$ and $\{\nu^-_{k,p_k}(\hat{\Omega})\}^{\infty}_{p_k=0}$.

We have now developed enough machinery to be able to obtain some very general oscillation properties possessed by the eigenfunctions of the system (1) (2) (B). These results are developed in [8] . To give one such result we make the further assumptions

(C) (i) $q_r > 0$, for all $x_r \in [0, 1]$, $r = 1, \ldots, k$

(ii) $\alpha_r \in (0, \pi/2]$, $\beta_r \in [\pi/2, \pi)$.

The conditions (B) (C) applied to the system (1) (2) give rise to the "left-definite" multiparameter eigenvalue problem (see [7]).

Returning to the problem (18) (19) we know that the eigenvalues of this system are precisely the abscissae of the points of intersection of the curves S_{p_k}, $p_k = 0, 1, \ldots$, with the line $\Omega = 1$. From lemma 1 we observe that for each $p_k \geq 0$, $\Omega_{p_k}(\nu_k) > 0$ in $(\nu^-_{k,p_k}(0), \nu^+_{k,p_k}(0))$ and $\Omega_{p_k}(\nu_k) < 0$ in $(-\infty, \nu^-_{k,p_k}(0))$ and $(\nu^+_{k,p_k}(0), \infty)$; $\dfrac{d\Omega_{p_k}(\nu_k)}{d\nu_k} > 0$ in $(-\infty, \nu^-_{k,p_k}(0)]$, $\dfrac{d\Omega_{p_k}(\nu_k)}{d\nu_k} < 0$ in

$[\nu^+_{k,p_k}(0), \infty)$ and $\lim\limits_{\nu_k \to \infty} \Omega_{p_k}(\nu_k) = \lim\limits_{\nu_k \to -\infty} \Omega\,(\nu_k) = -\infty$. Thus from the

analyticity of $\Omega_{p_k}(\nu_k)$ we conclude that $\Omega_{p_k}(\nu_k)$ attains its absolute maximum in

$(-\infty, \infty)$ in at most a finite number of points, all lying in $(\nu^-_{k,p_k}(0), \nu^+_{k,p_k}(0))$.

Let

$$\Omega^*_{p_k} = \sup_{\nu^-_{k,p_k}(0) \leq \nu_k \leq \nu^+_{k,p_k}(0)} \Omega_{p_k}(\nu_k); \tag{23}$$

then from Theorem 3 we have

$$0 < \Omega^*_0 < \Omega^*_1 < \ldots\ldots\ldots\ldots, \lim_{p_k \to \infty} \Omega^*_{p_k} = \infty.$$

If we let N be the infimum of the non-negative integers p_k for which $\Omega^*_{p_k} \geq 1$ then, using condition (C), it may be seen that $N = 0$ and for each $p_k \geq 0$, S_{p_k} has precisely two points of intersection with the line $\Omega = 1$. Furthermore one of these points has positive abscissa and the other negative abscissa.

Transforming back to the original parameters $(\lambda_1, \ldots, \lambda_k)$ via (15) we may summarize the above results in the following analogue of Theorem 1.

<u>Theorem 4</u>

<u>The eigenvalues of the system (1) (2) (B) and (C) form a countably infinite discrete set, lying in E^k, (Euclidean k-space), In particular if $(p_1, \ldots, p_k)$ is a k-tuple of non-negative integers, then there are precisely two eigenvalues of this set, say $\lambda^{*+}, \lambda^{*-} \in E^k$ such that if $\{y^{\pm}_r(x_r, \lambda^{*\pm})\}^k_{r=1}$ are corresponding sets of simultaneous solutions of (1) (2) (B) and (C) then $y^{\pm}_r(x_r, \lambda^{*\pm})$ has precisely p_r zeros in $0 < x_r < 1$, $r = 1, \ldots, k$.</u>

§5 <u>Completeness under conditions (B) and (C)</u>

In order to establish completeness of the eigenfunctions of the system (1) (2) we proceed in the following way. First, it may be shown [9] that the problem defined by (1) (2) can be formally replaced by the following system of eigenvalue problems

$$\sum_{s=1}^{k} \left(a^*_{sr} \frac{\partial^2 Y}{\partial x^2_s} - q_s\, a^*_{sr}\, Y\right) = -\lambda_r\, \Delta_k\, Y, \tag{24}$$

$r = 1, \ldots, k$ where $\Delta_k = \det \{a_{rs}\}_{r,s=1}^k$ and a^*_{sr} is the cofactor of a_{sr} in the determinant Δ_k. In (24) Y is a non-trivial solution satisfying boundary conditions of the form (2) on the sides of the k-dimensional cube I^k. Multiplication of (24) by μ_r and summation over $r = 1, \ldots, k$ leads to

$$\sum_{s=1}^{k} (h_s \frac{\partial^2 Y}{\partial x_s^2} - h_s q_s Y) = - \Lambda \Delta_k Y, \tag{25}$$

where

$$\Lambda = \sum_{r=1}^{k} \mu_r \lambda_r. \tag{26}$$

Because of condition (B*) it is clear that we now have an elliptic eigenvalue problem for Y. However the problem is not in the usual form since the coefficient of the spectral parameter, Λ, is not necessarily definite and so the standard theory is not applicable. If we had formulated (25) under the hypothesis (A) then, although Δ_k is positive, the left hand side of (25) would not in general be elliptic. The exception of course being the case $k = 2$.

In order to treat the boundary value problem (24) subject to the boundary conditions (2) a suitable Hilbert space is to be defined. The usual space consisting of those functions which are Lebesque measurable and in $L^2_{\Delta_k}[I^k]$ is clearly inappropriate since Δ_k is not necessarily definite. Instead a positive definite Dirichlet integral is to be associated with (25). Thus in the most general left definite case, where, for example, we assume the sufficient condition (C) to hold in addition to (B) the theory is developed in the Hilbert space H which is the completion of $C'(I^k)$ with respect to the inner product

$$D(u, v) = \int_{I^k} \sum_{s=1}^{k} (h_s \frac{\partial u}{\partial x_s} \frac{\partial \bar{v}}{\partial x_s} + h_s q_s u\bar{v}) d\chi$$
$$+ \sum_{s=1}^{k} \int_{I_s^k} h_s (\cot \alpha_s u\bar{v}\Big|_{x_s=0} - \cot \beta_s u\bar{v}\Big|_{x_s=1}) d\chi_s, \tag{27}$$

where I^k_s and $d\chi_s$ are as defined in (22).

<u>Remark</u>

Suppose the boundary ∂I^k of I^k consists of two parts Ω_1 and Ω_2, i.e. $\partial I^k = \Omega_1 \cup \Omega_2$ for which we have Robin or Neumann conditions on Ω_1 and Dirichlet conditions on Ω_2. Then we take the inner product $D(u, v)$ as defined in (27) but with boundary integrals only over Ω_1. The Hilbert space H is then the completion of the set $\{u \in C'(I^k);\ u = 0 \text{ on } \Omega_2\}$ with respect to this modified inner product.

The theory of Stummel [13] is now available; indeed we have the result.

Theorem 5

The spectrum of (25) consists of a countable real set, having no finite point of accumulation, of eigenvalues with finite multiplicities. Furthermore the eigenfunctions of (25) are complete in the space $H \ominus H(\infty)$, where $H(\infty)$ is the set $\{u \in H,\ \Delta_k u = 0$ and u satisfies the boundary conditions$\}$.

Let us denote the eigenvalues by Λ_n, $n = 1, 2, \ldots$, and the associated eigenfunctions by $\Phi_n(\underset{\sim}{x}, \Lambda_n)$. We now show how these eigenfunctions are related to the eigenfunctions of our original problem (1) (2).

Since not all the μ_r, $r = 1, \ldots, k$ in (B) are zero we may suppose for example that $\mu_k \neq 0$; then from (26) we can solve for λ_k to give

$$\lambda_k = \frac{\Lambda_n}{\mu_k} - \sum_{r=1}^{k-1} \frac{\mu_r}{\mu_k} \lambda_r. \tag{28}$$

Substituting this into the first $k - 1$ equation of (1) say, gives the new system

$$\frac{d^2 y_r}{dx_r^2} + \{ \sum_{s=1}^{k-1} (a_{rs} - a_{rk} \frac{\mu_s}{\mu_k}) \lambda_s + a_{rk} \frac{\Lambda_n}{\mu_k} - q_r \} y_r = 0, \tag{29}$$

$r = 1, \ldots, k - 1$. Thus we have generated a $(k - 1)$-parameter eigenvalue problem. As for (16) the determinant

$$\Delta_{k-1} \equiv \det \{a_{rs} - \frac{\mu_s}{\mu_k} a_{rk}\}_{r,s=1}^{k-1}$$

$$= \frac{1}{\mu_k} h_k. \tag{30}$$

Thus the problem defined by (29) and the boundary conditions (2) constitutes a $(k - 1)$-parameter eigenvalue problem for which the equivalent of hypothesis (A) holds. From Theorem 2 we know that the eigenfunctions of this system are complete in the space $K \equiv L^2_{\Delta_{k-1}}(I^{k-1})$.

Let the eigenfunctions be denoted by

$$E^m_{k-1} \equiv y^m_1(x_1;\ \tilde{\lambda}^m) \otimes y^m_2(x_2;\ \tilde{\lambda}^m) \otimes \ldots \otimes y^m_{k-1}(x_{k-1};\ \tilde{\lambda}^m), \tag{31}$$

where $\tilde{\lambda}^m = (\lambda^m_1, \lambda^m_2, \ldots, \lambda^m_{k-1})$ and $\lambda^m_r = \lambda^m_r(\Lambda_n)$, $r = 1, \ldots, k-1$. Now the eigenfunction Φ_n of (25) considered as a function of $x_1, \ldots, x_{k-1}$ is an element of K and so Φ_n can be expanded as a convergent series

$$\Phi_n(x_1, \ldots, x_k;\ \Lambda_n) = \sum_{\tilde{\lambda}^m} C_m(x_k)\, E^m_{k-1}, \tag{32}$$

convergence of course being in the sense of the norm induced by K. Using the orthogonality property of the eigenfunctions E^m_{k-1} we have

$$C_m(x_k)\, \| E^m_{k-1} \|^2_{\Delta_{k-1}} = \int_{I^{k-1}} \Phi_n\, E^m_{k-1}\, \Delta_{k-1}\, dx_1, \ldots, dx_{k-1}, \tag{33}$$

where the norm on the left hand side of (33) is the norm induced by the space K. The idea now proceeds by showing that $C_m(x_k)$ is an eigenfunction of the kth member of the system (1) (2) with the same eigenvalues $\lambda^m_1, \ldots, \lambda^m_k$, ($\lambda^m_k$ being determined from (26)) as the previous k - 1 members of the system; see [7]. We also note that the series (32) is a finite sum. This follows from the fact that the eigenvalues Λ_n have finite multiplicity which implies that only a finite number of the eigenfunctions $\{C_m E^m_{k-1}\}^\infty_{m=1}$ are linearly independent. Putting these results together we have the analogue of Theorem 2.

Theorem 6

Under hypotheses (B) and (C) the spectrum of the system (1) (2) consists of a countable set, having no finite point of accumulation of real eigenvalues with finite multiplicities. Furthermore the eigenfunctions of (1) (2) form an orthonormal set with respect to the D metric (27) and are complete in the space $H \ominus H(\infty)$.

Acknowledgement

I should like to emphasize that most of the results discussed in this lecture were obtained in collaboration with my colleague Anders Källström in the course of research supported by the Science Research Council.

References

[1] Atkinson, F.V. Multiparameter eigenvalue problems Vol. 1 Academic Press, New York and London (1972)

[2] Browne, P.J. A multiparameter eigenvalue problem, J. Math. Anal. Appl. 38 (1972) 553-568

[3] Faierman, M. The completeness and expansion theorems associated with the multiparameter eigenvalue problem in ordinary differential equations, J. diff. equations 5 (1969) 179-213

[4] Faierman, M. An oscillation theorem for a one-parameter ordinary differential equation of the second order, J. diff. equations 11 (1972) 10-37

[5] Ince, E.L. Ordinary differential equations, Dover, New York (1956)

[6] Källström, A. and Sleeman, B.D. A multiparameter Sturm-Liouville problem, Proceedings of the conference on the theory of ordinary and partial differential equations (1974). Lecture Notes in Mathematics. Springer-Verlag (to appear)

[7] Källström, A. and Sleeman, B.D. A left definite multiparameter eigenvalue problem in ordinary differential equations, Dundee University Mathematics report (1974)

[8] Källström, A. and Sleeman, B.D. Klein oscillation theorems for multiparameter eigenvalue problems in ordinary differential equations (in preparation)

[9] Sleeman, B.D. Multiparameter eigenvalue problems in ordinary differential equations, Bul. Inst. Poli. Jossy 17 (21) (1971) 51-60

[10] Sleeman, B.D. The two parameter Sturm-Liouville problem for ordinary differential equations, Proc. Roy. Soc. (Edin) (A) 69 (1971) 139-148

[11] Sleeman, B.D. The two parameter Sturm-Liouville problem for ordinary differential equations II, Proc. Amer. Math. Soc. 34 (1972) 165-170

[12] Sleeman, B.D. Completeness and expansion theorems for a two-parameter eigenvalue problem in ordinary differential equations using vatiational principles, J. Lond. Math. Soc. (2) 6 (1973) 705-712

[13] Stummel, F. Singular perturbations of elliptic sesquilinear forms, Proceedings of the conference on the theory of ordinary and partial differential equations (1972). Lecture Notes in Mathematics Vol. 280 155-180. Springer-Verlag.

Vol. 277: Séminaire Banach. Edité par C. Houzel. VII, 229 pages. 1972. DM 22,–

Vol. 278: H. Jacquet, Automorphic Forms on GL(2) Part II. XIII, 142 pages. 1972. DM 18,–

Vol. 279: R. Bott, S. Gitler and I. M. James, Lectures on Algebraic and Differential Topology. V, 174 pages. 1972. DM 20,–

Vol. 280: Conference on the Theory of Ordinary and Partial Differential Equations. Edited by W. N. Everitt and B. D. Sleeman. XV, 367 pages. 1972. DM 29,–

Vol. 281: Coherence in Categories. Edited by S. Mac Lane. VII, 235 pages. 1972. DM 22,–

Vol. 282: W. Klingenberg und P. Flaschel, Riemannsche Hilbertmannigfaltigkeiten. Periodische Geodätische. VII, 211 Seiten. 1972. DM 22,–

Vol. 283: L. Illusie, Complexe Cotangent et Déformations II. VII, 304 pages. 1972. DM 27,–

Vol. 284: P. A. Meyer, Martingales and Stochastic Integrals I. VI, 89 pages. 1972. DM 18,–

Vol. 285: P. de la Harpe, Classical Banach-Lie Algebras and Banach-Lie Groups of Operators in Hilbert Space. III, 160 pages. 1972. DM 18,–

Vol. 286: S. Murakami, On Automorphisms of Siegel Domains. V, 95 pages. 1972. DM 18,–

Vol. 287: Hyperfunctions and Pseudo-Differential Equations. Edited by H. Komatsu. VII, 529 pages. 1973. DM 40,–

Vol. 288: Groupes de Monodromie en Géométrie Algébrique. (SGA 7 I). Dirigé par A. Grothendieck. IX, 523 pages. 1972. DM 55,–

Vol. 289: B. Fuglede, Finely Harmonic Functions. III, 188. 1972. DM 20,–

Vol. 290: D. B. Zagier, Equivariant Pontrjagin Classes and Applications to Orbit Spaces. IX, 130 pages. 1972. DM 18,–

Vol. 291: P. Orlik, Seifert Manifolds. VIII, 155 pages. 1972. DM 18,–

Vol. 292: W. D. Wallis, A. P. Street and J. S. Wallis, Combinatorics: Room Squares, Sum-Free Sets, Hadamard Matrices. V, 508 pages. 1972. DM 55,–

Vol. 293: R. A. DeVore, The Approximation of Continuous Functions by Positive Linear Operators. VIII, 289 pages. 1972. DM 27,–

Vol. 294: Stability of Stochastic Dynamical Systems. Edited by R. F. Curtain. IX, 332 pages. 1972. DM 29,–

Vol. 295: C. Dellacherie, Ensembles Analytiques Capacités. Mesures de Hausdorff. XII, 123 pages. 1972. DM 18,–

Vol. 296: Probability and Information Theory II. Edited by M. Behara, K. Krickeberg and J. Wolfowitz. V, 223 pages. 1973. DM 22,–

Vol. 297: J. Garnett, Analytic Capacity and Measure. IV, 138 pages. 1972. DM 18,–

Vol. 298: Proceedings of the Second Conference on Compact Transformation Groups. Part 1. XIII, 453 pages. 1972. DM 35,–

Vol. 299: Proceedings of the Second Conference on Compact Transformation Groups. Part 2. XIV, 327 pages. 1972. DM 29,–

Vol. 300: P. Eymard, Moyennes Invariantes et Représentations Unitaires. II, 113 pages. 1972. DM 18,–

Vol. 301: F. Pittnauer, Vorlesungen über asymptotische Reihen. VI, 186 Seiten. 1972. DM 18,–

Vol. 302: M. Demazure, Lectures on p-Divisible Groups. V, 98 pages. 1972. DM 18,–

Vol. 303: Graph Theory and Applications. Edited by Y. Alavi, D. R. Lick and A. T. White. IX, 329 pages. 1972. DM 26,–

Vol. 304: A. K. Bousfield and D. M. Kan, Homotopy Limits, Completions and Localizations. V, 348 pages. 1972. DM 29,–

Vol. 305: Théorie des Topos et Cohomologie Etale des Schémas. Tome 3. (SGA 4). Dirigé par M. Artin, A. Grothendieck et J. L. Verdier. VI, 640 pages. 1973. DM 55,–

Vol. 306: H. Luckhardt, Extensional Gödel Functional Interpretation. VI, 161 pages. 1973. DM 20,–

Vol. 307: J. L. Bretagnolle, S. D. Chatterji et P.-A. Meyer, Ecole d'été de Probabilités: Processus Stochastiques. VI, 198 pages. 1973. DM 22,–

Vol. 308: D. Knutson, λ-Rings and the Representation Theory of the Symmetric Group. IV, 203 pages. 1973. DM 22,–

Vol. 309: D. H. Sattinger, Topics in Stability and Bifurcation Theory. VI, 190 pages. 1973. DM 20,–

Vol. 310: B. Iversen, Generic Local Structure of the Morphisms in Commutative Algebra. IV, 108 pages. 1973. DM 18,–

Vol. 311: Conference on Commutative Algebra. Edited by J. W. Brewer and E. A. Rutter. VII, 251 pages. 1973. DM 24,–

Vol. 312: Symposium on Ordinary Differential Equations. Edited by W. A. Harris, Jr. and Y. Sibuya. VIII, 204 pages. 1973. DM 22,–

Vol. 313: K. Jörgens and J. Weidmann, Spectral Properties of Hamiltonian Operators. III, 140 pages. 1973. DM 18,–

Vol. 314: M. Deuring, Lectures on the Theory of Algebraic Functions of One Variable. VI, 151 pages. 1973. DM 18,–

Vol. 315: K. Bichteler, Integration Theory (with Special Attention to Vector Measures). VI, 357 pages. 1973. DM 29,–

Vol. 316: Symposium on Non-Well-Posed Problems and Logarithmic Convexity. Edited by R. J. Knops. V, 176 pages. 1973. DM 20,–

Vol. 317: Séminaire Bourbaki - vol. 1971/72. Exposés 400–417. IV, 361 pages. 1973. DM 29,–

Vol. 318: Recent Advances in Topological Dynamics. Edited by A. Beck. VIII, 285 pages. 1973. DM 27,–

Vol. 319: Conference on Group Theory. Edited by R. W. Gatterdam and K. W. Weston. V, 188 pages. 1973. DM 20,–

Vol. 320: Modular Functions of One Variable I. Edited by W. Kuyk. V, 195 pages. 1973. DM 20,–

Vol. 321: Séminaire de Probabilités VII. Edité par P. A. Meyer. VI, 322 pages. 1973. DM 29,–

Vol. 322: Nonlinear Problems in the Physical Sciences and Biology. Edited by I. Stakgold, D. D. Joseph and D. H. Sattinger. VIII, 357 pages. 1973. DM 29,–

Vol. 323: J. L. Lions, Perturbations Singulières dans les Problèmes aux Limites et en Contrôle Optimal. XII, 645 pages. 1973. DM 46,–

Vol. 324: K. Kreith, Oscillation Theory. VI, 109 pages. 1973. DM 18,–

Vol. 325: C.-C. Chou, La Transformation de Fourier Complexe et L'Equation de Convolution. IX, 137 pages. 1973. DM 18,–

Vol. 326: A. Robert, Elliptic Curves. VIII, 264 pages. 1973. DM 24,–

Vol. 327: E. Matlis, One-Dimensional Cohen-Macaulay Rings. XII, 157 pages. 1973. DM 20,–

Vol. 328: J. R. Büchi and D. Siefkes, The Monadic Second Order Theory of All Countable Ordinals. VI, 217 pages. 1973. DM 22,–

Vol. 329: W. Trebels, Multipliers for (C, α)-Bounded Fourier Expansions in Banach Spaces and Approximation Theory. VII, 103 pages. 1973. DM 18,–

Vol. 330: Proceedings of the Second Japan-USSR Symposium on Probability Theory. Edited by G. Maruyama and Yu. V. Prokhorov. VI, 550 pages. 1973. DM 40,–

Vol. 331: Summer School on Topological Vector Spaces. Edited by L. Waelbroeck. VI, 226 pages. 1973. DM 22,–

Vol. 332: Séminaire Pierre Lelong (Analyse) Année 1971-1972. V, 131 pages. 1973. DM 18,–

Vol. 333: Numerische, insbesondere approximationstheoretische Behandlung von Funktionalgleichungen. Herausgegeben von R. Ansorge und W. Törnig. VI, 296 Seiten. 1973. DM 27,–

Vol. 334: F. Schweiger, The Metrical Theory of Jacobi-Perron Algorithm. V, 111 pages. 1973. DM 18,–

Vol. 335: H. Huck, R. Roitzsch, U. Simon, W. Vortisch, R. Walden, B. Wegner und W. Wendland, Beweismethoden der Differentialgeometrie im Großen. IX, 159 Seiten. 1973. DM 20,–

Vol. 336: L'Analyse Harmonique dans le Domaine Complexe. Edité par E. J. Akutowicz. VIII, 169 pages. 1973. DM 20,–

Vol. 337: Cambridge Summer School in Mathematical Logic. Edited by A. R. D. Mathias and H. Rogers. IX, 660 pages. 1973. DM 46,–

Vol: 338: J. Lindenstrauss and L. Tzafriri, Classical Banach Spaces. IX, 243 pages. 1973. DM 24,–

Vol. 339: G. Kempf, F. Knudsen, D. Mumford and B. Saint-Donat, Toroidal Embeddings I. VIII, 209 pages. 1973. DM 22,–

Vol. 340: Groupes de Monodromie en Géométrie Algébrique. (SGA 7 II). Par P. Deligne et N. Katz. X, 438 pages. 1973. DM 44,–

Vol. 341: Algebraic K-Theory I, Higher K-Theories. Edited by H. Bass. XV, 335 pages. 1973. DM 29,–

Vol. 342: Algebraic K-Theory II, "Classical" Algebraic K-Theory, and Connections with Arithmetic. Edited by H. Bass. XV, 527 pages. 1973. DM 40,-

Vol. 343: Algebraic K-Theory III, Hermitian K-Theory and Geometric Applications. Edited by H. Bass. XV, 572 pages. 1973. DM 40,-

Vol. 344: A. S. Troelstra (Editor), Metamathematical Investigation of Intuitionistic Arithmetic and Analysis. XVII, 485 pages. 1973. DM 38,-

Vol. 345: Proceedings of a Conference on Operator Theory. Edited by P. A. Fillmore. VI, 228 pages. 1973. DM 22,-

Vol. 346: Fučík et al., Spectral Analysis of Nonlinear Operators. II, 287 pages. 1973. DM 26,-

Vol. 347: J. M. Boardman and R. M. Vogt, Homotopy Invariant Algebraic Structures on Topological Spaces. X, 257 pages. 1973. DM 24,-

Vol. 348: A. M. Mathai and R. K. Saxena, Generalized Hypergeometric Functions with Applications in Statistics and Physical Sciences. VII, 314 pages. 1973. DM 26,-

Vol. 349: Modular Functions of One Variable II. Edited by W. Kuyk and P. Deligne. V, 598 pages. 1973. DM 38,-

Vol. 350: Modular Functions of One Variable III. Edited by W. Kuyk and J.-P. Serre. V, 350 pages. 1973. DM 26,-

Vol. 351: H. Tachikawa, Quasi-Frobenius Rings and Generalizations. XI, 172 pages. 1973. DM 20,-

Vol. 352: J. D. Fay, Theta Functions on Riemann Surfaces. V, 137 pages. 1973. DM 18,-

Vol. 353: Proceedings of the Conference on Orders, Group Rings and Related Topics. Organized by J. S. Hsia, M. L. Madan and T. G. Ralley. X, 224 pages. 1973. DM 22,-

Vol. 354: K. J. Devlin, Aspects of Constructibility. XII, 240 pages. 1973. DM 24,-

Vol. 355: M. Sion, A Theory of Semigroup Valued Measures. V, 140 pages. 1973. DM 18,-

Vol. 356: W. L. J. van der Kallen, Infinitesimally Central-Extensions of Chevalley Groups. VII, 147 pages. 1973. DM 18,-

Vol. 357: W. Borho, P. Gabriel und R. Rentschler, Primideale in Einhüllenden auflösbarer Lie-Algebren. V, 182 Seiten. 1973. DM 20,-

Vol. 358: F. L. Williams, Tensor Products of Principal Series Representations. VI, 132 pages. 1973. DM 18,-

Vol. 359: U. Stammbach, Homology in Group Theory. VIII, 183 pages. 1973. DM 20,-

Vol. 360: W. J. Padgett and R. L. Taylor, Laws of Large Numbers for Normed Linear Spaces and Certain Fréchet Spaces. VI, 111 pages. 1973. DM 18,-

Vol. 361: J. W. Schutz, Foundations of Special Relativity: Kinematic Axioms for Minkowski Space Time. XX, 314 pages. 1973. DM 26,-

Vol. 362: Proceedings of the Conference on Numerical Solution of Ordinary Differential Equations. Edited by D. Bettis. VIII, 490 pages. 1974. DM 34,-

Vol. 363: Conference on the Numerical Solution of Differential Equations. Edited by G. A. Watson. IX, 221 pages. 1974. DM 20,-

Vol. 364: Proceedings on Infinite Dimensional Holomorphy. Edited by T. L. Hayden and T. J. Suffridge. VII, 212 pages. 1974. DM 20,-

Vol. 365: R. P. Gilbert, Constructive Methods for Elliptic Equations. VII, 397 pages. 1974. DM 26,-

Vol. 366: R. Steinberg, Conjugacy Classes in Algebraic Groups (Notes by V. V. Deodhar). VI, 159 pages. 1974. DM 18,-

Vol. 367: K. Langmann und W. Lütkebohmert, Cousinverteilungen und Fortsetzungssätze. VI, 151 Seiten. 1974. DM 16,-

Vol. 368: R. J. Milgram, Unstable Homotopy from the Stable Point of View. V, 109 pages. 1974. DM 16,-

Vol. 369: Victoria Symposium on Nonstandard Analysis. Edited by A. Hurd and P. Loeb. XVIII, 339 pages. 1974. DM 26,-

Vol. 370: B. Mazur and W. Messing, Universal Extensions and One Dimensional Crystalline Cohomology. VII, 134 pages. 1974. DM 16,-

Vol. 371: V. Poenaru, Analyse Différentielle. V, 228 pages. 1974. DM 20,-

Vol. 372: Proceedings of the Second International Conference on the Theory of Groups 1973. Edited by M. F. Newman. VII, 740 pages. 1974. DM 48,-

Vol. 373: A. E. R. Woodcock and T. Poston, A Geometrical Study of the Elementary Catastrophes. V, 257 pages. 1974. DM 22,-

Vol. 374: S. Yamamuro, Differential Calculus in Topological Linear Spaces. IV, 179 pages. 1974. DM 18,-

Vol. 375: Topology Conference 1973. Edited by R. F. Dickman Jr. and P. Fletcher. X, 283 pages. 1974. DM 24,-

Vol. 376: D. B. Osteyee and I. J. Good, Information, Weight of Evidence, the Singularity between Probability Measures and Signal Detection. XI, 156 pages. 1974. DM 16.-

Vol. 377: A. M. Fink, Almost Periodic Differential Equations. VIII, 336 pages. 1974. DM 26,-

Vol. 378: TOPO 72 - General Topology and its Applications. Proceedings 1972. Edited by R. Alò, R. W. Heath and J. Nagata. XIV, 651 pages. 1974. DM 50,-

Vol. 379: A. Badrikian et S. Chevet, Mesures Cylindriques, Espaces de Wiener et Fonctions Aléatoires Gaussiennes. X, 383 pages. 1974. DM 32,-

Vol. 380: M. Petrich, Rings and Semigroups. VIII, 182 pages. 1974. DM 18,-

Vol. 381: Séminaire de Probabilités VIII. Edité par P. A. Meyer. IX, 354 pages. 1974. DM 32,-

Vol. 382: J. H. van Lint, Combinatorial Theory Seminar Eindhoven University of Technology. VI, 131 pages. 1974. DM 18,-

Vol. 383: Séminaire Bourbaki - vol. 1972/73. Exposés 418-435 IV, 334 pages. 1974. DM 30,-

Vol. 384: Functional Analysis and Applications, Proceedings 1972. Edited by L. Nachbin. V, 270 pages. 1974. DM 22,-

Vol. 385: J. Douglas Jr. and T. Dupont, Collocation Methods for Parabolic Equations in a Single Space Variable (Based on C^1-Piecewise-Polynomial Spaces). V, 147 pages. 1974. DM 16,-

Vol. 386: J. Tits, Buildings of Spherical Type and Finite BN-Pairs. IX, 299 pages. 1974. DM 24,-

Vol. 387: C. P. Bruter, Eléments de la Théorie des Matroïdes. V, 138 pages. 1974. DM 18,-

Vol. 388: R. L. Lipsman, Group Representations. X, 166 pages. 1974. DM 20,-

Vol. 389: M.-A. Knus et M. Ojanguren, Théorie de la Descente et Algèbres d' Azumaya. IV, 163 pages. 1974. DM 20,-

Vol. 390: P. A. Meyer, P. Priouret et F. Spitzer, Ecole d'Eté de Probabilités de Saint-Flour III - 1973. Edité par A. Badrikian et P.-L. Hennequin. VIII, 189 pages. 1974. DM 20,-

Vol. 391: J. Gray, Formal Category Theory: Adjointness for 2-Categories. XII, 282 pages. 1974. DM 24,-

Vol. 392: Géométrie Différentielle, Colloque, Santiago de Compostela, Espagne 1972. Edité par E. Vidal. VI, 225 pages. 1974. DM 20,-

Vol. 393: G. Wassermann, Stability of Unfoldings. IX, 164 pages. 1974. DM 20,-

Vol. 394: W. M. Patterson 3rd, Iterative Methods for the Solution of a Linear Operator Equation in Hilbert Space - A Survey. III, 183 pages. 1974. DM 20,-

Vol. 395: Numerische Behandlung nichtlinearer Integrodifferential- und Differentialgleichungen. Tagung 1973. Herausgegeben von R. Ansorge und W. Törnig. VII, 313 Seiten. 1974. DM 28,-

Vol. 396: K. H. Hofmann, M. Mislove and A. Stralka, The Pontryagin Duality of Compact O-Dimensional Semilattices and its Applications. XVI, 122 pages. 1974. DM 18,-

Vol. 397: T. Yamada, The Schur Subgroup of the Brauer Group. V, 159 pages. 1974. DM 18,-

Vol. 398: Théories de l'Information, Actes des Rencontres de Marseille-Luminy, 1973. Edité par J. Kampé de Fériet et C. Picard. XII, 201 pages. 1974. DM 23,-